Advances in Solid State Physics
Volume 46

Advances in Solid State Physics

Advances in Solid State Physics is a book series with a history of about 50 years. It contains the invited lectures presented at the Spring Meetings of the "Arbeitskreis Festkörperphysik" of the "Deutsche Physikalische Gesellschaft", held in March of each year. The invited talks are intended to reflect the most recent achievements of researchers working in the field both in Germany and worldwide. Thus the volume of the series represents a continuous documentation of most recent developments in what can be considered as one of the most important and active fields of modern physics. Since the majority of invited talks are usually given by young researchers at the start of their career, the articles can also be considered as indicating important future developments.
The speakers of the invited lectures and of the symposia are asked to contribute to the yearly volumes with the written version of their lecture at the forthcoming Spring Meeting of the Deutsche Physikalische Gesellschaft by the Series Editor. Colored figures are available in the online version for some of the articles.
Advances in Solid State Physics is addressed to all scientists at universities and in industry who wish to obtain an overview and to keep informed on the latest developments in solid state physics. The language of publication is English.

Series Editor

Prof. Dr. Rolf Haug

Abteilung Nanostrukturen
Institut für Festkörperphysik
Universität Hannover
Appelstr. 2
30167 Hannover
Germany
haug@nano.uni-hannover.de

Rolf Haug (Ed.)

Advances in Solid State Physics

46

With 171 Figures and 4 Tables

Springer

Prof. Dr. Rolf Haug (Ed.)
Abteilung Nanostrukturen
Institut für Festkörperphysik
Universität Hannover
Appelstr. 2
30167 Hannover
Germany
haug@nano.uni-hannover.de

Physics and Astronomy Classification Scheme (PACS): 60.00; 70.00; 80.00

ISSN print edition: 1438-4329
ISSN electronic edition: 1617-5034
ISBN-13 978-3-540-38234-8 Springer Berlin Heidelberg New York

Springer is a part of Springer Science+Business Media

springeronline.com

Printed in Germany

Typesetting by the authors using a Springer TEX macro package
Final processing: DA-TEX · Gerd Blumenstein · www.da-tex.de
Production: LE-TEX GbR, Leipzig, www.le-tex.de

Cover concept using a background picture by Dr. Ralf Stannarius, Faculty of Physics and Earth Sciences, Institute of Experimental Physics I, University of Leipzig, Germany
Cover design: WMXDesign GmbH, Heidelberg

Printed on acid-free paper 56/3180/YL 5 4 3 2 1 0

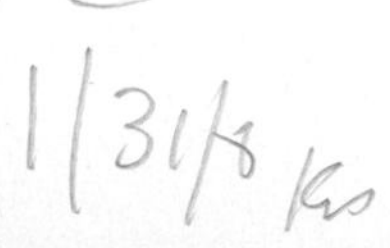

Preface

The 2006 Spring meeting of the Arbeitskreis Festkörperphysik of the Deutsche Physikalische Gesellschaft was held in Dresden between March 27 and March 31, 2006 in conjunction with the 21st General Conference of the European Physical Society, Condensed Matter Division. The number of participants reached 4500 with 3808 scientific contributions. These impressive numbers made it the largest solid-state physics meeting of the year in Europe and clearly show that this meeting was attractive to a large number of scientists from Germany and from all over Europe.

The present volume of the Advances in Solid State Physics contains the written version of a large number of the invited talks and gives a nice overview of the status of solid-state physics and of the most interesting subjects within it. Low-dimensional physics dominates the contributions. The themes ranged from zero-dimensional physics in quantum dots, molecules and nanoparticles through one-dimensional physics in nanowires and 1d systems to more applied subjects like optoelectronics and materials science in thin films. The contributions span the whole breadth of solid-state physics ranging from truly basic science to applications.

Rolf J. Haug

Contents

Part I Quantum Dots

Part III Nanowires and 1D Systems

Part I

Quantum Dots

Single Photons from Single Quantum Dots – New Light for Quantum Information Processing

Matthias Scholz[1], Thomas Aichele[2], and Oliver Benson[1]

[1] Institute of Physics, Humboldt-University Berlin,
Hausvogteiplatz 5–7, 10117 Berlin, Germany
matthias.scholz@physik.hu-berlin.de

[2] Equipe Nanophysique et Semiconducteurs, CEA / Université J. Fourier,
17 rue des Martyrs, 38054 Grenoble Cedex 9, France
thomas.aichele@cea.fr

Abstract. In this paper, we discuss the prospects of single quantum dots for single-photon generation and their implementation in quantum information experiments. We review the characterization of quantum dot emission and, as an example, report a recently demonstrated application in a quantum cryptography protocol.

1 Introduction

There has been considerable effort in the recent past to exploit light for quantum information processing. Photons are ideal tools to transmit quantum information over large distances. In 1984, *Bennett* and *Brassard* proposed a secret key-distribution protocol [1] that uses the single-particle character of a photon to avoid any possibility of eavesdropping on an encoded message (for a review see [2]). Also, the implementation of efficient quantum gates based on photons and linear optics were proposed [3, 4] and demonstrated [5, 6]. Proposals extend the role of photons as the information carriers in larger networks [7] between processing knots of stationary qubits, like ions [8–11], atoms [12, 13], quantum dots (QDs) [14, 15], and Josephson qubits [16, 17].

Linear optics applications in quantum information processing require the reliable deterministic generation of single- or few-photon states. However, due to their bosonic character, photons tend to appear in bunches. Thus, classical light sources provide a broad photon number distribution, as depicted for thermal and laser light in Figs. 1a and 1b, respectively. This characteristics hinders the implementation of classical sources particularly in quantum cryptographic systems since an eavesdropper may gain partial information by a beam splitter attack. Similar obstacles occur for linear optics quantum computation where photonic quantum gates [3], quantum repeaters [18], and quantum teleportation [19] require the preparation of single- or few-photon states on demand in order to obtain reliability and high efficiency. While an ideal non-classical single-photon source emits a sub-Poissonian photon number distribution with exactly one photon at a time, real sources have in-

R. Haug (Ed.): Advances in Solid State Physics,
Adv. in Solid State Phys. **46**, 3–14 (2008)

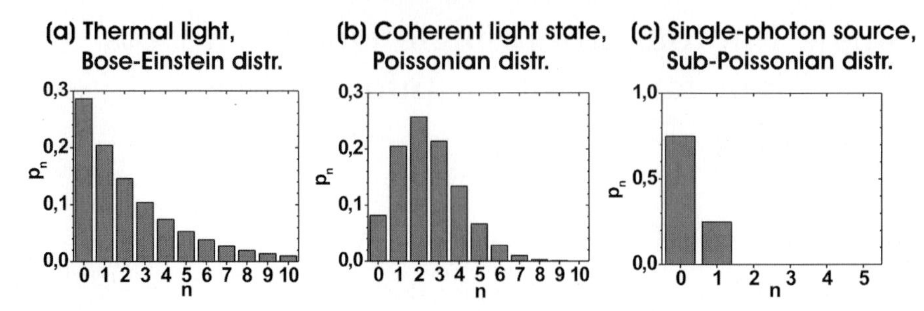

Fig. 1. Photon number distributions of (**a**) thermal light, (**b**) a coherent state, and (**c**) a single-photon source with 25 % efficiency

evitable losses due to scattering and absorption which lead to typical photon number distributions as shown in Fig. 1c [20].

A promising process for single-photon generation is the spontaneous emission from a single quantum emitter. Numerous emitters have been used to demonstrate single-photon emission [21]. Single atoms or ions are the most fundamental systems. They have been trapped and coupled to optical resonators to obtain single-mode emission [22, 23]. Other systems capable of single-photon generation are single molecules and single nanocrystals [24–26]. However, their drawback is their susceptibility for photo-bleaching and blinking [27]. Stable alternatives are nitrogen-vacancy defect centers in diamond [28, 29], but they show broad optical spectra together with comparably long lifetimes ($\approx 12\,\mathrm{ns}$).

In this article, we focus on single-photon generation from self-assembled single QDs. QDs are few-nanometer sized semiconductor structures which resemble features known from atoms, like discrete emission spectrum and electronic structure, and which are therefore often cited as artificial atoms. Most experiments with QDs have to be conducted at cryogenic temperature in order to reduce electron-phonon interaction and thermal ionization. High count rates can be obtained due to their short transition lifetimes, and their spectral lines are nearly lifetime-limited with material systems covering the ultraviolet, visible, and infrared spectrum. QDs also gain attractiveness by the possibility of electric excitation [30] and the implementation in integrated photonic structures [31].

Our paper is organized as follows. Section 2 provides an introduction to the electronic properties and radiative decay of single QDs. Section 3 covers the characterization and measurement of single-photon states. In Sect. 4, we show decay cascades in single QDs. Their application in a multiplexed quantum cryptography setup is demonstrated in Sect. 5. Section 6 concludes with a short summary.

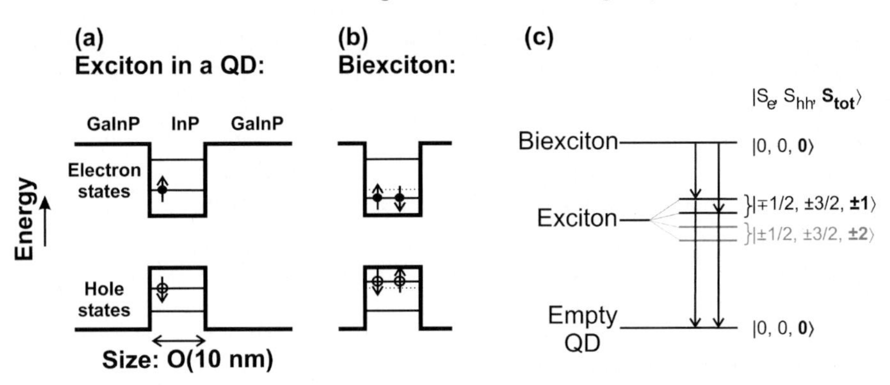

Fig. 2. Excitations in a QD: (**a**) An electron-hole pair forms an exciton. (**b**) Two electron-hole pairs form a biexciton, generally at an energy different from the exciton. (**c**) Schematic term scheme for the exciton and biexciton decay cascade

2 Single Quantum Dots

All experiments in this article were performed with self-assembled QDs fabricated by Stranski–Krastanov growth [32].

Overgrowth of QDs is required to obtain high-efficient and optically stable emission. As it is impossible to characterize the exact shape or material composition of an overgrown QD a priori, certain simplified models for the electronic structure are often used. Figure 2 shows a spherical potential which traps a single electron-hole pair (exciton, Fig. 2a) or two electron-hole pairs (biexciton, Fig. 2b). In Fig. 2c, the possible decay paths of a biexciton via the so-called bright excitons are depicted. A level splitting is shown for clarity. The recombination of a carrier pair leads to the emission of a single photon, generally at different wavelengths for exciton and biexciton due to Coulomb interaction.

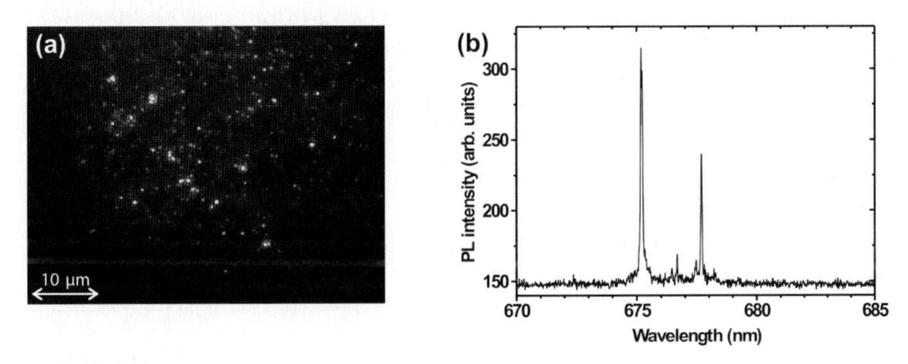

Fig. 3. (**a**) Micro-photoluminescence image of InP QDs in GaInP. (**b**) Spectrum of a single InP/GaInP QD with spectral lines of exciton and biexciton decay

The image of an ensemble of InP QDs is displayed in Fig. 3a. Figure 3b shows the spectrum of a single InP QD in a GaInP matrix with two dominant spectral lines originating from exciton and biexciton decay. This material system can generate single photons in the 620–750 nm region which perfectly fits the maximum detection efficiency of commercial silicon avalanche photo diodes (APD) with over 70 % at wavelengths around 700 nm.

3 Single-Photon Generation

3.1 Correlation Measurements

The single-photon character of the photoluminescence from a single QD can be tested by measuring the normalized second-order correlation function $g^{(2)}(t_1, t_2)$ via detecting the light intensity $\langle \hat{I}(t) \rangle$ at two points in time. For stationary fields, it reads

$$g^{(2)}(\tau = t_1 - t_2) = \frac{\langle : \hat{I}(0)\hat{I}(\tau) : \rangle}{\langle \hat{I}(0) \rangle^2} ,$$

where :: denotes normal operator ordering. For classical fields, $g^{(2)}(0) \geq 1$ and $g^{(2)}(0) \geq g^{(2)}(\tau)$ hold [33] which prohibits values smaller than unity. For thermal light sources, there is an increased probability to detect a photon shortly after another. This bunching phenomenon leads to $g^{(2)}(0) \geq 1$ (Fig. 4a). Coherent states show $g^{(2)}(\tau) = 1$ for all τ according to a Poissonian photon number distribution (Fig. 4b). A single-photon state shows the anti-bunching effect of a sub-Poissonian distribution with $g^{(2)}(0) \leq 1$ (Fig. 4c).

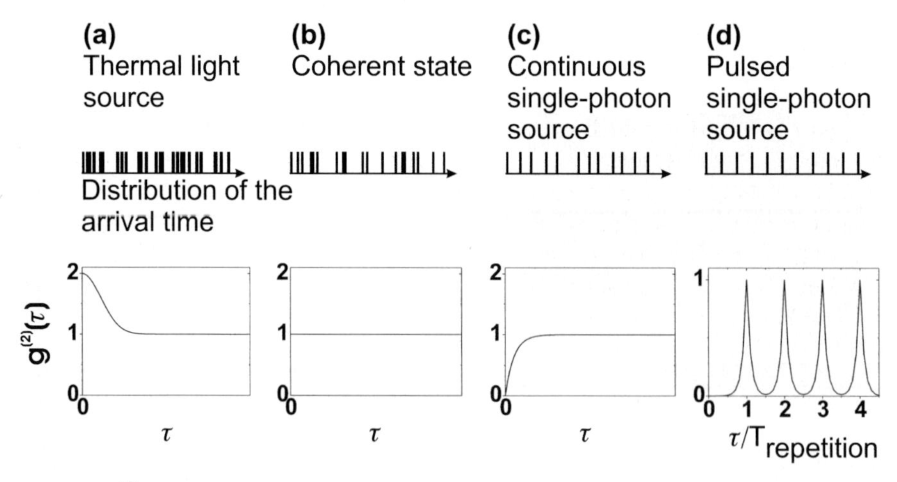

Fig. 4. $g^{(2)}$ function of (**a**) a thermal light source, (**b**) coherent light, (**c**) a continuously driven single-photon source, and (**d**) a pulsed single-photon source

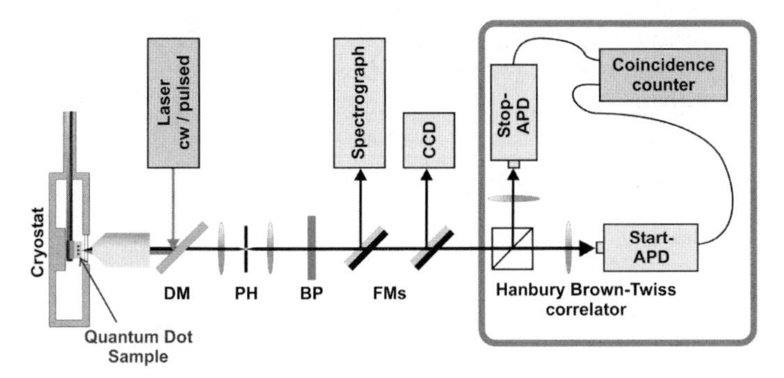

Fig. 5. Micro-PL setup (FM: mirrors on flip mounts, DM: dichroic mirror, PH: pinhole, BP: narrow bandpass filter, APD: avalanche photo diode)

For a Fock state $|n\rangle$, one has $g^{(2)}(0) = 1 - 1/n$ which yields $g^{(2)}(0) = 0$ for the special case of $|n = 1\rangle$. The pulsed excitation of a single-photon source leads to a peaked structure in $g^{(2)}(\tau)$ with a missing peak at zero delay indicating single-photon emission (Fig. 4d).

In order to circumvent the detectors' dead time (≈ 50 ns for APDs [34]), the second-order correlation function is measured in a *Hanbury Brown–Twiss* arrangement [35] as depicted in Fig. 5, consisting of two photo detectors and a $50:50$ beam splitter. A large number of time intervals between detection events is measured and binned together in a histogram.

3.2 Micro-Photoluminescence

Optical investigation of QDs is usually performed in a micro-photoluminescence (PL) setup which combines excellent spatial resolution with high detection efficiency. Samples with low densities of 10^8–10^{11} dots/cm^2 are required to isolate a single dot. In our setup, the sample is held at 4 K inside a continuous-flow liquid Helium cryostat (Fig. 5). The dots are excited either pulsed (Ti:Sa, pulse width 400 fs, repetition rate 76 MHz, frequency-doubled to 400 nm) or continuously (Nd:YVO_4, 532 nm). The microscope system (NA $= 0.75$) reaches a lateral resolution of 500 nm, and further spatial and spectral filtering selects a single transition of a single QD. The PL can be imaged on a CCD camera (see also Fig. 3), on a spectrograph, or can be sent to a Hanbury Brown–Twiss setup with a time resolution of about 800 ps to prove single-photon emission.

3.3 Single Photons from InP Quantum Dots

The sample used in our experiment was grown by metal-organic vapor phase epitaxy. An aluminum mirror was evaporated on the sample which was then thinned down to 400 nm and glued on a Si-substrate (see Fig. 6a). Figure 6b

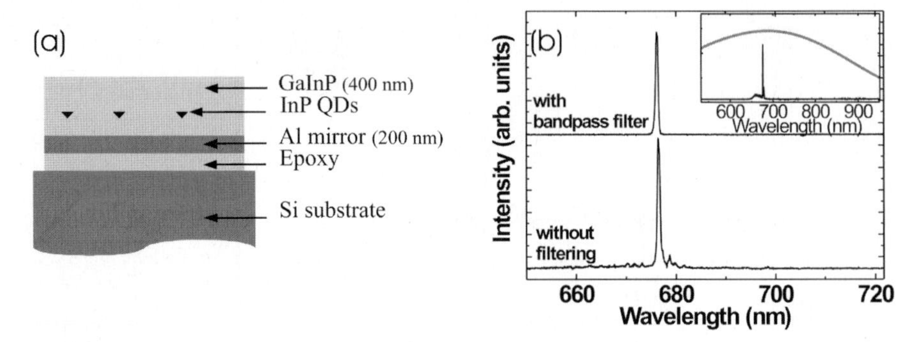

Fig. 6. (**a**) Structure of the InP/GaInP sample. (**b**) PL spectra from a single InP QD. An offset was added to separate the graphs. The *inset* shows a wider spectrum together with the APD detection efficiency

depicts the spectrum of a single InP QD on the sample under weak excitation showing only one dominant spectral line. Its linear dependance on the excitation power identifies it as an exciton emission. A 1-nm bandpass filter was used to further reduce the background.

Figure 7 shows the corresponding cw correlation measurement at a count rate of 1.1×10^5 and 10 K. The function is modeled as a convolution of the expected shape of the ideal correlation function $g^{(2)}(\tau) = 1 - \exp(-\gamma\,\tau)$ and a Gaussian distribution according to a 800 ps time resolution of the detectors. The width of the anti-bunching dip (2.3 ns) depends on both the transition lifetime and the excitation timescale. Equivalent measurements at pulsed excitation show a vanishing peak at zero delay (Fig. 8a) which again proves single-photon generation. With higher temperature, phonon interactions lead to an increased incoherent background with other spectral lines overlapping the filter transmission window (Fig. 8b). However, a characteristic anti-bunching was observed up to 50 K [36].

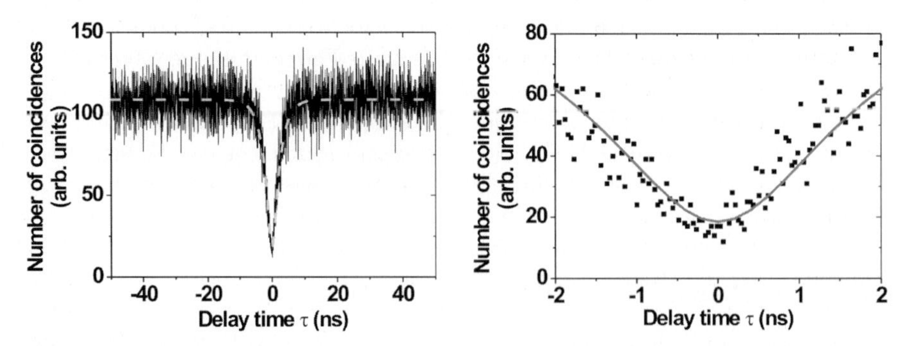

Fig. 7. $g^{(2)}$ function at continuous excitation. The fit function corresponds to an ideal single-photon source with limited time resolution. The *right image* shows a zoom into the central dip region

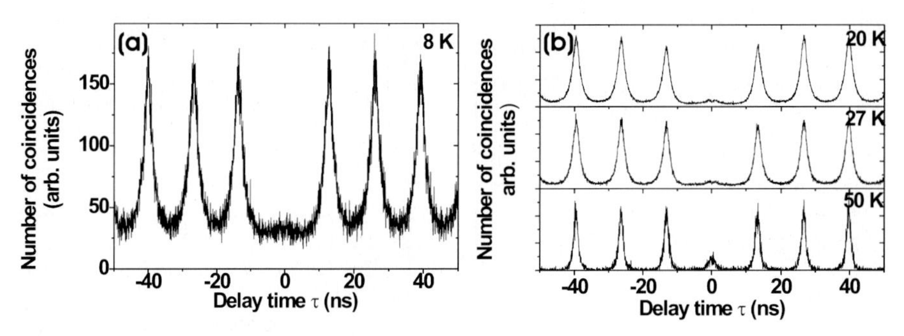

Fig. 8. $g^{(2)}$ function of a single InP QD at pulsed excitation: (**a**) at 8 K and (**b**) between 20 and 50 K

4 A Multi-Color Photon Source

A great advantage of QDs as photon emitters is their potential to create more complex states of light consisting of few photons. For example, by trapping two or more electron-hole pairs (compare Fig. 2), photon cascades can be produced. These cascades can be used to enhance the transmission rate of cryptographic systems usually limited by spontaneous lifetime. Previous experiments have studied cascaded decays in InAs [20, 37, 38] and II–VI QDs [39, 40]. Very recently, also entangled photon pair generation has been demonstrated [41, 42] following earlier proposals [43].

In order to study correlations between photons emitted in a cascade, the cross-correlation function

$$g^{(2)}_{\alpha\beta}(\tau) = \frac{\langle : \hat{I}_\alpha(t)\hat{I}_\beta(t+\tau) : \rangle}{\langle \hat{I}_\alpha(t)\rangle\langle \hat{I}_\beta(t)\rangle}$$

has to be measured with a Hanbury Brown–Twiss configuration modified by an interference filter in each arm. The filters can be tuned, e.g., to the exciton and biexciton transition lines in a two-photon cascade indicated by the index α and β in $g_{\alpha\beta}$, respectively. Cross-correlation measurements confirm assignments of spectral lines to exciton, biexciton, and triexciton transitions by the strong asymmetric shape of the correlation function (Fig. 9). If a biexciton decay starts the measurement and an exciton stops it, photon bunching occurs since an exciton photon is predominantly emitted shortly after a biexciton photon in a cascade. The opposite holds for switched start and stop channels. In this case, the detection of the exciton photon projects the QD to its empty (ground) state. The required re-excitation process manifests in an anti-bunching dip. Biexciton-triexciton correlations can be explained equivalently, but on different timescales. Since cross-correlation functions of two independent transitions show no (anti-)bunching, the results in Fig. 9 confirm a three-photon cascade from the triexciton to the ground state of a single QD. The dashed lines in Fig. 9 are a fit to the rate equation model described in [44, 45].

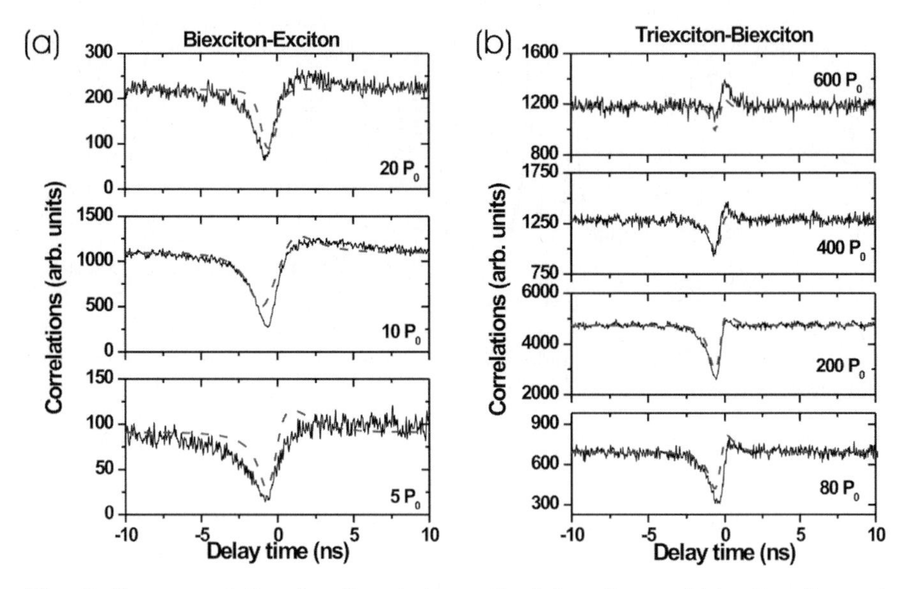

Fig. 9. Cross-correlation functions between the (**a**) exciton and biexciton line and (**b**) biexciton and triexciton line

5 Multiplexed Quantum Cryptography on the Single-Photon Level

5.1 A Single-Photon Add/Drop Filter

For quantum communication protocols, a single-photon source with high repetition rate and collection efficiency is desired. Beside passive elements like mirrors and solid immersion lenses, resonant techniques exploiting the Purcell effect can greatly enhance the emission rate [46, 47] if the source relies on the decay of an excited state.

A well-established technique from classical communication is multiplexing. Each signal is marked with a physical label like its wavelength and identified at the receiver using filters tuned to the carrier frequencies. For single photons, their wavelength can be used as their distinguishing label and polarization to encode quantum information [48]. If the two photons from a biexciton-exciton cascade pass a Michelson interferometer, constructive and destructive interference will be observed at the two output ports by proper choice of the relative path difference (Fig. 10). These photons are fed into an optical fiber each, and one of them is delayed by half the repetition rate of the excitation laser. Recombination at a beam splitter then leads to a photon stream with a doubled count rate. This is also reflected by autocorrelation measurements with a spacing of only 6.6 ns in the peak structure (see Fig. 10d) [48].

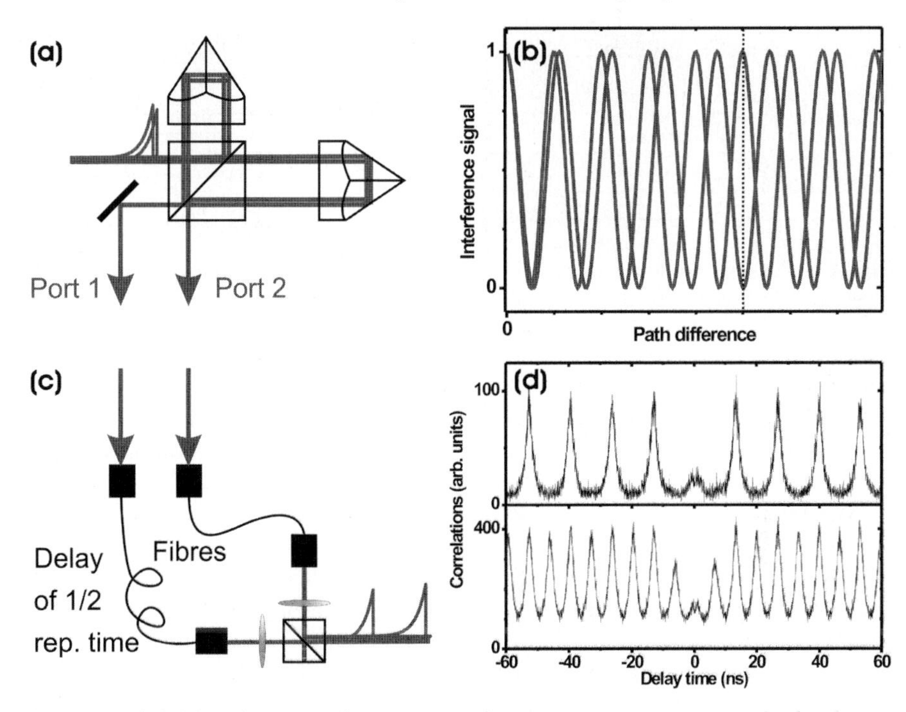

Fig. 10. (**a**) Michelson interferometer with two output ports as a single-photon add/drop filter. (**b**) Interference pattern for two distinct wavelengths versus the relative arm length. (**c**) Merging the separated photons with one path delayed by half the excitation repetition rate. (**d**) Intensity correlation of the exciton spectral line and the multiplexed signal

5.2 Application to Quantum Key Distribution

Multiplexing can enhance the bandwidth of quantum communication systems which use single-photon sources based on spontaneous emission. In our experiment, we implemented the BB84 protocol [1, 2] where the exact photon wavelength is unimportant. Quantum cryptography has been realized with single-photon states from diamond defect centers [49] and single QDs [50]. Figure 11 shows an implementation of interferometric multiplexing in the BB84 protocol using a cascaded photon source. Behind an exciton-biexciton add/drop filter, Alice randomly prepares the photons' polarization. Bob's detection consists of a second EOM, an analyzing polarizer, and an APD. In our experiment, all EOM and APD operation is performed automatically by a LabVIEW program. A visualization of a successful secret transmission of data is depicted in Fig. 12. After exchange of the key, Alice encrypts her data by applying an XOR operation between every image bit and the sequence of random bits (the secret key) which she has previously shared with Bob by performing the BB84 protocol. The encoded image (shown in Fig. 12b) is

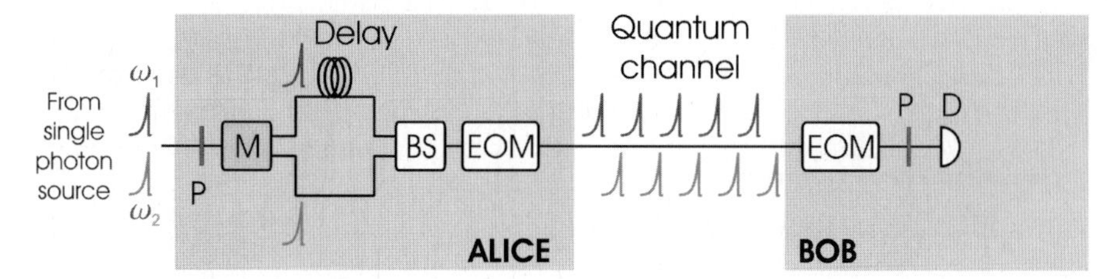

Fig. 11. Possible implementation of the multiplexer using the BB84 protocol. For a detailed description of the setup see text

transmitted to Bob over a distance of 1 m with 30 bits/s. Another XOR operation by Bob reveals the original image with an error rate of 5.5 %.

6 Summary

In this article, we have reviewed single-photon generation from single InP QDs and its application to quantum information processing. Single-photon statistics and cross-correlations of various transitions from multi-excitonic states in the visible spectrum around 690 nm have been demonstrated. We described a proof-of-principle experiment that shows how multi-photon generation from single QDs may find applications in quantum cryptography devices to enhance their maximum bandwidth. In this experiment, multiplexing on the single-photon level and its implementation in the BB84 quantum key distribution protocol has been realized for the first time.

Our experiments show that today's single-photon sources based on QDs have reached the level of sophistication to be ready-to-use non-classical light sources. The on demand character of the emission is extremely useful in all-solid state implementations of quantum information devices. Single QDs will therefore play an important role in the field of quantum cryptography and quantum computation, but also for realizing interfaces in larger quantum networks.

Fig. 12. Visualization of the quantum key distribution. Alice's original data (**a**), a photography taken out of our lab window in Berlin, is encrypted and sent to Bob (**b**). After decryption with his key, Bob obtains image (**c**)

Acknowledgements

The authors want to thank W. Seifert for providing the QD sample and V. Zwiller, G. Reinaudi, and J. Persson for valuable assistance. This work was supported by Deutsche Forschungsgemeinschaft (SFB 296) and the European Union (EFRE). M. Scholz acknowledges financial support from Ev. Studienwerk Villigst and T. Aichele by DAAD.

References

[1] C. M. Bennett and G. Brassard, in: *Proc. IEEE Conference on Computers, Systems, and Signal Processing in Bangalore, India* (IEEE, New York, 1984) p. 175
[2] N. Gisin, G. Ribordy, W. Tittel, H. Zbinden, Rev. Mod. Phys. **74**, 145 (2002)
[3] E. Knill, R. Laflamme, G. J. Milburn, Nature **409**, 46 (2001)
[4] D. Gottesman and I. L. Chuang, Nature **402**, 390 (1999)
[5] S. Gasparoni, J.-W. Pan, P. Walther, T. Rudolph, A. Zeilinger, Phys. Rev. Lett. **93**, 020504 (2004)
[6] N. Kiesel, C. Schmid, U. Weber, R. Ursin, H. Weinfurter, Phys. Rev. Lett. **95**, 210505 (2005)
[7] L. Tian, P. Rabl, R. Blatt, P. Zoller, Phys. Rev. Lett. **92**, 247902 (2004)
[8] J. I. Cirac and P. Zoller, Phys. Rev. Lett. **74**, 4091 (1995)
[9] J. I. Cirac and P. Zoller, Phys. Today **57**, 38 (2004)
[10] F. Schmidt-Kaler, H. Häffner, M. Riebe, S. Gulde, G. P. T. Lancaster, T. Deuschle, C. Becher, C. F. Roos, J. Eschner, R. Blatt, Nature **422**, 408 (2003)
[11] D. Leibfried, B. Demarco, V. Meyer, D. Lucas, M. Barrett, J. Britton, W. M. Itano, B. Jelenkovic, C. Langer, T. Rosenband, et al., Science **422**, 412 (2003)
[12] D. Kielpinski, C. Monroe, D. J. Wineland, Nature **417**, 709 (2002)
[13] J. I. Cirac and P. Zoller, Nature **404**, 579 (2000)
[14] D. Loss and D. P. DiVicenzo, Phys. Rev. A **57**, 120 (1998)
[15] W. G. van der Wiel, S. D. Franceschi, J. M. Elzerman, T. Fujisawa, S. Tarucha, L. P. Kouwenhoven, Rev. Mod. Phys. **75**, 1(2003)
[16] J. E. Mooij, T. P. Orlando, L. Levitov, L. Tian, C. H. van der Wal, S. Lloyd, Science **285**, 1036 (1999)
[17] Y. A. Pashkin, T. Yamamoto, O. Astafiev, Y. Nakamura, D. V. Averin, J. S. Tsai, Nature **421**, 823 (2003)
[18] H.-J. Briegel, W. Dür, J. I. Cirac, P. Zoller, Phys. Rev. Lett. **81**, 5932 (1998)
[19] D. Bouwmeester, J.-W. Pan, K. Mattle, M. Eibl, H. Weinfurter, A. Zeilinger, Nature **390**, 575 (1997)
[20] E. Moreau, I. Robert, L. Manin, V. Thierry-Mieg, J. M. Gérard, I. Abram, Phys. Rev. Lett. **87**, 183601 (2001)
[21] *New Journal of Physics* (special issue on single-photon sources) **6** (2004)
[22] A. Kuhn, M. Hennrich, G. Rempe, Phys. Rev. Lett. **89**, 067901 (2002)
[23] M. Keller, B. Lange, K. Hayasaka, W. Lange, H. Walther, Nature **431**, 1075 (2004)
[24] C. Brunel, B. Lounis, P. Tamarat, M. Orrit, Phys. Rev. Lett. **83**, 2722 (1999)

[25] B. Lounis and W. E. Moerner, Nature **407**, 491 (2000)
[26] P. Michler, A. Imamağlu, M. D. Mason, P. J. Carson, G. F. Strouse, S. K. Buratto, Nature **406**, 968 (2000)
[27] M. Nirmal, B. O. Dabbousi, M. G. Bawendi, J. J. Macklin, J. K. Trautman, T. D. Harris, L. E. Brus, Nature **383**, 802 (1996)
[28] C. Kurtsiefer, S. Mayer, P. Zarda, H. Weinfurter, Phys. Rev. Lett. **85**, 290 (2000)
[29] A. Beveratos, S. Kühn, R. Brouri, T. Gacoin, J.-P. Poizat, P. Grangier, Eur. Phys. J. D **18**, 191 (2002)
[30] Z. Yuan, B. E. Kardynal, R. M. Stevenson, A. J. Shields, C. J. Lobo, K. Copper, N. S. Beattie, D. A. Ritchie, M. Pepper, Science **295**, 102 (2002)
[31] A. Badolato, K. Hennessy, M. Atatre, J. Dreiser, E. Hu, P. M. Petroff, A. Imamoğlu, Science **308**, 1158 (2005)
[32] D. Bimberg, M. Grundmann, N. Ledentsov, *Quantum Dot Heterostructures* (Wiley, Chichester, UK, 1988)
[33] L. Mandel and E. Wolf: *Optical Coherence and Quantum Optics* (Cambridge University Press, 1995)
[34] *Single-photon counting module - SPCM-AQR series specifications*, Laser Components GmbH, Germany (2004)
[35] R. Hanbury Brown and R. Q. Twiss, Nature **178**, 1046 (1956)
[36] V. Zwiller, T. Aichele, W. Seifert, J. Persson, O. Benson, Appl. Phys. Lett. **82**, 1509 (2003)
[37] D. V. Regelman, U. Mizrahi, D. Gershoni, E. Ehrenfreund, W. V. Schoenfeld, P. M. Petroff, Phys. Rev. Lett. **87**, 257401 (2001)
[38] C. Santori, D. Fattal, M. Pelton, G. S. Salomon, Y. Yamamoto, Phys. Rev. B **66**, 045308 (2002)
[39] S. M. Ulrich, S. Strauf, P. Michler, G. Bacher, A. Forchel, Appl. Phys. Lett. **83**, 1848 (2003)
[40] C. Couteau, S. Moehl, F. Tinjod, J. M. Gérard, K. Kheng, H. Mariette, J. A. Gaj, R. Romestain, J. P. Poizat, Appl. Phys. Lett. **85**, 6251 (2004)
[41] R. M. Stevenson, R. J. Young, P. Atkinson, K. Cooper, D. A. Ritchie, A. J. Shields, Nature **439**, 179 (2006)
[42] N. Akopian, N. H. Lindner, E. Poem, Y. Berlatzky, J. Avron, D. Gershoni, B. D. Gerardot, P. M. Petroff, Phys. Rev. Lett. **96**, 130501 (2006)
[43] O. Benson, C. Santori, M. Pelton, Y. Yamamoto, Phys. Rev. Lett. **84**, 2513 (2000)
[44] J. Persson, T. Aichele, V. Zwiller, L. Samuelson, O. Benson, Phys. Rev. B **69**, 233314 (2004)
[45] T. Aichele, V. Zwiller, M. Scholz, G. Renaudi, J. Persson, O. Benson, Proc. SPIE **5722**, 30 (2005)
[46] G. Solomon, M. Pelton, Y. Yamamoto, Phys. Rev. Lett. **86**, 3903 (2001)
[47] J. M. Gérard, B. Sermage, B. Gayral, B. Legrand, E. Costard, V. Thierry-Mieg, Phys. Rev. Lett. **81**, 1110 (1998)
[48] T. Aichele, G. Reinaudi, O. Benson, Phys. Rev. B **70**, 235329 (2004)
[49] A. Beveratos, R. Brouri, T. Gacoin, A. Villing, J.-P. Poizat, P. Grangier, Phys. Rev. Lett. **89**, 187901 (2002)
[50] E. Waks, K. Inoue, C. Santori, D. Fattal, J. Vučković, G. S. Solomon, Y. Yamamoto, Nature **420**, 762 (2002)

Quantum-Dot Spin Qubit and Hyperfine Interaction

D. Klauser, W. A. Coish, and Daniel Loss

Department of Physics and Astronomy, University of Basel, Klingelbergstrasse 82, 4056 Basel, Switzerland

Abstract. We review our investigation of the spin dynamics for two electrons confined to a double quantum dot under the influence of the hyperfine interaction between the electron spins and the surrounding nuclei. Further we propose a scheme to narrow the distribution of difference in polarization between the two dots in order to suppress hyperfine induced decoherence.

1 Introduction

The fields of semiconductor physics and electronics have been successfully combined for many years. The invention of the transistor meant a revolution for electronics and has led to significant development of semiconductor physics and its industry. More recently, the use of the spin degree of freedom of electrons, as well as the charge, has attracted great interest [1]. In addition to applications for spin electronics (spintronics) in conventional devices, for instance based on the giant magneto-resistance effect [2] and spin-polarized field-effect transistors [3], there are applications that exploit the quantum coherence of the spin. Since the electron spin is a two-level system, it is a natural candidate for the realization of a quantum bit (qubit) [4]. A qubit is the basic unit of information in quantum computation. The confinement of electrons in semiconductor structures like quantum dots allows for better control and isolation of the electron spin from its environment. Control and isolation are important issues to consider for the design of a quantum computer.

Formally, a quantum computation is performed through a set of transformations, called gates [5]. A gate applies a unitary transformation U to a set of qubits in a quantum state $|\Psi\rangle$. At the end of the calculation, a measurement is performed on the qubits (which are in the state $|\Psi'\rangle = U\,|\Psi\rangle$). There are many ways to choose sets of universal quantum gates. These are sets of gates from which any computation can be constructed, or at least approximated as precisely as desired. Such a set allows one to perform any arbitrary calculation without inventing a new gate each time. The implementation of a set of universal gates is therefore of crucial importance. It can be shown that it is possible to construct such a set with gates that act only on one or two qubits at a time [6].

R. Haug (Ed.): Advances in Solid State Physics,
Adv. in Solid State Phys. **46**, 15–26 (2008)

The successful implementation of a quantum computer demands that some basic requirements be fulfilled. These are known as the *DiVincenzo* criteria [7] and can be summarized in the following:

1. *Information storage – the qubit:* We need to find some quantum property of a scalable physical system in which to encode our bit of information, that lives long enough to enable us to perform computations.
2. *Initial state preparation:* It should be possible to set the state of the qubits to 0 before each new computation.
3. *Isolation:* The quantum nature of the qubits should be tenable; this will require enough isolation of the qubit from the environment to reduce the effects of decoherence.
4. *Gate implementation:* We need to be able to manipulate the states of individual qubits with reasonable precision, as well as to induce interactions between them in a controlled way, so that the implementation of gates is possible. Also, the gate operation time τ_s has to be much shorter than the decoherence time T_2, so that $\tau_s/T_2 \ll r$, where r is the maximum tolerable error rate for quantum error correction schemes to be effective.
5. *Readout:* It must be possible to measure the final state of our qubits once the computation is finished, to obtain the output of the computation.

To construct quantum computers of practical use, we emphasize that the scalability of the device should not be overlooked. This means it should be possible to enlarge the device to contain many qubits, while still adhering to all requirements described above. It should be mentioned here that this represents a challenging issue in most of the physical setups proposed so far.

2 Quantum-Dot Spin Qubit

The requirement for scalability motivated the *Loss–DiVincenzo* proposal [4] for a solid-state quantum computer based on electron spin qubits.

The qubits of the Loss–DiVincenzo quantum computer are formed from the two spin states ($|\uparrow\rangle, |\downarrow\rangle$) of a confined electron. The considerations discussed in this proposal are generally applicable to electrons confined to any structure, such as atoms, molecules, etc., although the original proposal focuses on electrons localized in quantum dots. These dots are typically generated from a two-dimensional electron gas (2DEG), in which the electrons are strongly confined in the vertical direction. Lateral confinement is provided by electrostatic top gates, which push the electrons into small localized regions of the 2DEG (see Fig. 1). Initialization of the quantum computer can be achieved by allowing all spins to reach their thermodynamic ground state at low temperature T in an applied magnetic field B (i.e., virtually all spins will be aligned if the condition $|g\mu_B B| \gg k_B T$ is satisfied, with g-factor g, Bohr magneton μ_B, and Boltzmann's constant k_B). Single-qubit operations

Fig. 1. Two neighbouring electron spins confined to quantum dots, as in the Loss–DiVincenzo proposal. The lateral confinement is controlled by top gates. A time-dependent Heisenberg exchange coupling $J(t)$ can be pulsed high by pushing the electron spins closer, generating an appreciable overlap between the neighbouring orbital wave functions

can be performed, in principle, by changing the local effective Zeeman interaction at each dot individually. To do this may require large magnetic field gradients [8], g-factor engineering [9], magnetic layers, the inclusion of nearby ferromagnetic dots [4], polarized nuclear spins, or optical schemes.

In the Loss–DiVincenzo proposal, two-qubit operations are performed by pulsing the electrostatic barrier between neighboring spins. When the barrier is high, the spins are decoupled. When the inter-dot barrier is pulsed low, an appreciable overlap develops between the two electron wave functions, resulting in a non-zero Heisenberg exchange coupling J. The Hamiltonian describing this time-dependent process is given by

$$H(t) = J(t)\mathbf{S}_{\mathrm{L}} \cdot \mathbf{S}_{\mathrm{R}} \,. \tag{1}$$

This Hamiltonian induces a unitary evolution given by the operator $U = \mathcal{T} \exp\left\{-i \int H(t)\,\mathrm{d}t/\hbar\right\}$, where $\mathcal{T}$ is the time-ordering operator. If the exchange is pulsed on for a time τ_{s} such that $\int J(t)\,\mathrm{d}t/\hbar = J_0\tau_{\mathrm{s}}/\hbar = \pi$, the states of the two spins, with associated operators $\mathbf{S}_{\mathrm{L}}$ and $\mathbf{S}_{\mathrm{R}}$, as shown in Fig. 1, will be exchanged. This is the SWAP operation. Pulsing the exchange for the shorter time $\tau_{\mathrm{s}}/2$ generates the "square-root of SWAP" operation, which can be used in conjunction with single-qubit operations to generate the controlled-NOT (quantum XOR) gate [4]. The "square-root of SWAP" gate has recently been implemented in an experiment by *Petta* et al. [10] with a switching time $\tau_{\mathrm{s}} = 180\,\mathrm{ps}$. For scalability, and application of quantum error correction procedures in any quantum computing proposal, it is important to turn off inter-qubit interactions in the idle state. In the Loss–DiVincenzo proposal, this is achieved with exponential accuracy since the overlap of neighboring electron wave functions is exponentially suppressed with increasing separation. A detailed investigation of decoherence during gating due to a bosonic environment was performed in the original work of Loss and DiVincenzo. Since then, there have been many studies of leakage and decoherence in the context of the quantum-dot quantum computing proposal.

In addition to the interaction-based gate operations introduced above, it has been shown recently [11, 12] that it is also possible to generate the controlled-NOT based on partial Bell state (parity) measurements.

For both interaction-based and measurement-based quantum computation with the quantum-dot spin qubit, decoherence due to the coupling of the qubit to its environment is a major obstacle. There are two important sources of decoherence in GaAs quantum dots: spin-orbit coupling (interaction between spin and charge fluctuations) and hyperfine coupling (interaction between the electron spin and nuclear spins). In the case of spin-orbit interaction alone it has been shown that the decoherence time T_2 (which is the relevant timescale for quantum computing tasks) exceeds the relaxation time T_1 and is given by $T_2 = 2T_1$ to leading order in the spin-orbit coupling [13]. Since the T_1 obtained in measurements [14, 15] is on the order of ms, but the ensemble-averaged dephasing time T_2^* measured is $\sim$ 10 ns, spin-orbit interaction is not limiting for the dephasing time T_2^*. The limiting source of decoherence is the hyperfine interaction [10].

3 Hyperfine Interaction in Single and Double Dots

The hyperfine interaction between the electron spin and the nuclear spins present in all III-V semiconductors [16] leads to the strongest decoherence effect [10, 17–24]. Experiments [10, 25–27] have yielded values for the free-induction spin dephasing time T_2^* that are consistent with $T_2^* \sim \sqrt{N}/A \sim 10\,\mathrm{ns}$ [20–22] for $N = 10^6$, $\hbar = 1$, and $A = 90\,\mu\mathrm{eV}$ in GaAs, where N is the number of nuclei within one quantum dot Bohr radius and A characterizes the weighted average hyperfine coupling strength in GaAs [28]. This is to be contrasted with potential spin-echo envelope decay times, which may be much longer [23, 29–31]. With a two-qubit switching time of $\tau_s \sim 180\,\mathrm{ps}$ [10] this only allows $\sim 10^2$ gate operations within T_2^*, which falls short (by a factor of 10 to 10^2) of current requirements for efficient quantum error correction [32].

There are several ways to overcome the problem of hyperfine-induced decoherence, of which measurement and thus projection of the nuclear spin state may be the most promising [23]. Other methods include polarization [17, 22,23,33] of the nuclear spins and spin echo techniques [10,23,30]. However, in order to extend the decay time by an order of magnitude through polarization of the nuclear spins, a polarization of above 99% is required [23], but the best result so far reached is only $\sim$ 60% in quantum dots [25]. With spin-echo techniques, gate operations still must be performed within the single-spin free-induction decay time, which requires faster gate operations. A projective measurement of the nuclear spin state leads to an extension of the free-induction decay time for the spin. This extension is only limited by the ability to do a strong measurement since the longitudinal nuclear spin in a quantum dot is expected to survive up to the spin diffusion time, which is on the order of seconds for nuclear spins surrounding donors in GaAs [34].

A detailed analysis of the spin dynamics for one electron in a single quantum dot can be found in [23]. Here we concentrate on the case of two electrons in a double quantum dot. The spin $\mathbf{S}_l$ of an electron in quantum dot $l = 1, 2$, interacts with the surrounding nuclear spins $\mathbf{I}_k$ via the Fermi contact hyperfine interaction:

$$H_{\mathrm{hf}} = \mathbf{S}_l \cdot \mathbf{h}_l; \;\; \mathbf{h}_l = \sum_k A_k^l \mathbf{I}_k; \;\; A_k^l = v_0 A \left|\psi^l(\mathbf{r}_k)\right|^2 , \tag{2}$$

where v_0 is the volume of a crystal unit cell containing one nuclear spin. The effective Hamiltonian in the subspace of one electron on each dot is best written in terms of the sum and difference of electron spin and collective nuclear spin operators: $\mathbf{S} = \mathbf{S}_1 + \mathbf{S}_2, \delta\mathbf{S} = \mathbf{S}_1 - \mathbf{S}_2$ and $\mathbf{h} = \frac{1}{2}(\mathbf{h}_1 + \mathbf{h}_2), \delta\mathbf{h} = \frac{1}{2}(\mathbf{h}_1 - \mathbf{h}_2)$:

$$H_{\mathrm{eff}}(t) = \epsilon_{\mathrm{z}} S^z + \mathbf{h} \cdot \mathbf{S} + \delta\mathbf{h} \cdot \delta\mathbf{S} + \frac{J(t)}{2} \mathbf{S} \cdot \mathbf{S} - J(t) , \tag{3}$$

where $\epsilon_{\mathrm{z}} = g\mu_{\mathrm{B}} B$ is the Zeeman splitting induced by an applied magnetic field $\mathbf{B} = (0, 0, B), B > 0$. We assume that the Zeeman splitting is much larger than $\langle \delta\mathbf{h} \rangle_{\mathrm{rms}}$ and $\langle \mathbf{h}_i \rangle_{\mathrm{rms}}$, where $\langle \mathcal{O} \rangle_{\mathrm{rms}} = \langle \psi_I | \mathcal{O}^2 | \psi_I \rangle^{1/2}$ is the root-mean-square expectation value of the operator $\mathcal{O}$ with respect to the nuclear spin state $|\psi_I\rangle$. Under these conditions the relevant spin Hamiltonian becomes block diagonal with blocks labeled by the total electron spin projection along the magnetic field S^z. In the subspace of $S^z = 0$ the Hamiltonian can be written as ($\hbar = 1$) [24, 35]

$$H_0(t) = \frac{J(t)}{2} \tau^x - \frac{1}{2} \Omega \tau^z ; \;\; J(t) = J_0 + j \cos(\omega t), \;\; \Omega = 2(\delta h^z + \delta b^z). \tag{4}$$

Here, δb^z is the inhomogeneity of an externally applied classical static magnetic field with $\delta b^z \ll B$. The Pauli matrices τ^α, $\alpha = x, y, z$ are given in the basis of $|+\rangle \equiv |\tau^z = 1\rangle = |\downarrow\uparrow\rangle$ and $|-\rangle \equiv |\tau^z = -1\rangle = |\uparrow\downarrow\rangle$.

The dynamics of the two-electron spin states depends strongly on the initial state of the nuclear spin system. We denote by $|n\rangle$ the eigenstates of δh^z with $\delta h^z |n\rangle = \delta h^z_n |n\rangle$. If the initial state of the nuclear spin system is $\rho_I(0) = |n\rangle \langle n|$ and if we neglect spin-flip processes (as can be done for a large enough magnetic field B), then the initial spin state of the electron does not decay. Thus, if it is possible to prepare the nuclear spin system in an eigenstate $|n\rangle$, hyperfine-induced decoherence could be overcome. In general, however, the initial state of the nuclear spin system is not an eigenstate $|n\rangle$ but a general mixture:

$$\rho_I(0) = \sum_i p_i \left|\psi_I^i\right\rangle \left\langle\psi_I^i\right|; \ \left|\psi_I^i\right\rangle = \sum_n a_n^i \left|n\right\rangle, \tag{5}$$

where the a_n^i satisfy the normalization condition $\sum_n |a_n^i|^2 = 1$ and $\sum_i p_i = 1$. We denote by $\rho_I(n) = \sum_i p_i |a_n^i|^2$ the diagonal elements of the nuclear spin density operator.

For a large number of nuclear spins $N \gg 1$ which are in a superposition of δh^z-eigenstates $|n\rangle$, $\rho_I(n)$ describes a continuous Gaussian distribution of δh_n^z values, with mean $\overline{\delta h^z}$ and variance $\sigma^2 = \overline{\left(\delta h^z - \overline{\delta h^z}\right)^2}$. In the limit of large N, the approach to a Gaussian distribution for a sufficiently randomized nuclear system is guaranteed by the central limit theorem [23]. We perform the continuum limit according to

$$\sum_n \rho_I(n) f(n) \to \int \mathrm{d}x \rho_I(x) f(x); \ \rho_I(x) = \frac{1}{\sqrt{2\pi}\sigma} \exp\left(-\frac{(x - x_0)^2}{2\sigma_0^2}\right), \tag{6}$$

where $x = \delta h_n^z + \delta b^z$, $x_0 = \overline{\delta h^z} + \delta b^z$ and $\sigma_0^2 = \overline{x^2} - x_0^2$.

For the case of a static exchange interaction $J(t) = J_0$, the decay of the two-electron spin states in the $S^z = 0$ subspace due to the Gaussian distribution of nuclear spin states may be calculated in several interesting limits [24,35]. Assuming the initial state of the two-electron system is $\rho_e(0) = |+\rangle\langle+|$, the probability P^+ to measure the $|+\rangle$ state as a function of time is given by

$$P^+_{J=J_0}(t) = \int_{-\infty}^{\infty} \rho_I(x) \left(\frac{1}{2} + \frac{2x^2}{s(x)^2} + \frac{J^2}{2s(x)^2} \cos(s(x)t)\right) \tag{7}$$

with $s(x) = \sqrt{J^2 + 4x^2}$. In the limit $\sigma_0 \to 0$, which corresponds to one fixed eigenvalue, there is no decay. However, for $\sigma_0 > 0$ there is decay. For the regime $|x_0| \gg \sigma_0$ we have a Gaussian decay at short times with a decay time $t_0 \sim 1/\sigma_0$:

$$P^+_{J=J_0}(t) \approx \frac{1}{2} + \frac{2x_0^2}{\omega_0^2} + \left(\frac{1}{2} - \frac{2x_0^2}{\omega_0^2}\right) \exp\left(-\frac{t^2}{2t_0^2}\right) \cos(\omega_0 t), \tag{8}$$

$$\omega_0 = \sqrt{J^2 + 4x_0^2}, \ t_0 = \frac{\omega_0}{4|x_0|\sigma_0}; \ |x_0| \gg \sigma_0, \ t \ll \frac{\omega_0^{3/2}}{2J^2\sigma_0^2}. \tag{9}$$

Thus, decreasing σ_0 increases the coherence time t_0. Hence, the strategy to suppress hyperfine-induced decoherence is to narrow the Gaussian distribution of nuclear spin eigenvalues through a measurement of the eigenvalue of δh^z, i.e., of the difference in polarization between the two dots [23, 24]. It has also been proposed to measure the nuclear spin polarization using a phase estimation method [36]. In the ideal case, phase estimation yields one bit of information about the nuclear spin system for each perfectly measured electron. Optical methods have also been proposed [37]. The all-electrical method we propose here can be applied with current technology used in [10, 27].

4 Nuclear Spin State Narrowing

The general idea behind state narrowing is that the evolution of the two-electron system is dependent on the collective nuclear spin state and thus knowing the evolution of the two-electron system determines the nuclear spin state.

The eigenstates of the Hamiltonian H_0 are product states: if the nuclear spin system is in an eigenstate $|n\rangle$ of δh^z with $\delta h^z |n\rangle = \delta h^z_n |n\rangle$, we have $H|\psi\rangle = H_n |\psi_{\mathrm{e}}\rangle \otimes |n\rangle$, where in H_n the operator δh^z has been replaced by δh^z_n and $|\psi_{\mathrm{e}}\rangle$ is the electron spin part of the wave function. Thus, in the Hamiltonian for the evolution of the initial two-electron system, the parameter δh^z_n is determined by the state of the nuclear spin system. Initializing the two-electron system to the $|+\rangle$ state, i.e., $\rho_{\mathrm{e}}(0) = |+\rangle\langle+|$ and performing a measurement in the $|\pm\rangle$ basis at time t_{m} yields for the distribution of nuclear spin eigenvalues (which is the diagonal part of the nuclear spin density operator in the continuum limit) after the measurement [35]

$$\rho_{\mathrm{I}}^{(1,+,\omega)}(x) = \rho_{\mathrm{I}}(x)(1 - L_\omega(x))\frac{1}{P^+_\omega}\,, \tag{10}$$

$$\rho_{\mathrm{I}}^{(1,-,\omega)}(x) = \rho_{\mathrm{I}}(x)L_\omega(x)\frac{1}{P^-_\omega}\,, \tag{11}$$

where $\rho_{\mathrm{I}}(x)$ is the initial Gaussian distribution of nuclear spin eigenvalues (see (6)) and the probabilities $P^\pm$ for measuring $|\pm\rangle$ are given by

$$P^+_\omega = \int_{-\infty}^{\infty} \mathrm{d}x \rho_{\mathrm{I}}(x)(1 - L_\omega(x))\,, \tag{12}$$

$$P^-_\omega = \int_{-\infty}^{\infty} \mathrm{d}x \rho_{\mathrm{I}}(x)L_\omega(x)\,, \tag{13}$$

with

$$L_\omega(x) = \frac{1}{2}\frac{(j/4)^2}{(x - \frac{\omega}{2})^2 + (j/4)^2}\,. \tag{14}$$

To obtain this result we have assumed that the measurement is performed with low time resolution[1] $\Delta t \gg 1/j$ and that the parameters satisfy the requirements given in (15) below. The distribution of nuclear spin eigenvalues after the measurement depends on the result of the measurement

[1] This assumption is not necessary for our narrowing scheme. However, it does allow for the derivation of the analytical formulas in this section, which give insight into the mechanism of narrowing.

(whether $|+\rangle$ or $|-\rangle$ was measured) and on the driving frequency ω of the oscillating exchange interaction $J(t)$. In the case where the measurement outcome is $|+\rangle$, the initial distribution $\rho_{\mathrm{I}}(x)$ is multiplied by $1 - L_\omega(x)$ which causes a dip in $\rho_{\mathrm{I}}(x)$ at $x = \omega/2$. However, in the case where the result of the measurement was $|-\rangle$, the initial distribution $\rho_{\mathrm{I}}(x)$ is multiplied by $L(x)$. The full-width at half-maximum (FWHM) of $L_\omega(x)$ is $j/2$, i.e., half the amplitude of the applied oscillation exchange interaction $J(t)$. Thus, choosing $j < \sigma_0$, $\rho_{\mathrm{I}}^{(1,-,\omega)}(x)$ is dominated by the Lorentzian and therefore the FWHM of the initial nuclear spin distribution has been narrowed by a factor $\approx j/4\sigma_0$. The probability P_ω^- to measure $|-\rangle$ in the regime $j \ll \sigma_0$ is given by $P_\omega^- \approx (j/6\sigma_0)\exp((x_0 - \omega/2)^2/2\sigma_0^2)$ and the nuclear spin distribution after measuring $|-\rangle$ is centered around $\omega/2$. Thus, through such a measurement it is possible to choose the center of the nuclear spin distribution by choosing the driving frequency. However, the larger the difference $x_0 - \omega/2$, the smaller is the probability to have measurement outcome $|-\rangle$, which leads to narrowing.

4.1 Experimental Recipe

An experimental implementation of this scheme of course requires the ability to initialize to the state $|+\rangle$ and to read-out the states $|\pm\rangle$. This has recently been achieved in an experiment by *Petta* et al. [10] using adiabatic passage from the $S^z = 0$ singlet. What needs to be achieved in addition is to apply an external magnetic field gradient δb^z between the two dots in order to satisfy the requirements on the parameters of the system:

$$J_0 \ll x_0, \;\; j \ll x_0, \;\; \sigma_0 \ll x_0, \;\; j < \sigma_0 \,. \tag{15}$$

Typical values for the parameters satisfying these requirements are: $1/\sigma_0 = 10\,\mathrm{ns}$, $1/j = 100\,\mathrm{ns}$, $\omega = 2x_0 = 10^9$–$10^{10}\,\mathrm{Hz}$.

The pulse-sequence for one measurement is shown in Fig. 2. The parameter ϵ describes the detuning between the singlet state with two electrons on the right dot and the singlet state with one electron on each dot: $\epsilon = E_{S(0,2)} - E_{S(1,1)}$. First the system is set to the $S(1,1)$ from the $S(0,2)$ state by going from large positive to negative detuning ϵ (such that still $J \gg |x_0|$) as described in [10] (rapid adiabatic passage through $S(1,1) - T_+$ resonance). In the limit of $J \ll |x_0|$ and $x_0 > 0$, the ground state is $|+\rangle$ (for $x_0 < 0$, the ground state is $|-\rangle$ and $|\pm\rangle$ thus need to be interchanged in the following description) and initialization to $|+\rangle$ is thus possible by adiabatic passage from $S(1,1)$, i.e., by switching adiabatically to large negative detuning (such that $J \ll x_0$). Then the oscillating signal is applied to $J(t)$ for a time t_{m}. Finally we adiabatically switch back to $J \gg x_0$. With this the $|+\rangle$ state goes to the singlet $S(1,1)$, and the $|-\rangle$ state goes to the $S^z = 0$ triplet $T_0(1,1)$. Read-out of the singlet and triplet may then be achieved via switching to large positive detuning: the $S(1,1)$ state goes over to the $S(0,2)$

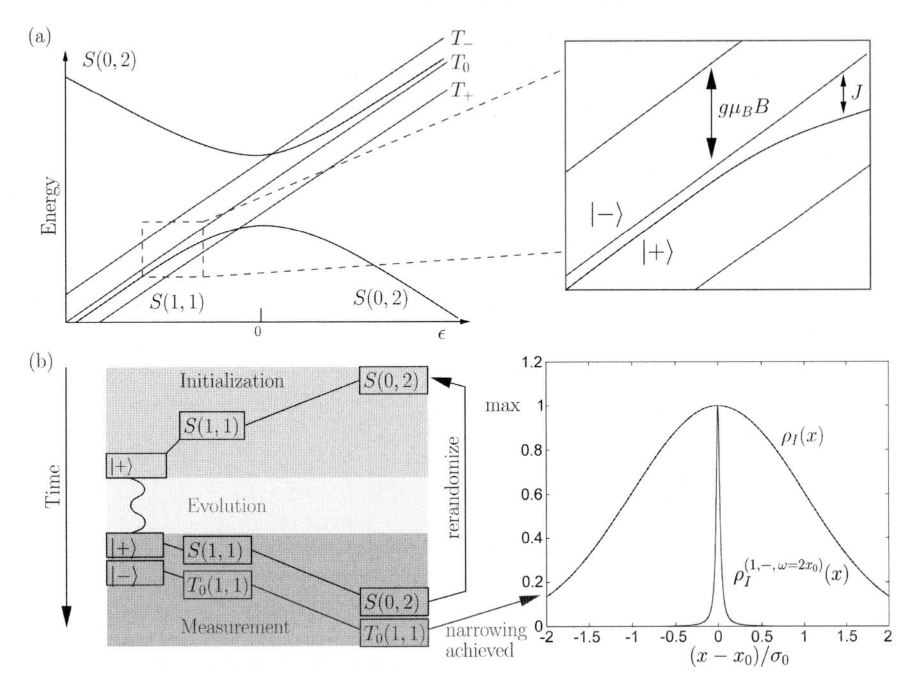

Fig. 2. In this figure the pulse-sequence for one measurement in the basis $|\pm\rangle$ is explained. (**a**) The level diagram for the two-electron spin states (sweeping $\epsilon = E_{S(1,1)} - E_{S(0,2)}$ with $E_{S(1,1)} + E_{S(0,2)}$ held constant). *Inset*: The splitting between the $S(1,1)$ and T_0 state is given by the exchange interaction J between the two dots. For $J \to 0$ the $|\pm\rangle$ states become eigenstates. (**b**) The change of the detuning ϵ during the course of the measurement is sketched: the position of the boxes corresponds to the value of ϵ. After applying the oscillating signal the system is in either one of $|\pm\rangle$, which results in a different state when switching back to positive detuning. If the system ends up in the T_0 state (which corresponds to measurement result $|-\rangle$) narrowing has been achieved, otherwise the nuclear system must be rerandomized and the measurement repeated

while the $T_0(1,1)$ does not since tunneling preserves spin. The number of electrons on the right dot can then be detected via a charge sensor (QPC). If the outcome of the measurement is $|-\rangle$, we have achieved narrowing. In the case where we have measured $|+\rangle$, the nuclear spin distribution is rerandomized and the measurement is repeated.

4.2 Adaptive Scheme

If measurements at several different driving frequencies can be performed, a systematic narrowing of the distribution can be achieved by an adaptive scheme. Such an adaptive scheme is more intricate than the one described above, but allows one to narrow by more than a factor $j/4\sigma_0$.

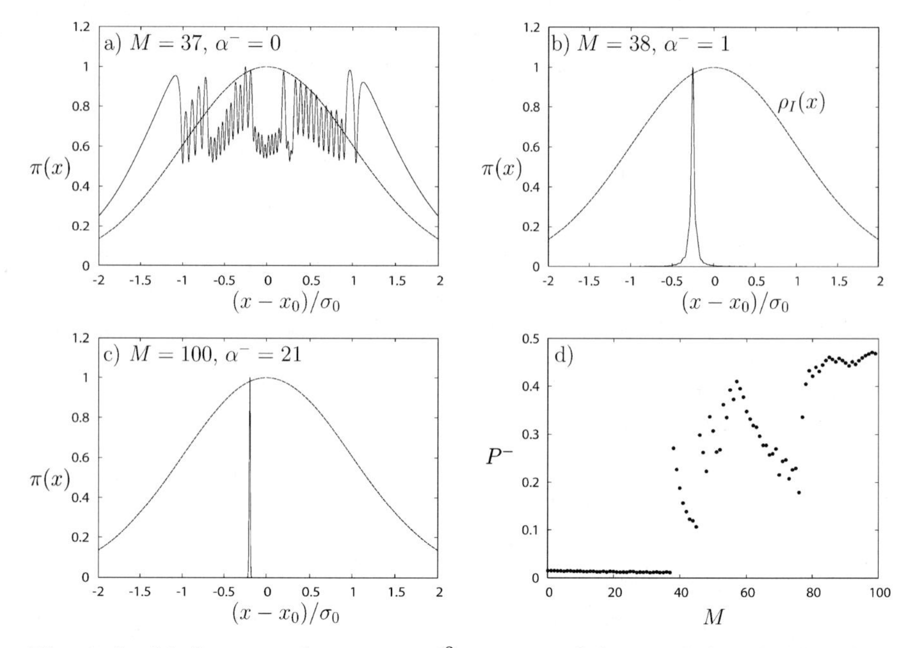

Fig. 3. In this figure we show a typical[3] sequence of the rescaled probability density of eigenvalues $\pi(x) = \rho_{\mathrm{I}}^{(M,\{\alpha_i^-\},\{\omega_i\})}(x)/\max\left(\rho_{\mathrm{I}}^{(M,\{\alpha_i^-\},\{\omega_i\})}(x)\right)$ for the adaptive scheme. Here, $\rho_{\mathrm{I}}^{(M,\{\alpha_i^-\},\{\omega_i\})}(x)$ is given in (16). We have $x = \delta h_n^z + \delta b^z$, $j/\sigma_0 = 1/10$, $\alpha^- = \sum_{i=1}^{M} \alpha_i^-$, and in (a)–(c) the initial Gaussian distribution (with FWHM $2\sigma_0\sqrt{2\ln 2} \approx 2\sigma_0$) is plotted for reference. (**a**) Up to $M = 37$ measurements the outcome is never $|-\rangle$ and thus each measurement "burns a hole" into the distribution where it previously had its maximum. (**b**) In the 38th measurement the outcome is $|-\rangle$ which leads to a narrowed distribution of nuclear spin eigenvalues (peak centered at ≈ -0.25) with a FWHM that is reduced by a factor $\approx j/4\sigma_0$. (**c**) Adapting the driving frequency ω to this peak, i.e., setting $\omega/2 = x_{\max}$ in subsequent measurements, leads to further narrowing every time $|-\rangle$ is measured. In this example the final FWHM is $\approx \sigma_0/100$, i.e., the distribution has been narrowed by a factor $\approx j/10\sigma_0$. (**d**) The probability P^- to measure $|-\rangle$ jumps up after the 38th measurement and after $|-\rangle$ is measured several more times, this probability saturates close to 1/2

[4]

The results of (10) and (11) may be generalized to the case of M subsequent measurements at different driving frequencies ω_i under the assumption, that the nuclear spin system is static between subsequent measurements:

$$\rho_{\mathrm{I}}^{(M,\{\alpha_i^-\},\{\omega_i\})}(x) = \frac{\rho_{\mathrm{I}}(x)}{Q(\{\alpha_i^-\},\{\omega_i\})} \prod_{i=1}^{M} L_{\omega_i}^{\alpha_i^-} (1 - L_{\omega_i})^{1-\alpha_i^-} , \tag{16}$$

[4] We have performed more than 60 runs of the simulation, varying M and j/σ_0.

where $Q(\{\alpha_i^-\}, \{\omega_i\})$ is the normalization factor, $\alpha_i^- = 1$ for measurement outcome $|-\rangle$ and $\alpha_i^- = 0$ for measurement outcome $|+\rangle$ in the i-th measurement with driving frequency ω_i. Further, $\{\omega_i\} = \{\omega_1, \ldots, \omega_M\}$ and $\{\alpha_i^-\} = \{\alpha_1^-, \ldots, \alpha_M^-\}$. As we have seen in the case of just one measurement, it is the measurement outcome $|-\rangle$ that leads to narrowing. Thus, before each measurement ω_i is chosen to maximize the probability $P_{\omega_i}^-$ to measure $|-\rangle$. The reason that ω_i must be adapted and that one should not keep measuring at the same driving frequency is that the measurement outcome $|+\rangle$ causes a dip in $\rho_{\mathrm{I}}(x)$ at the position where $L_{\omega_i}(x)$ has its peak and since $P_{\omega_i}^-$ is the overlap of $\rho_{\mathrm{I}}(x)$ and $L_{\omega_i}(x)$, this causes $P_{\omega_i}^-$ to diminish with each measurement.

To see what is a typical measurement history for such an adaptive scheme we have performed simulations. The results for a typical[5] sequence of measurements is shown in Fig. 3 (for another sequence see figure 2 in [35]).

5 Conclusions

We have reviewed our scheme that uses pseudospin measurements in the $S^z = 0$ subspace of two electron spin states in a double quantum dot to narrow the distribution of difference in nuclear polarization between the two dots. A successful experimental implementation of this scheme would allow to suppress hyperfine-induced decoherence and thus to reach the coherence times required for efficient quantum error correction.

Acknowledgements

We thank G. Burkard, M. Duckheim, J. Lehmann, F. H. L. Koppens, D. Stepanenko and, in particular, A. Yacoby for useful discussions. We acknowledge financial support from the Swiss NSF, the NCCR nanoscience, EU NoE MAGMANet, DARPA, ARO, ONR, JST ICORP, and NSERC of Canada.

References

[1] D. D. Awschalom, D. Loss, N. Samarth, *Semiconductor Spintronics and Quantum Computing* (Springer, Berlin, 2002).
[2] M. N. Baibich, J. M. Broto, A. Fert, F. Nguyen van Dau, F. Petroff, P. Etienne, G. Creuzet, A. Friederich, J. Chazelas, Phys. Rev. Lett. **61**, 2472 (1988).
[3] S. Datta and B. Das, Appl. Phys. Lett. **56**, 665 (1990).
[4] D. Loss and D. P. DiVincenzo, Phys. Rev. A **57**, 120 (1998).
[5] J. Preskill, http://theory.caltech.edu/people/preskill/ph229/ .

[5] We have performed more than 60 runs of the simulation, varying M and j/σ_0.

[6] A. Barenco, C. H. Bennett, R. Cleve, D. P. DiVincenzo, N. Margolus, P. Shor, T. Sleator, J. A. Smolin, H. Weinfurter, Phys. Rev. A **52**, 3457 (1995).
[7] D. P. DiVincenzo, D. Bacon, J. Kempe, G. Burkard, K. B. Whaley, Nature **408**, 339 (2000).
[8] L.-A. Wu, D. A. Lidar, M. Friesen, Phys. Rev. Lett. **93**, 030501 (2004).
[9] G. Medeiros-Ribeiro, E. Ribeiro, H. Westfahl, Appl. Phys. A-Mater. **77**, 725 (2003).
[10] J. R. Petta, A. C. Johnson, J. M. Taylor, E. A. Laird, A. Yacoby, M. D. Lukin, C. M. Marcus, M. P. Hanson, A. C. Gossard, Science **309**, 2180 (2005).
[11] H.-A. Engel and D. Loss, Science **309**, 586 (2005).
[12] C. W. Beenakker, D. P. Diqvincenzo, C. Emary, M. Kindermann, Phys. Rev. Lett. **93**, 020501 (2004).
[13] V. N. Golovach, A. V. Khaetskii, D. Loss, Phys. Rev. Lett. **93**, 016601 (2004).
[14] J. M. Elzerman, R. Hanson, L. H. Willems van Beveren, B. Witkamp, L. M. K. Vandersypen, L. P. Kouwenhoven, Nature **430**, 431 (2004).
[15] M. Kroutvar, Y. Ducommun, D. Heiss, M. Bichler, D. Schuh, G. Abstreiter, J. J. Finley, Nature **432**, 81 (2004).
[16] J. Schliemann, A. Khaetskii, D. Loss, J. Phys.: Condens. Matter **15**, 1809 (2003).
[17] G. Burkard, D. Loss, D. P. DiVincenzo, Phys. Rev. B **59**, 2070 (1999).
[18] S. I. Erlingsson, Y. V. Nazarov, V. I. Fal'ko, Phys. Rev. B **64**, 195306 (2001).
[19] S. I. Erlingsson and Y. V. Nazarov, Phys. Rev. B **66**, 155327 (2002).
[20] A. V. Khaetskii, D. Loss, L. Glazman, Phys. Rev. Lett. **88**, 186802 (2002).
[21] I. A. Merkulov, A. L. Efros, M. Rosen, Phys. Rev. B **65**, 205309 (2002).
[22] A. Khaetskii, D. Loss, L. Glazman, Phys. Rev. B **67**, 195329 (2003).
[23] W. A. Coish and D. Loss, Phys. Rev. B **70**, 195340 (2004).
[24] W. A. Coish and D. Loss, Phys. Rev. B **72**, 125337 (2005).
[25] A. S. Bracker, E. A. Stinaff, D. Gammon, M. E. Ware, J. G. Tischler, A. Shabaev, A. L. Efros, D. Park, D. Gershoni, V. L. Korenev, I. A. Merkulov, Phys. Rev. Lett. **94**, 047402 (2005).
[26] M. V. Dutt, J. Cheng, B. Li, X. Xu, X. Li, P. R. Berman, D. G. Steel, A. S. Bracker, D. Gammon, S. E. Economou, R.-B. Liu, L. J. Sham, Phys. Rev. Lett. **94**, 227403 (2005).
[27] F. H. L. Koppens, J. A. Folk, J. M. Elzerman, R. Hanson, L. H. W. van Beveren, I. T. Vink, H. P. Tranitz, W. Wegscheider, L. P. Kouwenhoven, L. M. K. Vandersypen, Science **309**, 1346 (2005).
[28] D. Paget, G. Lampel, B. Sapoval, V. I. Safarov, Phys. Rev. B **15**, 5780 (1977).
[29] R. de Sousa and S. Das Sarma, Phys. Rev. B **67**, 033301 (2003).
[30] N. Shenvi, R. de Sousa, K. B. Whaley, Phys. Rev. B **71**, 224411 (2005).
[31] W. Yao, R.-B. Liu, L. J. Sham, arXiv:cond-mat/0508441 (2005).
[32] A. M. Steane, Phys. Rev. A **68**, 042322 (2003).
[33] A. Imamoğlu, E. Knill, L. Tian, P. Zoller, Phys. Rev. Lett. **91**, 017402 (2003).
[34] D. Paget, Phys. Rev. B **25**, 4444 (1982).
[35] D. Klauser, W. A. Coish, D. Loss, arXiv:cond-mat/0510177 (Phys. Rev. B in press) (2005).
[36] G. Giedke, J. M. Taylor, D. D'Alessandro, M. D. Lukin, A. Imamoğlu, arXiv:quant-ph/0508144 (2005).
[37] D. Stepanenko, G. Burkard, G. Giedke, A. Imamoğlu, Phys. Rev. Lett. **96**, 136401 (2006).

Counting Statistics of Single Electron Transport in a Semiconductor Quantum Dot

S. Gustavsson[1], R. Leturcq[1], B. Simovič[1], R. Schleser[1], T. Ihn[1], P. Studerus[1], K. Ensslin[1], D. C. Driscoll[2], and A. C. Gossard[2]

[1] Solid State Physics Laboratory, ETH Zürich, 8093 Zürich, Switzerland

[2] Materials Departement, University of California, 93106 Santa Barbara, USA

Abstract. By using a quantum point contact as a charge detector, we show the measurement of current fluctuations in a semiconductor quantum dot by counting electrons tunneling through the system one by one. This method gives direct access to the full counting statistics of current fluctuations. In the sequential tunneling regime, we show the suppression of the noise compared to its classical Poissonian value, which is expected due to Coulomb blockade.

1 Introduction

In addition to the mean of the current, current fluctuations are very important in order to understand the transport mechanisms in a conductor [1]. In particular, they provide information on the involved charge. Many experiments have been concerned with measuring the shot noise, which is the variance of the current fluctuations. Not only the variance can be of interest, but also higher moments of current fluctuations could provide new information on the system, as it is widely used in quantum optics for probing photon entanglement [2]. For electronic systems, the third moment is of particular interest since it is not affected by the thermal noise, and could be used to determine the nature of the charge transport at high temperature [3, 4].

For independent particles, current fluctuations are expected to follow a Poissonian distribution. In the case of a quantum dot (QD) in the sequential tunneling regime, the noise is suppressed compared to the Poissonian distribution due to correlations between the electrons tunneling through the QD [5]: because of Coulomb blockade, an electron occupying the QD blocks the transport of the next electron. This suppression is maximum when the QD is symmetrically coupled to the leads, but vanishes for asymmetrically coupled QDs since the transport is limited by the weakly coupled contact. Few experiments on vertical quantum dots could measure a suppression of the noise [6–8], but measurements on lateral quantum dots are difficult due the very low current level involved.

An alternative way of measuring current fluctuations is to detect directly the charges traveling through a conductor. This method has been suggested

R. Haug (Ed.): Advances in Solid State Physics,
Adv. in Solid State Phys. **46**, 27–39 (2008)

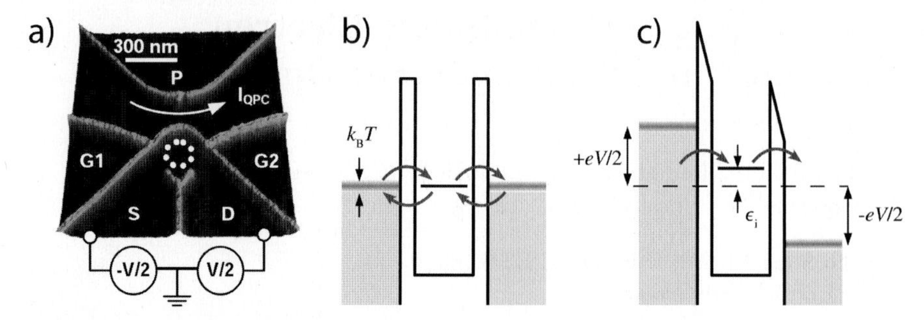

Fig. 1. (**a**) AFM micrograph of the oxide lines defining the nanostructure. The QD (*circle*) is connected to two leads S and D, and a nearby QPC is electrostatically coupled to the QD. Voltages applied on the lateral gates G1, G2 and P allow to tune respectively the coupling to source (S) and drain (D), and the conductance through the QPC. A bias voltage V is applied between S and D, and the conductance through the QPC is measured by applying a constant dc voltage and measuring the current I_{QPC}. (**b**) Energy diagram of the QD connected to the leads in the case where the level in the QD is aligned with the chemical potential in the leads, leading to equilibrium charge fluctuations in the QD. (**c**) Energy diagram of the QD in the case $eV/2 - \epsilon_i \gg k_{\mathrm{B}}T$, for which electrons tunnel into the QD from the source and tunnel out of the dot through the drain

by theories known as full counting statistics [9], and has been used since then as a theoretical tool to calculate current fluctuations in conductors. However, first attempts to measure the current by counting electrons could not achieve enough resolution in order to study the statistics of current fluctuations [10–12]. By using a quantum point contact (QPC) as a charge detector, we show here the direct measurement of the full distribution of current fluctuations in a semiconductor quantum dot [13].

2 Experimental Methods

The sample shown in Fig. 1a has been realized by local oxidation of the surface of a GaAs/AlGaAs heterostructure using an atomic force microscope. The oxide line obtained by scanning the biased AFM tip on top of the surface depletes the two-dimensional electron gas situated 34 nm below the surface, and allows to create high quality nanostructures [14,15]. Our sample consists in a quantum dot (QD) connected to two leads, source (S) and drain (D), and a nearby quantum point contact (QPC) capacitively coupled to the QD. The lateral gates G1 and G2 are used to tune the coupling of the QD to the leads, while the gate P controls the conductance of the QPC. The measurements have been done in a ^{3}He/^{4}He dilution refrigerator. The electronic temperature, measured by the Coulomb peak width, is 230 mK.

The strong dependence of the conductance of a QPC on the neighboring electrostatic potential makes it a very sensitive electrometer. A QPC can

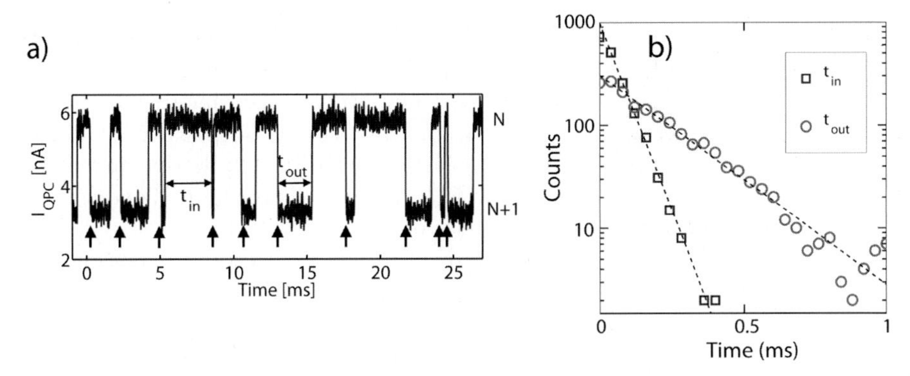

Fig. 2. (**a**) Typical time trace of the current measured through the QPC for a constant bias voltage applied on the QPC. The two levels correspond to zero (N) and one ($N+1$) excess electron in the QD. $\tau_{\rm in}$ is the time it takes for an electron to tunnel into the QD, and $\tau_{\rm out}$ the time it takes for an electron to tunnel out of the QD. (**b**) Probability density of the times $\tau_{\rm in}$ and $\tau_{\rm out}$. The *points* are experimental data, and the *lines* are fits with (13)

detect the charge state of a QD [16]. Time resolved detection of single electrons tunneling in and out of a QD has been used to detect spin states in a single [17] and double quantum dots [18], and to measure equilibrium charge fluctuations in a QD connected to a single lead [19, 20]. In these last experiments, the configuration was similar to the one depicted in Fig. 1b, in which thermal fluctuations induce hopping of electrons back and forth from one contact to the same contact (the second contact being pinched-off in these experiments). Here we propose to use the QPC to detect charge fluctuations in a QD connected symmetrically to two leads S and D. For a sufficiently large bias voltage applied between the two leads, the only possibility for an electron to tunnel through the QD is to come from the source contact, and to leave by the drain contact (see Fig. 1c). In this regime, the charge fluctuations measured in the QD are directly related to the fluctuations of the current through the QD, measured by detecting the tunneling of single electrons.

A typical time trace of the current measured through the QPC is shown in Fig. 2a. Fluctuations of charges in the QD are monitored by this current, which fluctuates between two states, corresponding to zero (upper states, N) and one (bottom state, $N+1$) excess electron in the QD. The trace of Fig. 2a shows the high signal-to-noise ratio in this set-up, allowing to measure single tunneling events with a short time resolution.

For a large bias voltage applied between S and D on the QD, a pair of one step downwards and one step upwards corresponds to one electron tunneling though the QD. The current through the QD can then be deduced from time traces similar to the one of Fig. 2a by counting the number of electron tunneling into the QD (corresponding to a step $N \to N+1$, see arrows in Fig. 2a, given that, due to Coulomb blockade, only one electron can enter the QD at a time. The same analysis on electrons tunneling out of the QD gives the same result. The current is simply given by the number of electrons tunneling through the QD in a given time interval. The bandwidth of the QPC circuit is 30 kHz, determined by the capacitance of the cables and the feedback resistor of the I–V converter. This bandwidth limits the current we can measure by counting electrons to 5 fA, while the lower limit is determined by the length of the time trace we take.

This way of counting electrons passing through a conductor to measure the current has been achieved in different configurations only very recently [10–12]. However, while these experiments were limited to measurement of the mean current, our measurement shows that we can measure not only the mean current, but also its fluctuations in time. In particular, we can deduce the current noise, which has been widely used to characterize mesoscopic systems [1]. But we can go even further, and measure the full distribution function of the current fluctuations.

3 Counting Statistics in the Sequential Tunneling Regime

The full counting statistics has been calculated for a QD in the sequential tunneling regime, and in this section we summarize the main results of [21] adapted to our way of counting events. The evolution of the occupancy of the QD, with 0 or 1 excess electron, can be describe by the rate equation:

$$\frac{\mathrm{d}}{\mathrm{d}t}\begin{pmatrix}0\\1\end{pmatrix} = \begin{pmatrix}-\Gamma_{\mathrm{in}} & \Gamma_{\mathrm{out}}\\ \Gamma_{\mathrm{in}} & -\Gamma_{\mathrm{out}}\end{pmatrix}\begin{pmatrix}0\\1\end{pmatrix}, \tag{1}$$

with

$$\Gamma_{\mathrm{in}} = g_{\mathrm{s}}\Gamma_{\mathrm{L}} f_{\mathrm{L}}(\epsilon_i) + g_{\mathrm{s}}\Gamma_{\mathrm{L}} f_{\mathrm{L}}(\epsilon_i) \tag{2}$$

$$\text{and} \quad \Gamma_{\mathrm{out}} = \Gamma_{\mathrm{L}}\left(1 - f_{\mathrm{L}}(\epsilon_i)\right) + \Gamma_{\mathrm{R}}\left(1 - f_{\mathrm{R}}(\epsilon_i)\right), \tag{3}$$

where f_{L} and f_{R} are the Fermi distributions in the left and right leads, ϵ_i is the energy of the level in the quantum dot, and g_{s} is the spin degeneracy of this level. These rates can be simplified in the case of large bias voltage, $|\pm eV/2 - \epsilon_i| \gg k_{\mathrm{B}}T$, for which the electron tunnels into the QD only through a single lead, the source (being either left or right lead, depending on the sign of the bias voltage), and tunnels out of the QD through the other contact, the drain:

$$\Gamma_{\mathrm{in}} = g_{\mathrm{s}}\Gamma_{\mathrm{L(R)}} = \Gamma_{\mathrm{s}} \qquad \text{and} \qquad \Gamma_{\mathrm{out}} = \Gamma_{\mathrm{R(L)}} = \Gamma_{\mathrm{D}}. \tag{4}$$

Here, Γ_{s} and Γ_{D} are effective tunneling rates which take into account possible spin degeneracy. Γ_{s} and Γ_{D} are assumed to be energy independent. In this limit, the model can be extended to the transport through multiple levels in the QD, the effective tunneling rates being the sum of the tunneling rates through individual levels.

To perform the counting statistics, we need to introduce a counting field $e^{i\chi}$ in the rate equation. In our case, we count electrons tunneling into the QD, and the matrix can be written:

$$M(\chi) = \begin{pmatrix} -\Gamma_{\text{in}} & \Gamma_{\text{out}} \\ \Gamma_{\text{in}} e^{i\chi} & -\Gamma_{\text{out}} \end{pmatrix} . \tag{5}$$

The distribution function of the number n of electrons tunneling through the quantum dot during a time t_0 can be calculated with the cumulant-generating function $S(\chi)$:

$$P(n) = \int_{-\pi}^{\pi} \frac{\mathrm{d}\chi}{2\pi} e^{-S(\chi)-n\chi} , \tag{6}$$

where $S(\chi)$ is given by the lowest eigenvalue of $M(\chi)$, $\lambda_0(\chi)$:

$$S(\chi) = -\lambda_0(\chi) t_0 = \frac{t_0}{2} \left[\Gamma_{\text{s}} + \Gamma_{\text{D}} - \sqrt{(\Gamma_{\text{s}} - \Gamma_{\text{D}})^2 + 4\Gamma_{\text{s}}\Gamma_{\text{D}} e^{-i\chi}} \right] . \tag{7}$$

From the distribution function $P(n)$, one can calculate all central moments characterizing the current fluctuations. The three first central moments μ_i, in which we are interested in the following, coincide with the cumulants C_i. They can then be deduced from the cumulant-generating function. The mean current is given by the mean, or the first cumulant C_1, of the distribution:

$$I = \frac{ie}{t_0} C_1 = \frac{ie}{t_0} \left(\frac{\mathrm{d}S}{\mathrm{d}\chi} \right)_{\chi=0} = -e \frac{\Gamma_{\text{s}}\Gamma_{\text{D}}}{\Gamma_{\text{s}} + \Gamma_{\text{D}}} . \tag{8}$$

The symmetrized shot noise is given by the variance, or the second cumulant C_2, of the distribution:

$$S_I = \frac{2e^2}{t_0} C_2 = \frac{2e^2}{t_0} \left(\frac{d^2 S}{d\chi^2} \right)_{\chi=0} , \tag{9}$$

... from which we can calculate the Fano factor:

$$F_2 = \frac{S_I}{2eI} = \frac{C_2}{iC_1} = \frac{\Gamma_{\text{s}}^2 + \Gamma_{\text{D}}^2}{(\Gamma_{\text{s}} + \Gamma_{\text{D}})^2} = \frac{1}{2} \left(1 + a^2 \right) , \tag{10}$$

where $a = (\Gamma_{\text{s}} - \Gamma_{\text{D}})/(\Gamma_{\text{s}} + \Gamma_{\text{D}})$ is the asymmetry of the coupling. This result recovers the earlier calculations for the shot noise in a quantum dot [5], and shows the reduction of the noise by a factor 1/2 for a QD symmetrically coupled to the leads, while the Poissonian limit, $F_2 = 1$, is reached for an asymmetrically coupled QD.

Finally we are also interested in the third central moment, or third cumulant C_3, of the fluctuations, which characterizes the asymmetry of the distribution (skewness):

$$S_I^3 = \frac{ie^3}{t_0} C_3 = \frac{ie^3}{t_0} \left(\frac{d^3 S}{d\chi^3} \right)_{\chi=0} , \tag{11}$$

which can also be characterized by its normalized value:

$$F_3 = \frac{C_3}{C_1} = \frac{\Gamma_\mathrm{s}^4 - 2\Gamma_\mathrm{s}^3\Gamma_\mathrm{D} + 6\Gamma_\mathrm{s}^2\Gamma_\mathrm{D}^2 - 2\Gamma_\mathrm{s}\Gamma_\mathrm{D}^3 + \Gamma_\mathrm{D}^4}{(\Gamma_\mathrm{s} + \Gamma_\mathrm{D})^4} = \frac{1}{4}\left(1 + 3a^4\right) . \tag{12}$$

This results shows the strong reduction of the third moment, by a factor 1/4, for a symmetrically coupled QD, and the Poissonian limit, $F_3 = 1$, for an asymmetrically coupled quantum dot.

4 Determination of the Individual Tunneling Rates

In order to make a quantitative comparison of the experimental results with the theory, it is important to determine the tunneling rates Γ_s and Γ_D, which are the only parameters of our model. In the time trace of Fig. 2a, the time τ_in represents the time an electron needs to tunnel into the QD, while the time τ_out is the time an electron needs to tunnel out of the QD. In the case where tunneling events are uncorrelated, the probability densities are expected to follow exponential functions:

$$P_{\tau_\mathrm{in}} \propto \Gamma_\mathrm{in} \exp(-\Gamma_\mathrm{in}\tau_\mathrm{in}) \qquad \text{and} \qquad P_{\tau_\mathrm{out}} \propto \Gamma_\mathrm{out} \exp(-\Gamma_\mathrm{out}\tau_\mathrm{out}) . \tag{13}$$

Figure 2b shows that these relations are well followed experimentally, proving that the tunneling events are uncorrelated. The tunneling rates Γ_in and Γ_out can be determined either by fitting the data in Fig. 2b with (13), or by taking the averages $\Gamma_\mathrm{in} = 1/\langle\tau_\mathrm{in}\rangle$ and $\Gamma_\mathrm{out} = 1/\langle\tau_\mathrm{out}\rangle$.

At large bias voltage, $| \pm eV/2 - \epsilon_i| \gg k_\mathrm{B}T$, the only possibility for an electron to tunnel in is to come from the source contact, and the only possibility to tunnel out is to go to the drain contact (see Fig. 1c). In this case, the tunneling rates from source and to drain can be directly calculated from the tunneling times considering that $\Gamma_\mathrm{s} = \Gamma_\mathrm{in}$ and $\Gamma_\mathrm{D} = \Gamma_\mathrm{out}$. As we emphasized in the previous section, this model can be extended to several levels.

5 Distribution Function of Current Fluctuations

In order to determine the distribution function of current fluctuations from the measurement, a time trace of length $T = 0.5\,\mathrm{s}$ is divided into intervalles of length t_0, during which we count the number n of electrons tunneling into the QD. The distribution function of the number of events in this time intervalle t_0 is shown in Fig. 3a, and characterizes directly the current fluctuations. In order to compare the experimental result with the theory presented in Sect. 3, the tunneling rates are first determined as explained in Sect. 4, and then included in (7) and (6) in order to determine the probability distribution $P_{t_0}(n)$. The result is shown as a plain line in Fig. 3a. The agreement with the experimental data is striking, knowing that no adjustable parameters are used.

From the time traces, we can also calculate the central moments given by $\mu = \langle n \rangle$ and $\mu_i = \langle n^i - \langle n \rangle^i \rangle$. For $i \leq 3$, the central moments are equal to the cumulants. For this reason we will use the same notation C_i for both, but experimentally only the central moments are calculated.

An important parameter for the analysis is the time t_0 in which we count the events. In Fig. 3b we show the values of the second and third central moments determined for several values of t_0. For small time $t_0 \ll 1/\Gamma_{\mathrm{S}}, 1/\Gamma_{\mathrm{D}}$, we expect to measure either 0 or 1 event, meaning that the resulting distribution will tend to a Bernoulli distribution, with a probability of measuring one event being:

$$p = \frac{\langle n_{\mathrm{total}} \rangle t_0}{T} = t_0 \frac{\Gamma_{\mathrm{S}} \Gamma_{\mathrm{D}}}{\Gamma_{\mathrm{S}} + \Gamma_{\mathrm{D}}} \, . \tag{14}$$

$\langle n_{\mathrm{total}} \rangle$ is the average total number of events expected within the time T. The central moments of this Bernoulli distribution are:

$$C_1 = p \quad , \quad C_2 = p(1-p) \quad \text{and} \quad C_3 = p(1-p)(1-2p) \, . \tag{15}$$

Since p tends to 0 when t_0 tends to zero, the normalized central moments $C_2/C_1 = (1-p)$ and $C_3/C_1 = (1-p)(1-2p)$ both tend to 1 when t_0 tends to zero in Fig. 3b. For this reason it is important to choose a time $t_0 > 1/\Gamma_{\mathrm{S}}, 1/\Gamma_{\mathrm{D}}$, corresponding to an average number of events measured during t_0 larger than one, $\langle n \rangle > 1$. For all analysis, $\langle n \rangle$ is kept close to 3.

6 Coulomb Diamonds Measured by Counting Electrons

By changing the gate voltage V_{G1} and the bias voltage V, we can map out the charge stability diagram of the QD (so called Coulomb diamonds). We have measured the distribution function for each point (V,V_{G1}), and the three first central moments are shown in Fig. 4. We point out that our method gives the current through the QD only under the condition $|\pm eV/2 - \epsilon_i| \gg k_{\mathrm{B}}T$, i.e., far from the conduction edges of the Coulomb diamonds. In particular, the enhancement of C_1 along the edges of the Coulomb diamonds is due to equilibrium fluctuation of charges between the QD and the leads (see Fig. 1b) and is not related to the current.

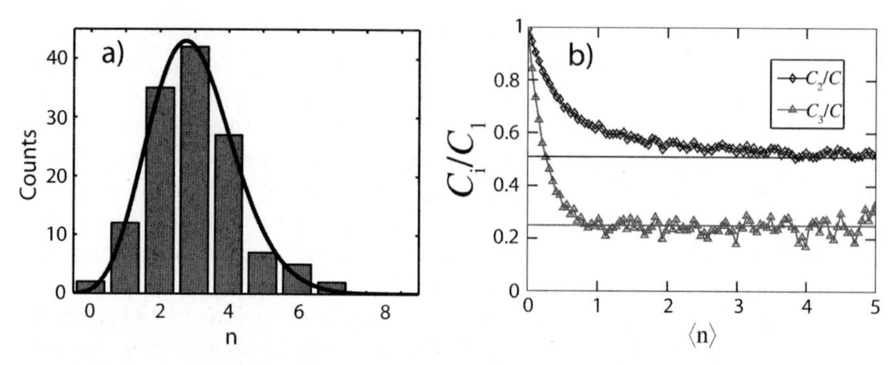

Fig. 3. (**a**) Distribution function of the number of electrons tunneling into the QD during a time t_0, obtained for $V_{G1} = -44\,\text{mV}$ and at large bias voltage. In this configuration, the tunneling rates are $\Gamma_s = 1748\,\text{Hz}$ and $\Gamma_D = 1449\,\text{Hz}$, determined as described in Sect. 4. The *red line* is calculated from (6) with the same tunneling rates. (**b**) Normalized central moments determined for different values of the time t_0, represented in the x-axis by the average number of event measured during the time t_0, $\langle n \rangle = t_0 \Gamma_s \Gamma_D / (\Gamma_s + \Gamma_D)$. The *horizontal lines* are the expected values

Figure 4a is very similar to conventional Coulomb diamonds measured by transport in quantum dots, and show well resolved excited states. However, the currents measured here are lower than 1 fA, and are well below what could be achieved with any conventional current measurement. The second central moment shows features that are very similar to the mean, and we have shown that excited states can also be resolved [13, 22], as it has been also observed for noise measurements in QDs formed in carbon nanotubes [23].

In most of the regions, the central moments correspond to a sub-Poissonian noise, with $C_2/C_1 < 1$ and $C_3/C_1 < 1$. This reduction of the noise is characteristic of a quantum dot in the Coulomb blockade regime, and is compatible with the model of sequential tunneling transport through a single level, or through multiple independent levels. This situation is however not true in the region encircled in Figs. 4b and 4c, corresponding to the chemical potential of one lead being aligned with the chemical potential in the dot ($eV/2 = \epsilon_i$). In this region, the noise is clearly super-Poissonian, with $C_2/C_1 > 1$ and $C_3/C_1 > 1$. This situation is of great interest since it corresponds to bunching of electrons, which is not expected for the transport of independent fermions. It is also not expected for interacting electrons in the model of sequential tunneling presented in Sect. 3, which can be extended to the situation where $eV/2 = \epsilon_i$. Using an extended model, we can show that this bunching is due to transport through two states with very different tunneling rates, given that the excited state has a long relaxation time [22].

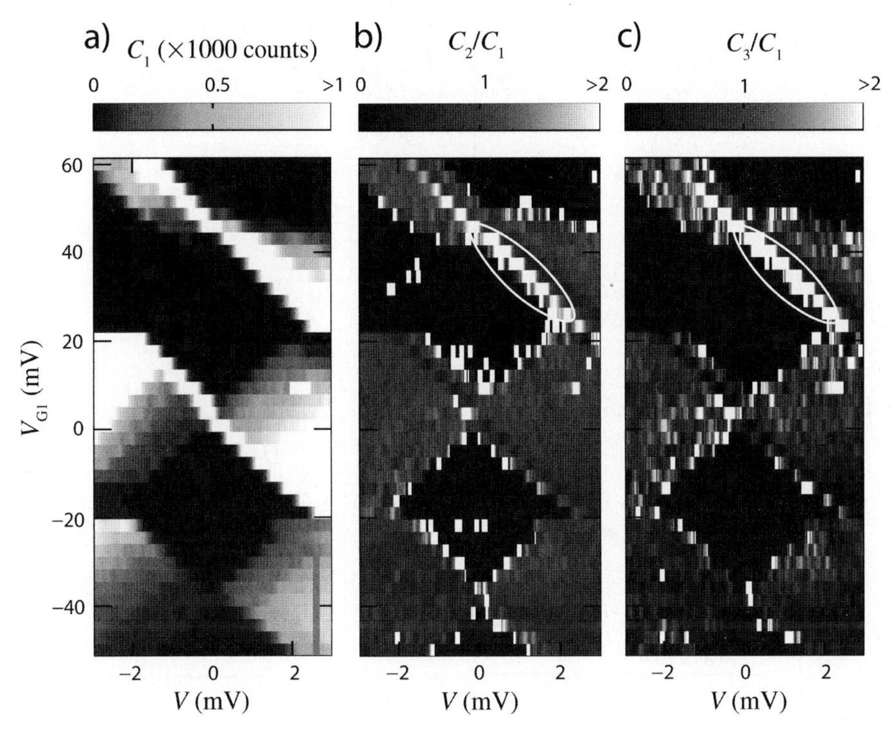

Fig. 4. Measurement of the charge stability diagram of the QD as a function of the voltage on gate G1 (used as plunger gate) and the bias voltage V. (**a**) First central moment, proportional to the mean current; (**b**) normalized second central moment, which is the Fano factor of the shot noise; (**c**) normalized third central moment. The *gray* (*green* online) *vertical bar* in (a) represents the range over which the analysis of Fig. 5 is done

7 Counting Statistics and Sub-Poissonian Noise

Equations (10) and (12) show that the second and third normalized cumulants depend on the tunneling rates only through the asymmetry $a = (\Gamma_{\mathrm{S}} - \Gamma_{\mathrm{D}})/(\Gamma_{\mathrm{S}} + \Gamma_{\mathrm{D}})$. To further check how the theory applies to our system, we have measured the distribution function for different values of the asymmetry a. Changing the asymmetry is achieved by changing the voltage on the gate G1: in addition to acting as a plunger gate and changing the number of electrons in the QD, as shown in Fig. 4, this gate also modifies the coupling of the source lead. This is shown in particular following the gray (green online) vertical bar in Fig. 4a, for which the asymmetry changes from −0.5 to 1, as shown in the inset of Fig. 5.

In Fig. 5, the second and third central moments are plotted as a function of V_{G1} and a. To increase the resolution, each data point correspond to an average over 50 time traces at a given gate voltage V_{G1} and in a bias window

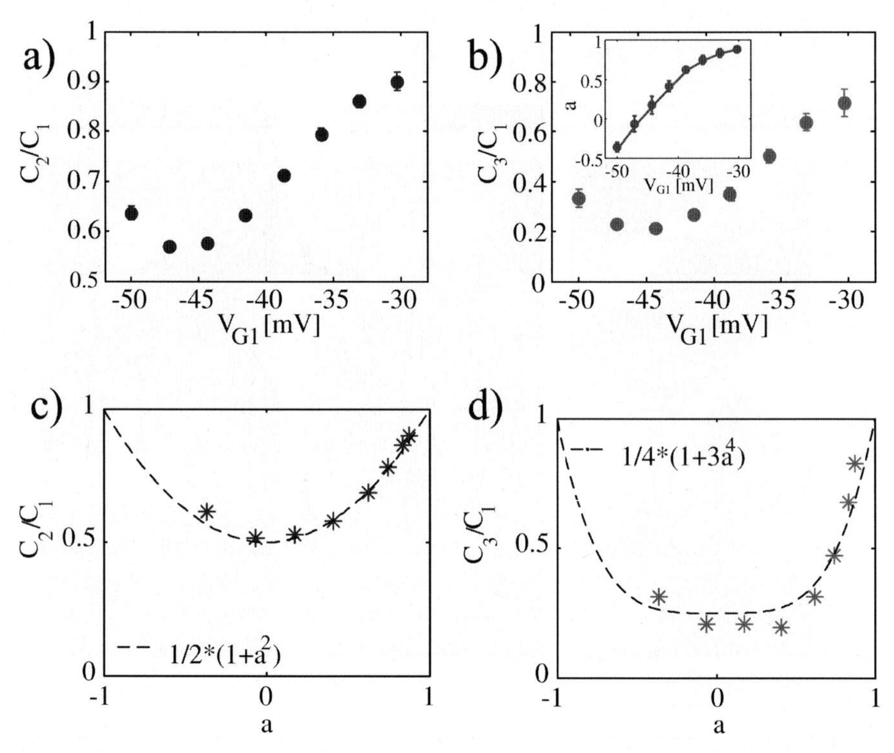

Fig. 5. (**a**,**b**) Second and third normalized central moments as a function of the gate voltage V_{G1}. (**c**,**d**) Same data as a function of the asymmetry of the tunneling rates. The points are experimental data, averaged over several traces in a bias voltage window of $1.5 < V < 3\,\mathrm{mV}$. The lines are the theoretical predictions given by (10) and (12). Inset: asymmetry of the tunneling rates $a = (\Gamma_{\mathrm{s}} - \Gamma_{\mathrm{D}})/(\Gamma_{\mathrm{s}} + \Gamma_{\mathrm{D}})$ vs. the gate voltage V_{G1}

$1.5 < V < 3\,\mathrm{mV}$, for which a does not change. We have plotted in Figs. 5c and 5d the theoretical predictions for the cumulants given by (10) and (12). Here again, the agreement between experiment and theory is very good, given that there is no adjustable parameter [13].

8 Equilibrium Charge Fluctuations

As pointed out before, the measurement of charge fluctuations in the QD are equivalent to current fluctuations only at large bias voltage, i.e., $| \pm eV/2 - \epsilon_i| \gg k_{\mathrm{B}}T$. This condition is shown in Fig. 6a presenting the central moments as a function of the bias voltage for a fixed gate voltage. While the equivalence between charge and current fluctuations is not valid at small bias voltage (gray regions in Fig. 6a), the theory of counting statistics presented in Sect. 3 can still be applied in this case. Taking into account the full expression

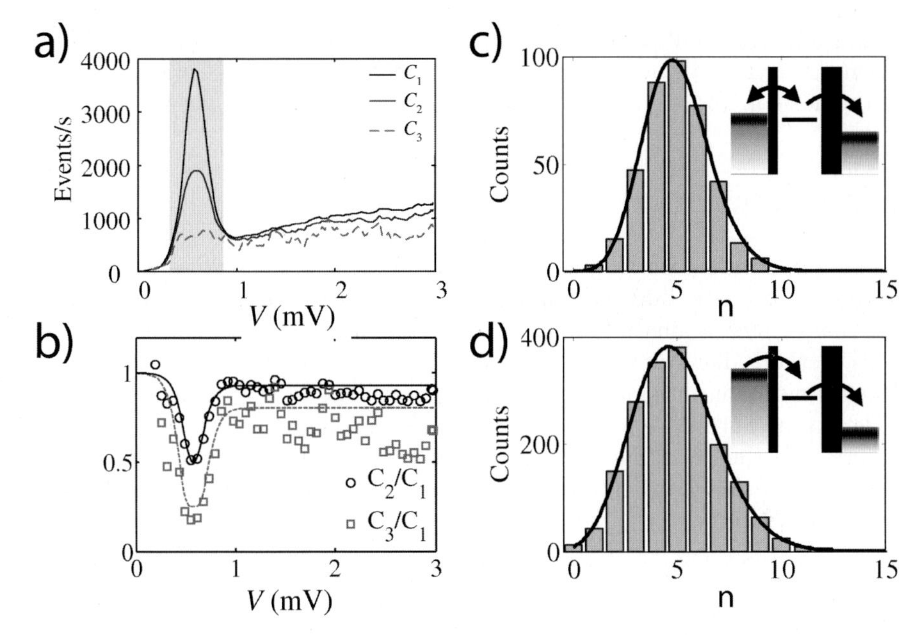

Fig. 6. (**a**) Central moments measured as a function of the bias voltage V for a fixed gate voltage $V_{\mathrm{G1}} = -2\,\mathrm{mV}$. The gray region represents the region where the condition $|\pm eV/2 - \epsilon_i| \gg k_{\mathrm{B}}T$ is not valid, and the charge fluctuations in the QD are not directly related to the current fluctuations through the QD. (**b**) Second and third normalized central moments in the same conditions. *Lines* are given by the theory of Sect. 3 extended to take into account equilibrium fluctuations. (**c**) Distribution function of the charge fluctuations at $V = 0.6\,\mathrm{mV}$, corresponding to the scheme in the *inset*. In this case, the chemical potential of the source lead is aligned with the chemical potential of the QD, and the equilibrium fluctuations dominate the charge fluctuations. The effective tunneling rates used for the fit are $\Gamma_{\mathrm{in}} = 7800\,\mathrm{Hz}$ and $\Gamma_{\mathrm{out}} = 7900\,\mathrm{Hz}$. (**d**) Distribution function at large bias voltage, resulting from an averaging of data at large bias voltage $2.5 < V < 3\,\mathrm{mV}$. The effective tunneling rates used for the fit are $\Gamma_{\mathrm{in}} = \Gamma_{\mathrm{s}} = 19500\,\mathrm{Hz}$ and $\Gamma_{\mathrm{out}} = \Gamma_{\mathrm{D}} = 1300\,\mathrm{Hz}$

for the tunneling rates in (2) and (3), including the Fermi distributions in the leads, we can calculate the distribution function of charge fluctuations in the QD (see Fig. 6b).

Qualitatively, the strong reduction of the second and third normalized moments at the conduction edge observed in Fig. 6b can be understood as follows. For $eV/2 \approx \epsilon_i$, the effective tunneling rates are functions of the value of the Fermi distribution at ϵ_i. When increasing the bias voltage, this value changes from 0 to 1, and the effective tunneling rate of the left lead (i.e., $\Gamma_{\mathrm{L}} f_{\mathrm{L}}(\epsilon_i)$ in (2)) changes from 0 to Γ_{L}. Considering the tunneling rate of the right lead constant in this region, and with the condition $\Gamma_{\mathrm{R}} \ll \Gamma_{\mathrm{L}}$, the asymmetry of the coupling will then change continuously from -1 to 0 and 1, giving this strong reduction of the normalized central moments to 1/2 and 1/4 when the asymmetry is close to zero. This effect is similar to the suppression of the noise in vertical quantum dots near the conduction edges [6–8].

9 Conclusion

Using a quantum point contact as a charge detector, we have measured the charge fluctuations in a quantum dot at large bias voltage. In this regime, the charge fluctuations are directly related to current fluctuations in the QD. This method allows to measure current and noise levels that could not be reached with conventional current measurements, and is of great interest in order to measure the shot noise in lateral semiconductor quantum dots. In addition to the mean current and the shot noise, this method gives direct access to the distribution function of current fluctuations, known as the full counting statistics.

References

[1] Y. M. Blanter, M. Büttiker, Phys. Rep. **336**, 1–166 (2000)
[2] A. Zeilinger, G. Weihs, T. Jennewein, M. Aspelmeyer, Nature (London) **433**, 230–238 (2005)
[3] B. Reulet, J. Senzier, D. E. Prober, Phys. Rev. Lett. **91**, 196601 (2003)
[4] Y. Bomze, G. Gershon, D. Shovkun, L. S. Levitov, M. Reznikov, Phys. Rev. Lett. **95**, 176601 (2005)
[5] S. Hershfield, J. H. Davies, P. Hyldgaard, C. J. Stanton, J. W. Wilkins, Phys. Rev. B **47**, 1967–1979 (1993)
[6] H. Birk, M. J. M. de Jong, C. Schönenberger, Phys. Rev. Lett. **75**, 1610–1613 (1995)
[7] A. Nauen, I. Hapke-Wurst, F. Hohls, U. Zeitler, R. J. Haug, K. Pierz, Phys. Rev. B **66**, 161303(R) (2002)
[8] A. Nauen, F. Hohls, N. Maire, K. Pierz, R. J. Haug, Phys. Rev. B **70**, 033305 (2004)
[9] L. S. Levitov, H. Lee, G. B. Lesovik, J. Math. Phys. **37**, 4845–4866 (1996)
[10] W. Lu, Z. Ji, L. Pfeiffer, K. W. West, A. J. Rimberg, Nature (London) **423**, 422–425 (2003)
[11] T. Fujisawa, T. Hayashi, Y. Hirayama, H. D. Cheong, Y. H. Jeong, Appl. Phys. Lett. **84**, 2343–2345 (2004)
[12] J. Bylander, T. Duty, P. Delsing, Nature (London) **434**, 361–364 (2005)
[13] S. Gustavsson, R. Leturcq, B. Simovič, R. Schleser, T. Ihn, P. Studerus, K. Ensslin, D. C. Driscoll, A. C. Gossard, Phys. Rev. Lett. **96**, 076605 (2006)
[14] R. Held, S. Lüscher, T. Heinzel, K. Ensslin, W. Wegscheider, Appl. Phys. Lett. **75**, 1134–1136 (1999)
[15] A. Fuhrer, A. Dorn, S. Lüscher, T. Heinzel, K. Ensslin, W. Wegscheider, M. Bichler, Superlattice Microst.. **31**, 19–42 (2002)

[16] M. Field, C. G. Smith, M. Pepper, D. A. Ritchie, J. E. F. Frost, G. A. C. Jones, D. G. Hasko, Phys. Rev. Lett. **70**, 1311–1314 (1993)
[17] J. M. Elzerman, R. Hanson, L. H. Willems van Beveren, B. Witkamp, L. M. K. Vandersypen, L. P. Kouwenhoven, Nature (London) **430**, 431–435 (2004)
[18] J. R. Petta, A. C. Johnson, J. M. Taylor, E. A. Laird, A. Yacoby, M. D. Lukin, C. M. Marcus, M. P. Hanson, A. C. Gossard, Science **309**, 2180–2184 (2005)
[19] R. Schleser, E. Ruh, T. Ihn, K. Ensslin, D. C. Driscoll, A. C. Gossard, Appl. Phys. Lett. **85**, 2005–2007 (2004)
[20] L. M. K. Vandersypen, J. M. Elzerman, R. N. Schouten, L. H. Willems van Beveren, R. Hanson, L. P. Kouwenhoven, Appl. Phys. Lett. **85**, 4394–4396 (2004)
[21] D. A. Bagrets, Y. V. Nazarov, Phys. Rev. B **67**, 085316 (2003)
[22] S. Gustavsson, R. Leturcq, B. Simovič, R. Schleser, T. Ihn, P. Studerus, K. Ensslin, D. C. Driscoll, A. C. Gossard (unpublished)
[23] E. Onac, F. Balestro, B. Trauzettel, C. F. J. Lodewijk, L. P. Kouwenhoven, Phys. Rev. Lett. **96**, 026803 (2006)

Size-Tunable Exchange Interaction in InAs/GaAs Quantum Dots

Udo W. Pohl, Andrei Schliwa, Robert Seguin, Sven Rodt,
Konstantin Pötschke, and Dieter Bimberg

Institut für Festkörperphysik, Technische Universität Berlin,
Hardenbergstr. 36, 10623 Berlin, Germany
pohl@physik.tu-berlin.de

Abstract. Single epitaxial quantum dots are promising candidates for the realization of quantum information schemes due to their atom-like electronic properties and the ease of integration into optoelectronic devices. Prerequisite for realistic applications is the ability to control the excitonic energies of the dot. A major step in this direction was recently reached by advanced self-organized quantum-dot growth, yielding ensembles of equally shaped InAs/GaAs dots with a multimodal size distribution. The well-defined sizes of spectrally well separated subensembles enable a direct correlation of structural and excitonic properties, representing an ideal model system to unravel the complex interplay of Coulomb interaction and the quantum dot's confining potential that depends on size, shape, and composition. In this paper we focus on the exciton-biexciton system with emphasis on the excitonic fine-structure splitting. Across the whole range of size variations within our multimodal quantum dot distribution a systematic trend from +520 μeV to −80 μeV is found for decreasing dot size. To identify the underlying effects calculations of the fine-structure splitting are performed. A systematic variation of the structural and piezoelectric properties of the modeled quantum dots excludes shape anisotropy and tags piezoelectricity as a key parameter controlling the fine-structure splitting in our quantum dots.

Introduction

Methods for secure data transmission are of particular importance in times of distributed information processing. Classic encryption schemes rely on undisclosed transmission of the crypto key, because eavesdropping may hardly be detected. This uncertainty is evaded by applying quantum information schemes for key distribution (e.g., [1]). Here the key may be transmitted via insecure channels, because eavesdropping will be revealed by statistical analysis based on quantum mechanics. The realization of such a cryptographic scheme relies on novel light sources, in particular single-photon emitters showing controllable polarized emission or polarization-entangled photon pairs. Utilizing bulky solid state or atomic lasers, first demonstrator devices have been realized by attenuated pulses of polarization-controlled photons and down-converted pulses for entangled photon pairs. Strongly improved performance and high wall-plug efficiency is expected when using quantum dots

R. Haug (Ed.): Advances in Solid State Physics,
Adv. in Solid State Phys. **46**, 41–54 (2008)

as sources of the single photons. Schemes were proposed to utilize pairs of polarization-entangled photons originating from the radiative decay of the biexciton → exciton → vacuum-state cascade [2]. However, in quantum dots the degeneracy of the intermediate exciton level was found to be lifted [3], thereby leading to solely classically correlated photon pairs. A measure that decides on the state of entanglement is the so-called fine-structure splitting of the bright exciton state. If it is small compared to the homogeneous line width of the transition entanglement is to be expected. The control over polarization and over the state of entanglement is gained via control over the fine-structure splitting of the bright exciton ground state. As will be demonstrated in this work, such control is feasible by tailoring the structural properties of the quantum dots.

1 Ground State of Excitons Confined in a Quantum Dot

The fine-structure splitting is caused by the exchange interaction, which couples the spins of electrons and holes, and their Zeeman interaction with internal and external magnetic fields [4, 5]. The Zeeman contribution is generally negligible in absence of an external magnetic field and will not be considered here. We focus on the exciton ground state in crystals with zincblende structure (like InAs and GaAs), which have T_d point group symmetry.

The exciton is made up of an electron in the first conduction band and a hole in the uppermost valence band; for simplification we consider only heavy-hole states, though our calculations presented in Sect. 3.2 include light-hole and split-off hole contributions. The ground state is then composed of angular-momentum projections $s_z = \pm 1/2$ and $j_z = \pm 3/2$. The resulting four exciton states therefore have momentum projections of M_z equal to ± 1 and ± 2. In absence of Coulomb interaction and neglecting band- and spin-coupling effects, the exciton ground state is hence fourfold degenerate due to four combinations of products $e_i h_j$ with $i, j \in \{1, 2\}$. The two-particle Hamiltonian can be expanded into a basis of antisymmetrized product wavefunctions

$$e_i(\mathbf{r}_1)\, h_j(\mathbf{r}_2) \rightarrow \frac{1}{\sqrt{2}} \left(e_i(\mathbf{r}_1)\, h_j(\mathbf{r}_2) - h_j(\mathbf{r}_1)\, e_i(\mathbf{r}_2) \right) \quad i, j \in \{1, 2\} , \tag{1}$$

leading to a representation within the four degenerate exciton states of M_z $(|1\rangle, |-1\rangle, |2\rangle, |-2\rangle)$ as depicted in Fig. 1. The degeneracy is lifted by the Coulomb interaction, which consists of direct and exchange terms and reads in this basis

$$\left(C \cdot \mathbf{1} + \begin{pmatrix} 0 & \Delta_\mathrm{B} & 0 & 0 \\ \Delta_\mathrm{B} & 0 & 0 & 0 \\ 0 & 0 & -K & \Delta_\mathrm{D} \\ 0 & 0 & \Delta_\mathrm{D} & -K \end{pmatrix} \right) \mathbf{u} = E_i \mathbf{u} \, . \tag{2}$$

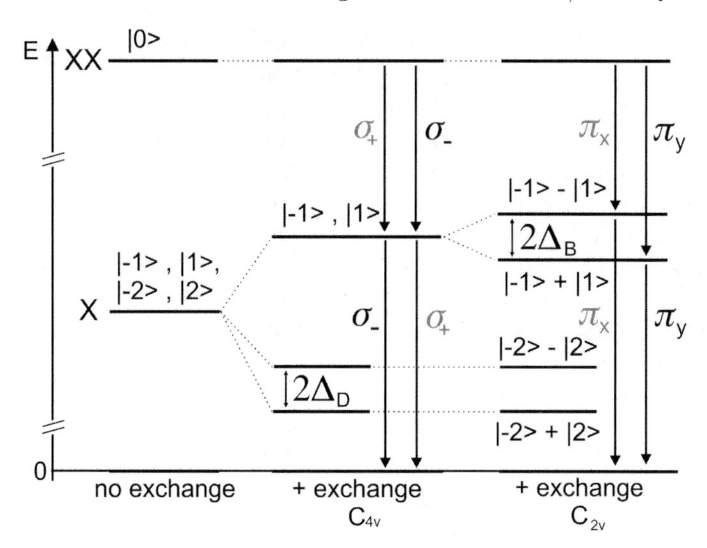

Fig. 1. Energy level scheme of exciton and biexciton states as a function of exchange interaction and symmetry of the confining potential. $2\Delta_D$ and $2\Delta_B$ mark the splittings of dark ($|2\pm\rangle$, D) and bright ($|1\pm\rangle$, B) states, respectively, $\sigma_\pm(\pi_{x/y})$ denote circular (linear) polarization

Here C is the direct Coulomb term $\langle e_i h_j | H_{\text{Coul}} | e_i h_j \rangle$ and K the diagonal exchange term $\langle e_i h_j | H_{\text{Coul}} | h_j e_i \rangle$. Solving this equation results in the eigenvectors and eigenstates

$$(E_1, E_2, E_3, E_4) = (\underbrace{C - K - \Delta_D, C - K + \Delta_D}_{D}, \underbrace{C - \Delta_B, C + \Delta_B}_{B}). \qquad (3)$$

The states labeled D comprise the states with $|M_z| = 2$. Since these states do not couple to photon fields and are usually not intermixed with the $|M_z| = 1$ states in absence of magnetic fields, they generally are not observed in luminescence experiments and hence denoted dark states. They are split by $E_2 - E_1 = 2\Delta_D$. The dark–dark splitting is accompanied by the formation of symmetric and antisymmetric linear combinations $|2\rangle + |-2\rangle$ and $|2\rangle - |-2\rangle$ of the pure spin states $|\pm 2\rangle$. The B states $|1\rangle + |-1\rangle$ and $|1\rangle - |-1\rangle$ are referred to as bright states. The difference $E_4 - E_3 = 2\Delta_B$ is called bright splitting or fine-structure splitting. Δ_B is equal 0 and the wave functions are pure spin states $|\pm 1\rangle$, if the confining potential has a high symmetry (at least C_{4v}). This is not the case for an exciton confined in a quantum dot, even if the shape is that of a square-based pyramid like those considered in this work. Piezoelectric effects lead to a reduction of the symmetry to C_{2v} [6, 7]. Indeed, the experimentally observed splitting presented in Sect. 3.1 demonstrates that the actual dot symmetry is C_{2v}. Additional potential reasons for the symmetry lowering from C_{4v} to C_{2v} are discussed in Sect. 3.2. The symmetry lowering also causes the pure spin states in C_{4v} to mix, and the corresponding circularly polarized transitions become linearly polarized in C_{2v}. The two directions of polarization are perpendicular to each other and reflect the symmetry characteristics of the confining potential.

The biexciton state corresponds to $M_z = 0$ and is therefore not split by exchange interaction. Since the single exciton is the final state in the XX $\rightarrow$ X transition, this emission also reflects the fine-structure splitting and hence the mixing of the spin states of the bright exciton.

2 Dot Ensembles with Multimodal Size Distribution

To study the effect leading to the fine-structure splitting experimentally, quantum dots with well-defined structure are required. In case of self-organized In(Ga)As/GaAs quantum dots such an investigation is not trivial as the quantum dots will differ simultaneously in size, shape, and composition. To overcome this problem we have developed a novel growth regime that results in an ensemble of self-similar quantum dots, which can be size-selectively addressed by their ground-state recombination energy.

Minimization of the total elastic strain is the driving force for the self-organized formation of QDs using the widely applied Stranski-Krastanow (SK) growth mode [8]. The equilibrium shape of free-standing coherently strained QDs in zincblende materials was modeled to adopt either the form of a peaked or a truncated pyramid with flat top and bottom interfaces on (001)-oriented substrates [9, 10], in agreement with experimental findings [11–13]. For capped QDs a large variety of shapes and compositions is reported, due to the strong impact of the non-equilibrium overgrowth procedure. Using specific growth conditions, the flat top and bottom facets of InAs/GaAs QDs could be preserved due to kinetic effects [14]. The dots of such ensembles keep their shape even during evolution to larger sizes, and the discrete variation of size within the ensemble leads to a multimodal distribution. These dots consist of pure InAs with heights varying in steps of complete InAs monolayers, and dots with a common height represent subensembles with small inhomogeneous broadening.

2.1 Kinetic Description of Multimodal Dot-Ensemble Formation

The formation of QDs with a multimodal size distribution is theoretically well described by a kinetic approach [15, 16]. The model outlined in Fig. 2 assumes strained QDs of truncated pyramidal shape being surrounded by an adatom sea, which represents the InAs wetting layer. Dot growth and dissolution occurs by adatom attachment and detachment showing a kinetics, which is largely controlled by the stress concentration at the base perimeter. The elastic energy density depicted in Fig. 2 creates a barrier for the nucleation at the side facets that increases with QD height. Growth hence proceeds essentially by adding new layers on the top facet, and Gibbs free energy has local

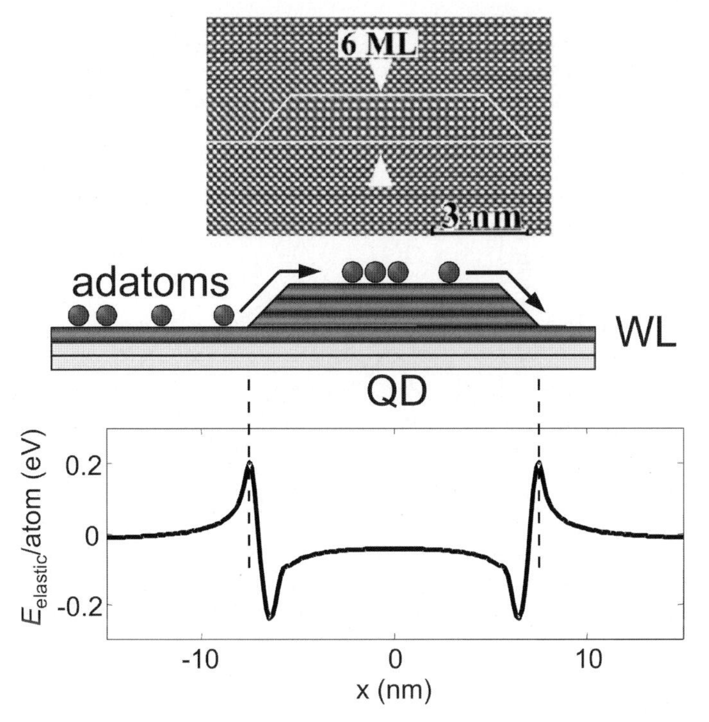

Fig. 2. *Top*: Fourier-filtered cross section transmission electron micrograph of an InAs/GaAs QD from an ensemble featuring a multi-modal size distribution. *Bottom*: Model InAs QD on 1 ML wetting layer surrounded by an adatom sea, and elastic energy density of the QD

minima each time a top layer is completed [14]. The interplay between the chemical potential of the adatom sea and Gibbs free energy of the QDs leads to a continuous shift of the optimum QD height towards larger values during a growth interruption. The material for QD growth is provided from dissolution of smaller QDs, leaving the thickness of the wetting layer unchanged as proved experimentally [14].

For numerical modeling, adatom exchange between sea and dot is described by rate equations [15, 16]. An Arrhenius dependence in the probability of QD-height increase or decrease accounts for the strain-dependent barrier. Materials properties enter by the elastic-energy change occurring at nucleation of a new layer and the step energy of the nucleus; for calculation of InAs/GaAs dots the parameters were adopted from continuum elasticity and density functional theory, respectively. Results of the numerical solution at various stages of the evolution are given in Fig. 3a. The histograms show that the ensemble initially consists of 1 monolayer high dots and some minor contributions of higher dots. The distribution in base length is rather broad [16], thereby representing the situation of a rough layer also found

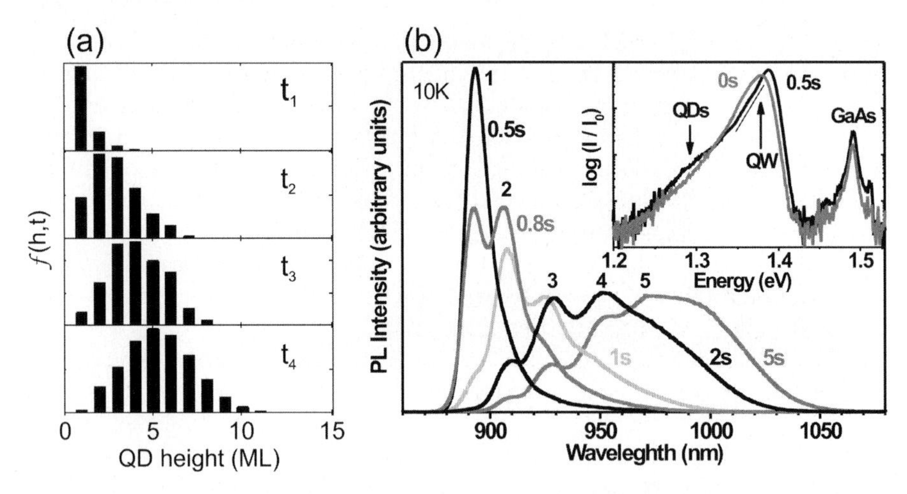

Fig. 3. (**a**) Normalized QD-height distribution $f(h,t)$ function for various stages t of the QD-ensemble evolution. (**b**) PL spectra of QD samples grown with varied growth interruptions given in seconds, numbers at the curves without unit denote QD height in monolayers assigned to maxima. *Inset*: Semilogarithmic PL spectra of samples grown without and with 0.5 s growth interruption; QW and GaAs denote emission from an InAs quantum well and the GaAs matrix, respectively

in the experiment described below. At later stages the maximum gradually shifts towards higher dots by dissolution of more shallow dots.

2.2 Epitaxy of Multimodal InAs/GaAs Quantum Dots

Multimodal QD ensemble formation was reported for InAs dots in both, GaAs [14, 17] and InP matrix [18, 19]. The dots studied here were obtained from InAs layers with a thickness near 1.8 monolayers, deposited on GaAs using metalorganic vapor phase epitaxy at typ. 490 °C. The thickness just exceeds the critical value for surface faceting in the SK growth mode. In situ studies demonstrated that a smooth growth front is maintained during strained layer deposition [20]. Consequently a rough pseudomorphic InAs/GaAs quantum well (QW) forms if a GaAs cap layer is deposited immediately after InAs growth [21, 22]. The corresponding PL given in the inset of Fig. 3b shows an exponential low-energy tail due to potential fluctuations from thickness variations. QDs form self-organized during a growth interruption (GRI) applied prior to cap-layer deposition. Fig. 3b shows that the PL of such QD ensembles exhibits a pronounced modulation due to the decomposition into subensembles. Furthermore a red shift of the PL is observed for prolonged growth interruption due to an increase of the average QD size, which is accompanied by a decrease of the QD areal density [14].

The effect of the GRI on the PL shows how the multimodal dot ensemble forms from the initial rough QW. A short interruption (0.5 s) leads to a blue shift of the QW-related PL and a weak broad emission around 1.3 eV, originating from thinning of the QW and enhanced appearance of locally thicker QW regions, respectively. The blue shift saturates for longer interruptions ($\geq$ 0.8 s), and the emission of a multimodal QD ensemble with individual subensemble peaks evolves from the shallow localizations in the density-of-states tail of the rough QW. The material of the initial InAs deposition hence partially concentrates at some QD precursors located on a wetting layer of constant thickness. The existence of such a constant, 1 monolayer thick wetting layer consisting of pure InAs was proved by PL excitation studies and confirmed by TEM investigations [14].

The multimodal nature of the ensemble size-distribution primarily concluded from optical spectra was confirmed by transmission-electron micrographs and cross-sectional scanning tunneling images [23]. Analysis of such images (e.g., Fig. 2) showed that both, the base and the top layers of the QDs are flat and that the dots consist always of plain, continuous InAs. QDs hence differ in height by integer numbers of InAs monolayers. The side facets are steep, giving the QDs the shape of truncated pyramids. Since the wetting layer of one monolayer InAs remains unchanged during the evolution of the ensemble, QD ripening occurs by dissolution of smaller QDs. This is in line with a small blue shift of subensemble PL peaks, indicating lateral shrinking [14, 21].

3 Size Dependence of the Fine-Structure Splitting

3.1 Single-Quantum-Dot Spectra

To exclude influences of inhomogeneous broadening due to remaining lateral size variations of QDs within a subensemble, we measure single QDs using spatially resolved cathodoluminescence. An optically opaque nearfield Au shadow mask is evaporated onto the sample surface to reduce the number of simultaneously probed QDs [24]. For $\sim$ 100 nm apertures and typically 4×10^{10} dots/cm^2 about three QDs are detected simultaneously. Further selection is achieved spectrally, and an unambiguous assignment of spectral lines to a single specific QD is provided by exploiting the effect of spectral diffusion [25]. To identify the corresponding electronic transitions, excitation- and polarization-dependent measurements are performed. The single-excitonic and biexcitonic emissions consist of a cross-polarized doublet structure. The former shows a linear increase of intensity as a function of excitation density and the latter a quadratic one.

Two polarized spectra showing exciton and biexciton emissions are displayed in Fig. 4. In both spectra the doublet structure of the emission lines and the reversed order of polarization is clearly revealed. Most interestingly the energetic order of the polarization direction within the excitonic and the

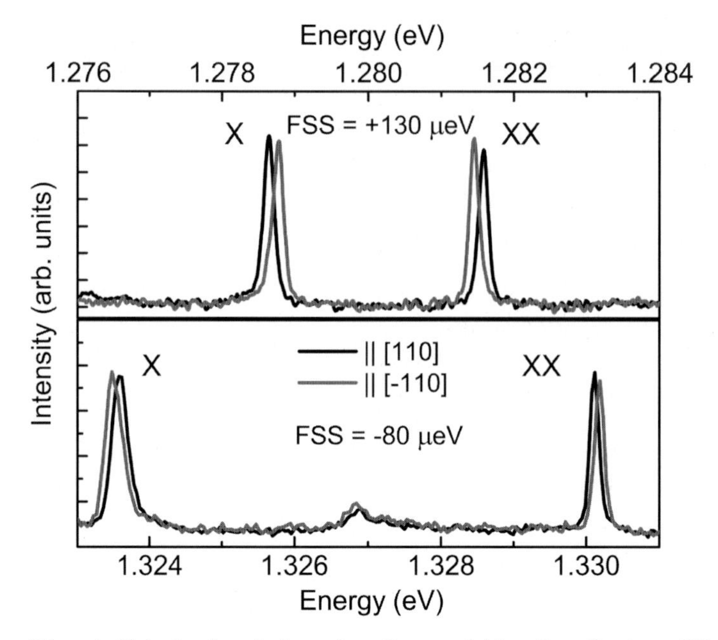

Fig. 4. Polarized emission of exciton and biexciton for two different single QDs. *Black* and *grey* spectra refer to linear polarization along the [110] and $[\bar{1}10]$ direction, respectively. Note, that the order of polarization changes between the two plots

biexcitonic doublet is not fixed. In the upper spectrum the exciton line polarized along [110] appears at lower energy, while in the lower spectrum it is at higher energy. For quantifying this, the fine-structure splitting is defined to be positive when the [110] polarization appears at lower energy in the exciton doublet and negative if vice versa. Accordingly positive and negative fine-structure splittings can be observed in Fig.4.

Figure 5 compiles fine-structure splittings as a function of the single-exciton recombination energy – or the quantum dot size – respectively. The exciton fine-structure splitting increases from small negative values for small QDs to positive values as large as 520 μeV for large QDs within the multimodal dot ensemble, as indicated by bars on top. Such large splittings exceed those reported for InAs/GaAs QDs by other groups (e.g., [5, 26–28]) by a factor of three and cannot be understood in the framework of generally discussed models and materials parameters [29].

3.2 Modeling of the Fine-Structure Splitting

The measured fine-structure splitting and the linear polarization of the radiative exciton decay presented in Sect. 3.1 clearly confirm that the effective dot symmetry is C_{2v} with symmetry axes along $\langle 110 \rangle$. Symme-

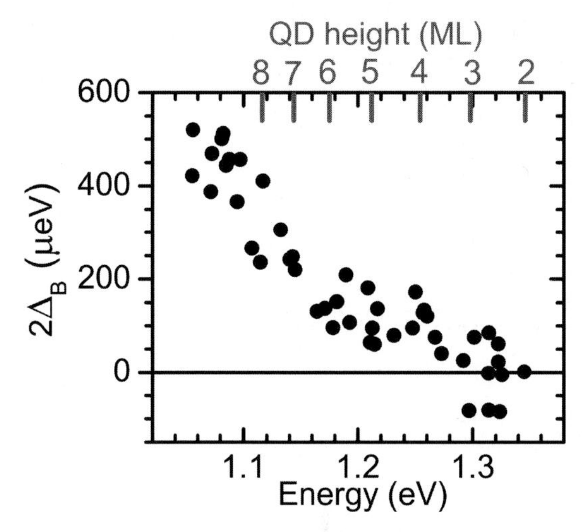

Fig. 5. Exciton fine-structure splitting $2\Delta_B$ for a large number of individual dots. *Grey bars* on top mark PL maxima of subensembles labeled by their height in monolayers

try lowering from C_{4v} can originate from a geometrical elongation of QDs along $[\bar{1}10]$ [30, 31], in addition to effects connected to the lack of inversion symmetry in the zincblende lattice. The latter leads to the atomistic symmetry anisotropy (ASA), which accounts for non-equivalent side facets of a pyramid on an atomic scale and can be further enhanced by strain [32]. Finally, the strain present in InAs QDs leads to an additional anisotropic potential originating from piezoelectricity [6]. Fig. 6 schematically displays the impact of these three sources on the orientation of the electron and hole wave-functions. Structural dot elongation and ASA [33] enforce both wave functions to point in the same direction, whereas the piezoelectric charges cause the orbitals to be perpendicular to each other. The effects of these qualitatively different sources of anisotropy are discussed below, when elongated QDs without piezoelectricity are compared to those having a square base and the piezoelectric effect included.

We calculated electron and hole states using the eight-band $\mathbf{k}\cdot\mathbf{p}$ method. This method provides a reliable link between the QD's structural and electronic properties. Since this theory is basically a continuum model, in contrast to, e.g., the empirical pseudopotential model, atomistic details like interface anisotropies and their impact on the electronic states are beyond its scope. But the probably dominating effects of elongation and piezoelectricity can be easily compared and assessed with respect to the fine-structure splitting. It should be emphasized that the pure interface asymmetry is expected to produce only minor splittings below 10 μeV [32]. The single particle orbitals are derived from a strain-dependent eight-band $\mathbf{k}\cdot\mathbf{p}$ Hamiltonian, thus exceeding the simple effective-mass approach of (1), and accounting for CB-VB

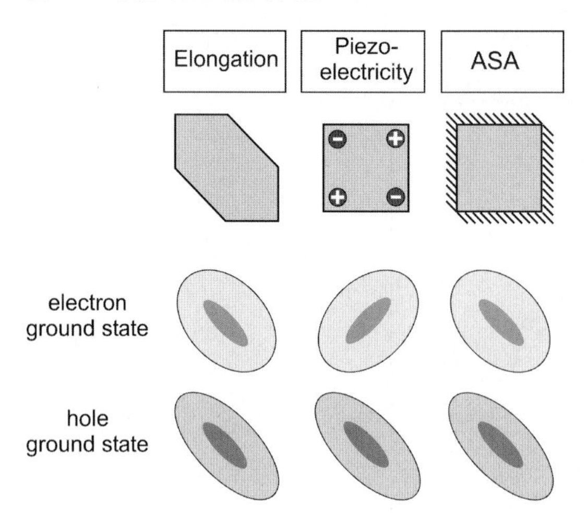

Fig. 6. Scheme of the three sources for laterally anisotropic electron and hole wave-functions

band coupling and heavy-hole, light-hole and split-off hole band mixing effects, following our previous work [7, 34]. A crucial part in the calculations is the evaluation of the Coulomb integral

$$\left\langle \Psi^a \Psi^b \right| C \left| \Psi^c \Psi^d \right\rangle = \frac{e^2}{4\pi\varepsilon} \int \frac{(\Psi^a \overline{\Psi^c})(\mathbf{r_1})\,(\Psi^b \overline{\Psi^d})(\mathbf{r_2})}{|\mathbf{r_1} - \mathbf{r_2}|}\, dV_1 dV_2 \,, \tag{4}$$

where the single particle wavefunction $\Psi(\mathbf{r})$ is given by

$$\Psi(\mathbf{r}) = \sum_R \sum_{i=1}^{8} \psi_i(\mathbf{R}) w_i(\mathbf{r} - \mathbf{R}) \,. \tag{5}$$

ψ is the 8-dimensional (i.e., 4 (e, lh, hh, split-off h)×2 (spin orientations)) envelope part determining the amplitude of Ψ at the locations $\mathbf{R}$ of the atoms, and w_i is the Wannier part featuring the S or P character. Two particle exciton energies are calculated using the configuration interaction (CI) method, including excited state configurations in the basis set [35]. Modeling thus accounts for direct Coulomb interaction, correlation and exchange. For the evaluation of (4) a multipole expansion of the $1/r$ term of the Coulomb integral is employed. As a result only monopole and dipole-dipole terms are expected to give sizable contributions. The results presented below consider solely the monopole term, since numeric implementation of the dipole term is presently not fully accomplished.

To distinguish between different influences on the fine-structure splitting $2\Delta_{\mathrm{B}}$, we compared the effects of piezoelectricity and elongation. The effect of piezoelectricity was calculated for a series of square-based QDs ranging from 10.2 nm base length and 3 monolayer height to 15.8 nm/13 monolayers, so

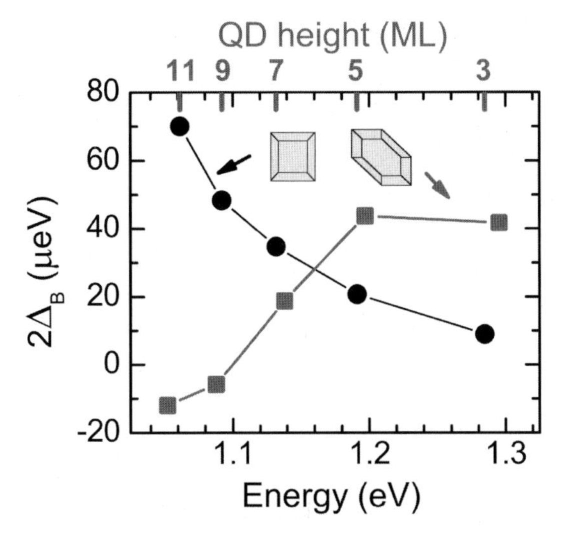

Fig. 7. Exciton fine-structure splitting $2\Delta_B$ calculated for a series of square-based QDs including piezoelectricity (*black circles*) and a series of QDs elongated along $[\bar{1}10]$ excluding piezoelectricity (*grey squares*). The *grey scale* on top gives the height of the modeled QDs. From [22]

as to match the PL spectra [36]. A piezoelectric constant extrapolated from strained InGaAs quantum wells to pure strained InAs was used as input parameter [29]. Calculations of the elongation effect consider QDs of the same volume and height to obtain comparable exciton energies, a lateral aspect ratio of 2 and absence of any piezoelectric fields. The results given in Fig. 7 show that the experimentally observed trend given in Fig. 5 is well reproduced by the calculation including piezoelectricity. The fine-structure splitting decreases monotonously for smaller QDs [29] due to a scaling of piezoelectricity being essentially proportional to QD size [6]. In contrast, the trend calculated for elongation along the $[\bar{1}10]$ direction is opposite to the experimental data, ruling out this effect as an origin of the observed splitting. It should be noted that elongation along [110] just reverses the sign of the data and hence also fails; moreover, no major elongation is seen in plan view TEM images [14]. The experimentally observed negative values might be attributed to some residual shape anisotropy of small dots; ASA and piezoelectricity are negligible in such QDs.

The calculations pinpoint piezoelectricity as the key parameter governing the exciton fine-structure splitting. However, the size of the fine-structure splitting is still too small when compared to the experiment. Recent investigations of the dipole term neglected so far suggest a fine-structure splitting that could exceed 1 meV using the same input parameters [37].

4 Conclusion

Self-similar InAs quantum dots in GaAs matrix with a well-defined multimodal size distribution were employed to study the origin of the bright-exciton fine-structure splitting. We established a systematic dependence of the magnitude on the size of the dots, ranging from small and even negative values for dots with 2 ML height to more than 500 μeV for dots higher than 9 ML. Realistic model calculations using 8-band $\mathbf{k} \cdot \mathbf{p}$ theory and CI rule out structural elongation of the dots as a cause of the splitting. The experimentally observed size-dependent trend is well explained by piezoelectricity, and quantitative agreement may be achieved in near future by accounting for dipole-dipole interaction. The size of a QD was recently demonstrated to affect also other excitonic properties, e.g., the binding energy of confined charged and neutral exciton complexes; here the change of the biexciton energy from anti-binding to binding for increasing dot size was shown to originate from correlation by a gradually increased number of bound states [22,34]. The unique properties of InAs dots with a multimodal size distribution provide an excellent model system to study confined excitons and to taylor their properties for developing novel applications in the field of single photon control.

Acknowledgements

This work was supported by the Deutsche Forschungsgemeinschaft in the framework of SFB 296 and the SANDiE Network of Excellence of the European Commission. Electronic-structure calculations were performed at HLRN within project bep00014. Growth modeling by M. B. Lifshits and V. A. Shchukin, Ioffe Institute St. Petersburg, and TEM studies by N. D. Zakharov and P. Werner, MPI Halle, are highly appreciated.

References

[1] C. H. Bennett and G. Brassard, in *Proc. IEEE Int. Conf. on Computers, Systems, and Signal Processing*, Bangalore India, p. 175 (1984).
[2] O. Benson, C. Santori, M. Pelton, and Y. Yamamoto, Phys. Rev. Lett. **84**, 2513 (2000).
[3] D. Gammon, E. S. Snow, B. V. Shanabrook, D. S. Katzer, and D. Park, Phys. Rev. Lett. **76**, 3005 (1996).
[4] D. Bimberg, Advances in Sol. State Phys. **17**, 195 (1977).
[5] M. Bayer, G. Ortner, O. Stern, A. Kuther, A. A. Gorbunov, A. Forchel, P. Hawrylak, S. Fafard, K. Hinzer, T. L. Reinecke, S. N. Walck, J. P. Reithmaier, F. Klopf, and F. Schäfer, Phys. Rev. B **65**, 195315 (2002).
[6] M. Grundmann, O. Stier, and D. Bimberg, Phys. Rev. B **52**, 11969 (1995).
[7] O. Stier, M. Grundmann, and D. Bimberg, Phys. Rev. B **59**, 5688 (1999).

[8] V. A. Shchukin, N. N. Ledentsov, P. S. Kop'ev, and D. Bimberg, Phys. Rev. Lett. **75**, 2968 (1995).
[9] N. Moll, M. Scheffler, and E. Pehlke, Phys. Rev. B **58**, 4566 (1998).
[10] Q. K. K. Liu, N. Moll, M. Scheffler, and E. Pehlke, Phys. Rev. B **60**, 17008 (1999).
[11] G. Costantini, C. Manzano, R. Songmuang, O. G. Schmidt, and K. Kern, Appl. Phys. Lett. **82**, 3194 (2003).
[12] S. Guha, A. Madhukar, and K. C. Rajkumar, Appl. Phys. Lett. **57**, 2110 (1990).
[13] M. Tabuchi, S. Noda, and A. Sasaki, in S. Namba, C. Hamaguchi, and T. Ando (Eds.), *Science and Technology of Mesoscopic Structures*, Springer, Tokyo 1992.
[14] U. W. Pohl, K. Pötschke, A. Schliwa, F. Guffarth, D. Bimberg, N. D. Zakharov, P. Werner, M. B. Lifshits, V. A. Shchukin, and D. E. Jesson, Phys. Rev. B **72**, 245332 (2005).
[15] M. B. Lifshits, V. A. Shchukin, D. Bimberg, and D. E. Jesson, *Proc. 13th Int. Symposium on Nanostructures: Physics and Technology, St. Petersburg, Russia 2005 (Ioffe Physico-Technical Institute*, St. Petersburg, 2005) pp. 308–309.
[16] U. W. Pohl, K. Pötschke, A. Schliwa, M. B. Lifshits, V. A. Shchukin, D. E. Jesson, and D. Bimberg, Physica E **32**, 9 (2006).
[17] M. Colocci, F. Bogani, L. Carraresi, R. Mattolini, A. Bosacchi, S. Franchi, P. Frigeri, M. Rosa-Clot, and S. Taddei, Appl. Phys. Lett. **70**, 3140 (1997).
[18] A. Gustafsson, D. Hessmann, L. Samuelson, J. F. Carlin, R. Houdré, and A. Rudra, J. Crystal Growth **147**, 27 (1995).
[19] S. Raymond, S. Studenikin, S.-J. Cheng, M. Pioro-Ladrière, M. Ciorga, P. J. Poole, and M. D. Robertson, Semicond. Sci. Technol. **18**, 385 (2003).
[20] U. W. Pohl, K. Pötschke, I. Kaiander, J.-T. Zettler, and D. Bimberg, J. Crystal Growth **272**, 143 (2004).
[21] U. W. Pohl, K. Pötschke, M. B. Lifshits, V. A. Shchukin, D. E. Jesson, and D. Bimberg, Appl. Surf. Sci. (2006), in print.
[22] U. W. Pohl, R. Seguin, S. Rodt, A. Schliwa, K. Pötschke, and D. Bimberg, Physica E (2006), in print.
[23] R. Timm, A. Lenz, H. Eisele, T.-Y. Kim, F. Streicher, K. Pötschke, U. W. Pohl, D. Bimberg, and M. Dähne, Physica E **32**, 25 (2006).
[24] S. Rodt, R. Heitz, A. Schliwa, R. L. Sellin, F. Guffarth, and D. Bimberg, Phys. Rev. B **68**, 035331 (2003).
[25] V. Türck, S. Rodt, O. Stier, R. Heitz, R. Engelhardt, U. W. Pohl, D. Bimberg, and R. Steingrüber, Phys. Rev. B **61**, 9944 (2000).
[26] K. Kowalik, O. Krebs, A. Lemaître, S. Laurent, P. Senellart, P. Voisin, and J. A. Gaj, Appl. Phys. Lett. **86**, 041907 (2005).
[27] A. S. Lenihan, M. V. Gurudev Dutt, D. G. Steel, S. Gosh, and P. K. Bhattacharya, Phys. Rev. Lett. **88**, 223601 (2002).
[28] R. J. Young, R. M. Stevenson, A. J. Shields, P. Atkinson, K. Cooper, D. A. Ritchie, K. M. Groom, A. I. Tartakovskii, and M. S. Skolnick, Phys. Rev. B **72**, 113305 (2005).
[29] R. Seguin, A. Schliwa, S. Rodt, K. Pötschke, U.W. Pohl, and D. Bimberg, Phys. Rev. Lett. **95**, 257402 (2005).

[30] V. D. Kulakovskii, G. Bacher, R. Weigand, T. Kümmell, A. Forchel, E. Borovitskaya, K. Leonardi, and D. Hommel, Phys. Rev. Lett. **82**, 1780 (1999).
[31] R. Songmuang, S. Kiravittaya, and O. G. Schmidt, J. Crystal Growth **249**, 416 (2003).
[32] G. Bester, S. Nair, and A. Zunger, Rev. B **67**, 161306(R) (2003).
[33] G. Bester and A. Zunger, Phys. Rev. B **71**, 045318 (2005).
[34] S. Rodt, A. Schliwa, K. Pötschke, F. Guffarth, and D. Bimberg, Phys. Rev. B **71**, 155325 (2005).
[35] O. Stier, R. Heitz, A. Schliwa, and D. Bimberg, Phys. Stat. Sol. (a) **190**, 477 (2002).
[36] R. Heitz, F. Guffarth, K. Pötschke, A. Schliwa, D. Bimberg, N. D. Zakharov, and P. Werner, Phys. Rev. B **71**, 045325 (2005).
[37] R. Zimmermann, A. Schliwa et al., to be published.

Quantum Dots in Planar Cavities – Single and Entangled Photon Sources

Robert Young[1], Mark Stevenson[1], Paola Atkinson[2], Ken Cooper[2], David Ritchie[2], and Andrew Shields[1]

[1] Toshiba Research Europe Ltd., Cambridge Science Park, CB4 0WE Cambridge, UK
robert.young@crl.toshiba.co.uk
[2] Cavendish Laboratory, Cambridge University, JJ Thomson Avenue, CB3 0HE Cambridge, UK

Abstract. We show that single InAs quantum dots embedded in a planar cavity, formed by mismatched sets of GaAs/AlAs distributed Bragg reflectors, can be a useful source of triggered single photons as well as polarisation-entangled photon pairs. The former is demonstrated with a second order correlation function under 0.1 and the latter with a fidelity exceeding 70 %. Such quantum dot devices may be useful in quantum communications and quantum information processing.

1 Introduction

Non-classical light is essential for quantum information applications [1]; classical sources of light, such as laser pulses, contain a distribution in the number of photons per pulse which obeys Poissonian statistics. Thus no matter how much a laser pulse is attenuated there will always be a finite chance of finding multiple photons in the pulse. The security of quantum cryptography is limited by pulses which contain more than one photon, allowing an eve dropper to attempt a photon number splitting attack [2].

In many physical ways single quantum dots are analogous to single atoms, three dimensional confinement of excitons within a dot results in an energy spectrum consisting of discrete levels, like those of an atom. Each exciton complex has a unique energy due to the Coulomb interaction between the confined carriers. Single photon emission is possible from any exciton complex which is optically active, following photon emission there is a finite time delay before re-excitation is possible and so simultaneous emission of multiple photons is greatly suppressed. The left-hand side of Fig. 1 shows a simplified energy level diagram for a quantum dot, including only neutral exciton and biexciton levels. The exciton transition first used to demonstrate single photon emission [3–5] from quantum dots is indicated with a solid arrow. Single photon emission has also been demonstrated from other exciton complexes, including the biexciton and charged exciton [4].

Emission cascades in quantum dots can be employed to produce multiple photons, emitted at different energies, in a single excitation cycle. The biexciton cascade is a well studied example of this, as the biexciton decays to the

R. Haug (Ed.): Advances in Solid State Physics,
Adv. in Solid State Phys. **46**, 55–65 (2008)

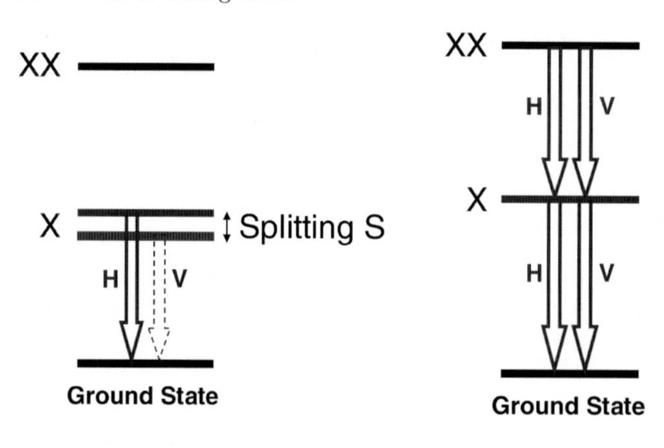

Fig. 1. Schematics showing simplified energy level diagrams for a quantum dot, including only neutral exciton (X) and biexciton (XX) levels. In (**a**) the decay scheme for a quantum dot as a source of polarised single photons from the horizontally polarised exciton level is shown, and in (**b**) the decay scheme for a quantum dot with a degenerate exciton level a source of polarisation-entangled photons. *Blue (Red) arrows (online color)* correspond to horizontally (vertically) linearly polarised photons

groundstate via the intermediate exciton level two photons are emitted [6]. Similarly the cascade from the tri-exciton state has been shown to produce three photons per excitation [7].

The bright exciton is a doublet as it can have a total angular momentum of either ± 1, in general the anisotropic exchange interaction [8, 9] hybridises these states, making them couple to linearly polarised photons as illustrated and split in energy. With this splitting present it has been shown that the biexciton cascade can be used as a source of polarisation-correlated photons [10–12].

Entangled photons also have applications in quantum key distribution, allowing long distance quantum communication using quantum repeaters [13]. They can also greatly reduce the resources required for linear optical quantum computing [14]. For these applications however, the number of photon pairs generated per cycle is of critical importance, since emission of multiple photon pairs introduces errors due to the possibility of two individual photons not being entangled. The most widely used technique to generate entangled photon pairs is currently parametric down conversion [15], producing only a probabilistic numbers of photons pairs per excitation cycle.

The biexciton decay in a single quantum dot, shown on the right hand side of Fig. 1, was proposed to provide a source of entangled photon pairs [16] if the splitting in the intermediated exciton level could be reduced to within the homogenous linewidth of the transition. As for single photon emission from an exciton complex, the biexciton cascade can decay no more than once per excitation cycle and thus generate no more than two photons. Such a device could find future applications in quantum optics therefore, with the added benefit that it might be realised in a simple LED structure [17, 18].

In this paper we will show optical measurements from a single quantum dot with no exciton level splitting embedded in a semiconductor planar cavity. Emission from a quantum dot is measured with a microscope objective which is only capable of collecting light emitted into a small solid angle. Embedding quantum dots in a mismatched planar cavity, designed to preferentially channel emission in the direction of the collection optics increases the collection efficiency by an order of magnitude [19]. The cavity has an optical mode with finite width which, when centred on the dot emission, helps to limit the background light collected, as it is emitted by layers other than the quantum dot and usually at a different energy. In previous work background emission from the InAs wetting layer was the major limiting factor in the proportion of both the amount of entangled light measured from the biexciton cascade and the suppression of multiple photon emission from the exciton decay in a quantum dot with a degenerate exciton level. We show that by limiting the amount of such emission being collected using a planar cavity we are able to greatly increase the degree of entanglement in the expected $(|H_{XX}H_X> + |V_{XX}V_X>)/\sqrt{2}$ state, to over 70 %, and decrease the second order correlation function for the exciton transition to under 0.1.

2 Experimental Details and Sample Design

Samples were fabricated by molecular beam epitaxy (MBE); a low density ($\sim 1/\mu m^2$) layer of InAs quantum dots was grown in a λ-cavity above fourteen pairs of AlAs/GaAs distributed Bragg reflector (DBR's) [20] and below two repeats of the same DBR mirror structure. The sample design is illustrated in Fig. 2.

A striking dependence on the fine structure splitting with the recombination energy of the exciton has been discovered [21] and remarkably the splitting is found to pass through zero at ~ 1.4 eV. This allows selection of quantum dots with no exciton level splitting within the homogeneous linewidth of the transition to be selected. Further to this it has also been reported that the exciton level splitting of some quantum dots can be tuned to zero with a magnetic field [22].

In samples studied previously [4] photoluminescence from the InAs wetting layer forms a broad intense peak to higher energy and is the main source of background light emission measured. To reduce this unwanted background the growth temperature of the InAs layer was nominally increased by 20°C, to increase In/Ga intermixing and in doing so increase the emission energy of the wetting layer.

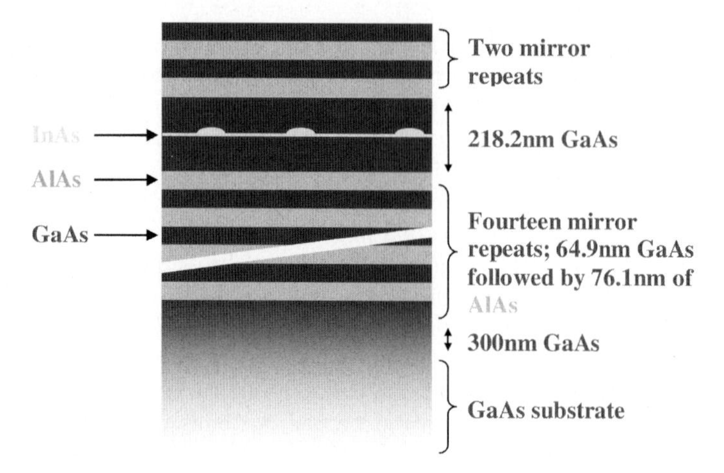

Fig. 2. The epitaxial layer structure used to form a planar cavity around a layer of InAs quantum dots. Fourteen mirrors, each comprising a $\lambda/4$ (64.9 nm) layer of GaAs (n $\simeq$ 3.5) followed by a $\lambda/4$ (76.1 nm) layer of AlAs (n $\simeq$ 3) were grown beneath the dot layer and two mirror repeats above. A 1λ GaAs cavity (218. nm) surrounds the dot layer

Photoluminescence (PL) from this sample was measured at $\sim$ 10 K, with excitation provided non-resonantly using a red laser diode emitting 100 ps pulses with an 80 MHz repetition rate. A microscope objective lens focussed the laser onto the surface of the sample, and collected the emitted light.

To isolate single quantum dots, a metal shadow mask containing circular apertures around 2 µm in diameter was defined on the surface of the sample. Figure 3 shows high resolution polarised photoluminescence spectra from a single quantum dot. The spectra is dominated by emission from the neutral exciton (X) and biexciton (XX) states, despite the high emission energy very little background from the InAs wetting layer is evident as it is suppressed by the stop-band of the cavity. The cavity also gives a notable enhancement in the PL collected from the quantum dots, around ten-fold.

3 Single Photon Emission

To measure the statistics of the photons emitted following exciton decay a Hanbury-Brown and Twiss arrangement was used. A spectrometer was first employed to select photons from the exciton transition, a linearly polarising 50/50 beam splitter then divided the emission between a pair of thermo-electrically cooled silicon avalanche photodiodes (APD's). The time delay (τ) between counts in the two detectors integrated over the dot emission cycle was recorded by a time interval analyzer, a histogram of the distribution in emission of the pairs is shown in Fig. 4.

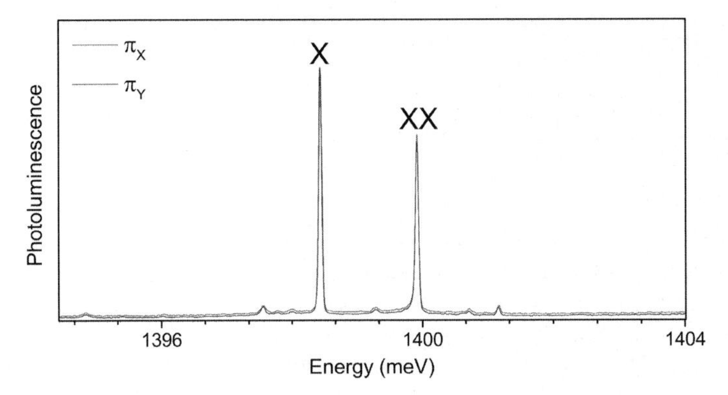

Fig. 3. High resolution photoluminescence spectra taken from a single quantum dot in a planar cavity. *Red (blue) lines (online color)* show horizontally (vertically) polarised spectra, the dominant features correspond to radiative recombination from the exciton (X) and biexciton (XX) states

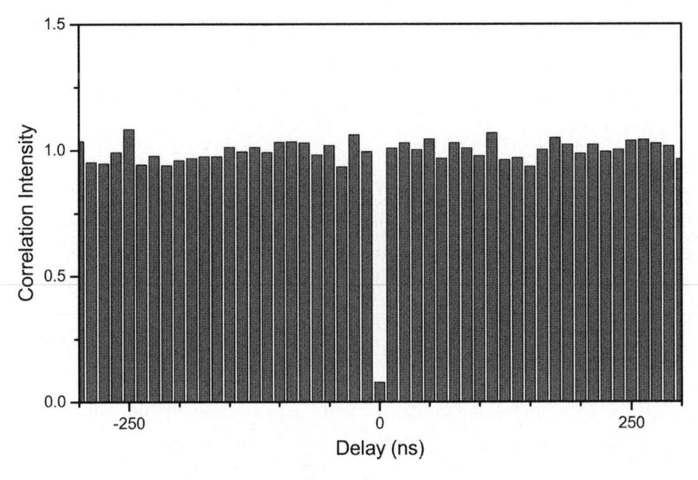

Fig. 4. Second order correlation of exciton emission from a quantum dot. The strong suppression of coincidences at zero time delay demonstrates close to ideal single photon emission from the quantum dot source

The strong suppression measured at zero delay is expected and characteristic of single-photon emission. The area of the zero delay peak indicates that the exciton level is emitting $\sim$ 14 times less multi-photon pulses than a Poissonian source of the same intensity. There should be no pairs with zero delay, its finite area can be attributed to two mechanisms; detector dark counts and emission from other exciton-confining layers in the sample at the same energy as this exciton. The narrow mode widths of microcavities help directly to reduce the latter and the increased collection efficiencies they afford reduce the relative contribution of the former, enabling better zero delay suppressions [23].

4 Entangled Photon Emission

The emission lines corresponding to neutral exciton and biexciton emission shown in Fig. 3 are broadened with line widths of $\sim 50\,\mu\text{eV}$. Fluctuations in the charge distribution surrounding the dot are thought to be the cause of this inhomogeneous broadening. Fitting Lorenzian line shapes to the measured peaks and averaging the splitting measured in the exciton and biexciton photons allows the fine structure splitting of single dots to be measured with high precision. This process is detailed elsewhere [21] and allows, so called 'degenerate' quantum dots with fine structure splittings which are smaller than the homogeneous linewidth ($\sim 1.5\,\mu\text{eV}$) of the exciton transition to be located. The spectra in Fig. 3 are taken from such a dot. No polarisation splitting is visible in the figure.

To analyse the properties of photon pairs emitted by a selected quantum dot, we measure the polarisation and time dependent correlations between the biexciton and exciton photons. The exciton photon was selected with a spectrometer in the same way as detailed in the previous section. A second spectrometer was introduced to the arrangement to select the biexciton photon. The biexciton photons were then measured by a third APD after passing through a linear polariser. The insertion of appropriately oriented quarter-wave or half-wave plates preceding each of the two spectrometers allows any polarisation measurement basis to be selected. The time intervals between detection events on different APD's were measured using a pair of single photon counting modules, to determine the second order correlation functions. This arrangement is illustrated for clarity in Fig. 5.

In Fig. 6a second order correlation as a function of the time delay between horizontally polarised biexciton photons and horizontally (vertically) polarised exciton photons is shown in red (blue). There is significant increase (decrease) in the area of the peak at zero delay above (below) the non-zero time delay peak average, as would be expected for a source emitting rectilinearly polarised correlated photons as well as an entangled photon source. This result has been reported previously in quantum dots with large exciton level splittings [10–12]. Strikingly however for this degenerate dot a similar result is found in panel (b) which shows second order correlation in the diagonal basis, a result which is expected for degenerate dots emitting entangled photons. Further proof that our degenerate dot is emitted photon pairs in the entangled basis $(|\mathrm{H_{XX}H_X}> + |\mathrm{V_{XX}V_X}>)/\sqrt{2}$ is presented in panel (c) which shows second order correlation measured in the circular basis. The expected entangled state is exactly equivalent to $(|\mathrm{R_{XX}L_X}> + |\mathrm{L_{XX}R_X}>)/\sqrt{2}$, hence correlation between right-hand polarised biexciton photons and left-hand po-

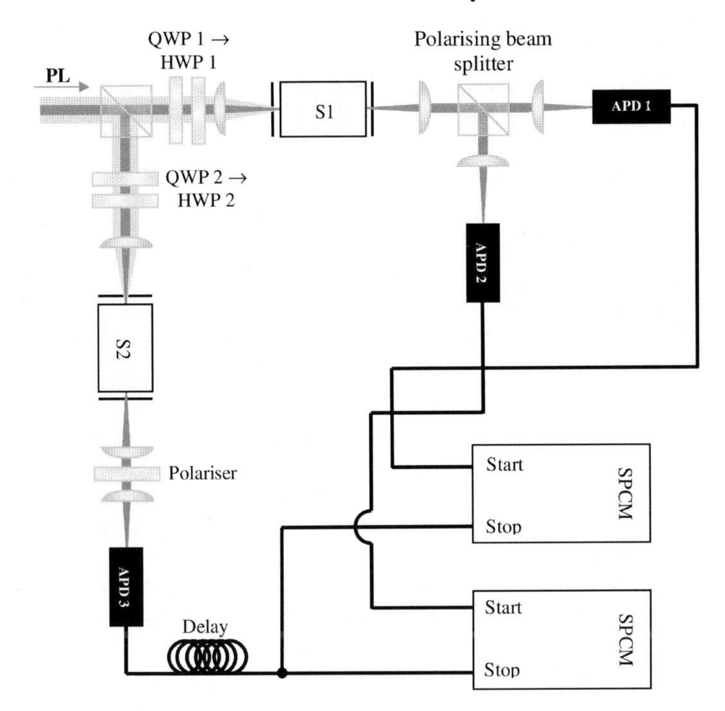

Fig. 5. A schematic of the experimental setup used to measure polarised cross correlation between the photoluminescence (PL) from the exciton and biexciton states, which differ in emission energy. Quarter-wave plates (QWP 1 and 2) and half-wave plates (HWP 1 and 2) were inserted in front of the entrance slits to the two spectrometers (S1 and S2). The avalanche photodiodes are labelled APD 1–3 and single photon counting modules SPCM

larised exciton photons and vice versa is expected and this is shown in the panel.

For cross correlations between biexciton and exciton photons, the probability of detecting a pair of coincident photons, relative to the probability of detecting photons separated by a number of excitation cycles, is proportional to the inverse of the probability of generating an exciton photon per cycle. Thus, the relative probability of detecting coincident photons is dependent on the excitation rate, which fluctuates during the integration time of our experiments. However, the correlations of the biexciton detection channel with each of the orthogonally polarised exciton detection channels are measured simultaneously with the same excitation conditions, and thus can be compared directly. Additionally, we verify that the time averaged biexciton and exciton emission from this quantum dot is unpolarised within experimental error, so the number of coincident photon pairs can be normalised relative to the average number of photon pairs separated by at least one cycle, which compensates for the different detection efficien-

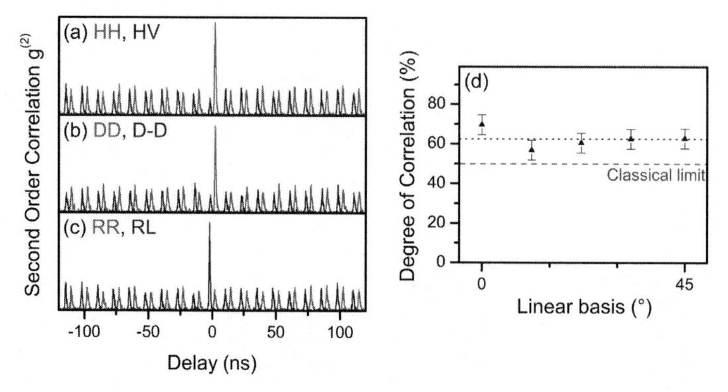

Fig. 6. Biexciton-exciton polarization correlations for a quantum dot with exciton level splitting S ≈ 0. The degree of polarization correlation is defined as the difference between the normalized polarization correlated and anti-correlated coincidences, divided by their sum. The second-order correlation as a function of the time delay between the detected photons measured in the rectilinear (Horizontal H and Vertical V) (**a**), diagonal (D and orthodiagonal $\overline{D}$) (**b**) and circular bases (Right- R and Left- L hand) (**c**). The degree of correlation is shown in (**d**) as a function of the rotation of the linear detection basis by a half-wave plate. The *green dashed line (online color)* shows the upper average limit for a classical source of photon pairs

cies of the measurement system. The degree of polarisation correlation is defined as $(g^{(2)}_{\mathrm{XX,X}} - g^{(2)}_{\mathrm{XX,\overline{X}}})/(g^{(2)}_{\mathrm{XX,X}} + g^{(2)}_{\mathrm{XX,\overline{X}}})$, where the second order correlation functions, $g^{(2)}_{\mathrm{XX,X}}$ and $g^{(2)}_{\mathrm{XX,\overline{X}}}$ are the simultaneously measured, normalised coincidences of the biexciton photon with the co-polarised exciton and orthogonally-polarised exciton photons respectively.

Figure 6d shows the degree of correlation as the function of the angle of a single half wave plate, placed directly after the microscope objective. It is found that the degree of polarisation correlation is approximately independent of the half wave plate angle. This is an expected result for photon pairs being emitted in the entangled $(|\mathrm{H_{XX}H_X}> + |\mathrm{V_{XX}V_X}>)/\sqrt{2}$ state; the linear polarisation measurement of the first photon defines the linear polarisation of the second photon. For classical polarisation correlated photon pairs, for example a quantum dot with an exciton splitting which is larger than the homogeneous linewidth of the transition, the degree of correlation varies sinusoidally with wave-plate angle, contrary to the emission of this dot. In fact, the average linear correlation measured is $62.4 \pm 2.4\,\%$ for this dot, is five standard deviations above the $50\,\%$ limit for classical pairs of photons, which proves that the degenerate quantum dot does emits polarisation entangled photon pairs. The noise on the data is caused by the statistical errors associated with the finite number of counts, from which the error in the degree of correlation can be estimated to be typically < 0.05.

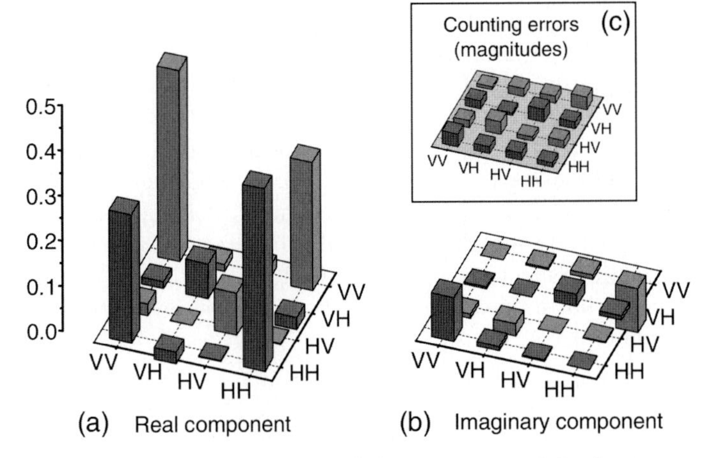

Fig. 7. Real (**a**) and imaginary (**b**) components of the density matrix for a quantum dot with bright exciton splitting $< 1.5\,\mu\mathrm{eV}$. The *inset* (**c**) shows the magnitudes of the counting errors for the sixteen components of the density matrix. A scale for all elements, including the errors, is shown on the *left hand side*

To fully characterise the two photon state emitted by the dot, the two photon density matrix can be constructed from correlation measurements, using quantum state tomography [24]. The measurements pairs required are the combinations of the V, H, L, and D biexciton polarisations, with the rectilinear, diagonal and circular polarised exciton detection bases. The resulting density matrix representing the emission from the quantum dot is shown in Fig. 7, with real and imaginary components shown in (a) and (b), respectively.

The strong outer diagonal elements in the real matrix demonstrate the high probability that the photon pairs have the same linear polarisation. The inner diagonal elements represent the probability of detecting oppositely linearly polarised photons, which is greatly suppressed over previous measurements. The residual average value of these elements is 0.085, of which we estimate 42 % is due to background light. The remaining contribution is likely to be due to scattering of the exciton spin states. This too seems to be suppressed compared to previous measurements, this is likely a direct result of reducing the density of wetting layer states resonant with the dots.

A direct consequence of the reduction of the background light is that the matrix more closely resembles the state of photon pairs generated by the dot itself. The outer off-diagonal elements in the real matrix are several times stronger than previously measured [25], and are clear indicators of entanglement. Small imaginary off-diagonal elements are additionally seen, which indicates that there may be a small phase difference between the $|HH\rangle$ and $|VV\rangle$ components of the entangled state [26]. All other elements are close to zero given the errors associated with the procedure, which are determined from the number of coincidences, and potted in Fig. 4c.

The measured two photon density matrix projects onto the expected $(|\mathrm{H_{XX}H_X}> +|\mathrm{V_{XX}V_X}>)/\sqrt{2}$ state with fidelity 0.702 ± 0.022. This proves that the photon pairs we detect are entangled, since for pure or mixed classical states, the fidelity cannot exceed 0.5. Other tests for entanglement include the tangle [27] which must be > 0 to prove entanglement, it is measured to be 0.194 ± 0.026 for the density matrix shown and the *Peres* test [28] which must be < 0 to prove entanglement, we measure -0.219 ± 0.021, again unambiguously demonstrating that a quantum dot with no exciton level splitting does emit entangled photons.

5 Conclusions

In conclusion, we have demonstrated that quantum dots are a useful source of non-classical light, including light with significant sub-Poissonian statistics and on-demand polarisation entangled photons. Semiconductor quantum dots integrate into device structures such as the ones studied in this paper are compact and robust, making them extremely promising candidates as inexpensive elements in future quantum information processing applications. We show that embedding a planar cavity into such devices increases the collection efficiency through standard microscope objective by an order of magnitude, the cavity's stop band has the additional effect of helping to suppress unwanted background emission. Another key advantage of the use of a planar cavity is that is requires no further processing steps over traditional cavity-free samples, as the cavity can be grown in-situ by MBE.

Acknowledgements

This work was partially funded by the EU projects QAP and SANDiE, and by the EPSRC.

References

[1] D. F. Walls and G. J. Bilburn: *Quantum Optics* (Springer, Berlin 1994)
[2] D. Bouwmeester, A. K. Ekert, A. Zeilinger: *The Physics of Quantum Information* (Springer, Berlin 2000)
[3] P. Michler, A. Imamoglu, M. D. Mason, P. J. Carson, G. F. Strouse, S. K. Buratto: Nature **406**, 968 (2000)
[4] R. M. Thompson, R. M. Stevenson, A. J. Shields, I. Farrer, C. J. Lobo, D. A. Ritchie, M. L. Leadbeater, M. Pepper: Phys. Rev. B(R) **64**, 201302 (2001)
[5] V. Zwiller, T. Aichele, W. Seifert, J. Persson, O. Benson: Appl. Phys. Lett. **82**, 1509 (2003)

[6] E. Moreau, I. Robert, L. Manin, V. Thierry-Mieg, J. M. Gérard, I. Abram: Phys. Rev. Lett. **87** 183601 (2001)
[7] J. Persson, T. Aichele, V. Zwiller, L. Samuelson, O. Benson: Phys. Rev. B **69**, 233314 (2004)
[8] H. W. van Kasteren, E. C. Cosman, W. A. J. A. van der Poel, C. T. B. Foxon: Phys. Rev. B **41**, 5283 (1990)
[9] E. Blackwood, M. J. Snelling, R. T. Harley, S. R. Andrews, C. T. B. Foxon: Phys. Rev. B **50**, 14246 (1994)
[10] R. M. Stevenson, R. M. Thompson, A. J. Shields, I. Farrer, B. E. Kardynal, D. A. Ritchie, M. Pepper: Phys. Rev. B(R) **66**, 081302 (2002)
[11] C. Santori, D. Fattal, M. Pelton, G. S. Solomon, Y. Yamamoto: Phys. Rev. B **66**, 045308 (2002)
[12] S. M. Ulrich, S. Strauf, P. Michler, G. Bacher, A. Forchel: Appl. Phys. Lett. **83**, 1848 (2003)
[13] H.-J. Briegel, W. Dür, J. I. Cirac, P. Zoller: Phys. Rev. Lett. **81** 5932 (1998)
[14] T. B. Pittman, B. C. Jacobs, J. D. Franson: Phys. Rev. A **64**, 062311 (2001)
[15] Y. H. Shih, C. O. Alley: Phys. Rev. Lett. **61**, 2921 (1988)
[16] O. Benson, C. Santori, M. Pelton, T. Yamamoto: Phys. Rev. Lett. **84**, 2513 (2000)
[17] Z. Yuan, B. E. Kardynal, R. M. Stevenson, A. J. Shields, C. J. Lobo, K. Cooper, N. S. Beattie, D. A. Ritchie, M. Pepper: Science **295**, 102 (2002)
[18] A. J. Bennett, D. C. Unitt, P. See, A. J. Shields, P. Atkinson, K. Cooper, D. A. Ritchie: Appl. Phys. Lett. **86**, 181102 (2005)
[19] H. Benisty, H. De Neve, C. Weisbuch: IEEE J. Quantum Electrin, **34**, 1612 (1998)
[20] M. Born, E. Wolf: *Principles of Optics (7th Ed.)* (Cambridge University Press 2002)
[21] R. J. Young, R. M. Stevenson, A. J. Shields, P. Atkinson, K. Cooper, D. A. Ritchie, K. M. Groom, A. I. Tartakovskii, M. S. Skolnick: Phys. Rev. B **72** 113305 (2005)
[22] R. M. Stevenson, R. J. Young, P. See, D. Gevaux, K. Cooper, P. Atkinson, I. Farrer, D. A. Ritchie, A. J. Shields: Phys. Rev. B **73**, 033306 (2006)
[23] A. J. Bennett, D. C. Unitt, P. Atkinson, D. A. Ritchie, A. J. Shields: Optics Express, **13**, 7772 (2005)
[24] D. F. V. James, P. G. Kwiat, W. J. Munro, A. G. White: Phys. Rev. A **64**, 052312 (2001)
[25] R. M. Stevenson, R. J. Young, P. Atkinson, K. Cooper, D. A. Ritchie: Nature **439**, 179 (2006)
[26] R. J. Young, R. M. Stevenson, P. Atkinson, K. Cooper, D. A. Ritchie: New J. Phys. **8**, 29 (2006)
[27] V. Coffman, J. Kundu, W. K. Wooters: Phys. Rev. A **64**, 052306 (2000)
[28] A. Peres: Phys. Rev. Lett. **77**, 1413 (1996)

Part II

Molecules and Nanoparticles

Periodic Structure Formation in Polymer Films with Embedded Gold Nanoparticles

Katrin Loeschner, Andreas Kiesow, and Andreas Heilmann

Fraunhofer Institute for Mechanics of Materials,
Heideallee 19, 06120 Halle/Saale, Germany
andreas.heilmann@iwmh.fraunhofer.de

Abstract. Plasma polymer films containing gold nanoparticles were irradiated with linearly polarized femtosecond laser pulses. The laser-induced periodic modification of the nanoparticle assemblies was studied by using scanning and transmission electron microscopy. The influence of laser intensity and laser wavelength on the structure formation process was investigated. Optical spectra of the irradiated films reveal a laser-induced dichroism. Optical modelling based on Rayleigh–Gans theory was applied to explain the anisotropic optical properties.

1 Introduction

The spontaneous formation of periodic surface gratings or so called "ripples" or "LIPS" (laser induced periodic structures) due to laser irradiation with a single laser beam was reported for the first time by *Birnbaum* in 1965 after ruby-laser irradiation of various semiconductor surfaces [1]. Further investigations have shown that LIPS formation is a quite general phenomenon which is practically found on nearly all kinds of material surfaces, e.g., semiconductors, metals, dielectrics, and polymers, within certain ranges of laser parameters [2, 3]. It is generally accepted that the effect originates from the interference of the incident laser light with the scattered light near the surface resulting in an intensity modulation on the irradiated material. Theoretical considerations concerning LIPS have been made by several authors [4, 5]. To the best of our knowledge, up to now, the LIPS formation effect was not used to modify metal nanoparticles and to arrange them in ordered structures within a polymer matrix. In [6] LIPS formation was observed after irradiation of thin silica films containing separated silver nanoparticles with nanosecond laser pulses.

In previous studies, a possibility to structure metal/polymer nanocomposites using linearly polarized femtosecond laser pulses was discovered, which is probably based on the LIPS formation mentioned above [7, 8]. Here, line-like periodic metal nanoparticle arrangements embedded in thin polymer films result after laser irradiation (Fig. 1). To understand the complex process of evolution of these periodic structures, different parameters concerning the laser irradiation and the metal/polymer film structure itself have to be investigated. Although, we observed this effect on different metals (Au, Ag, Cu),

R. Haug (Ed.): Advances in Solid State Physics,
Adv. in Solid State Phys. **46**, 69–82 (2008)

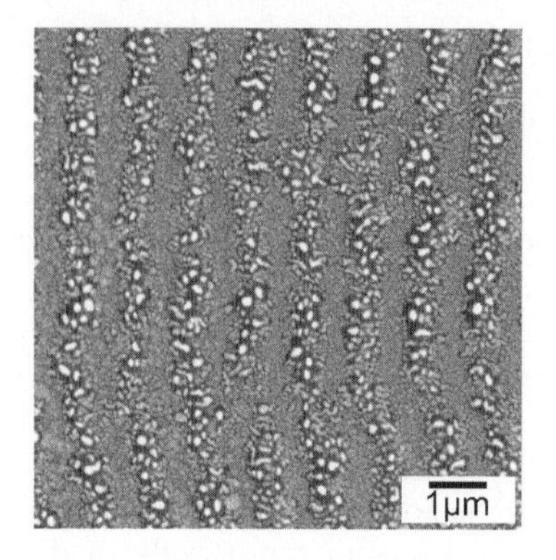

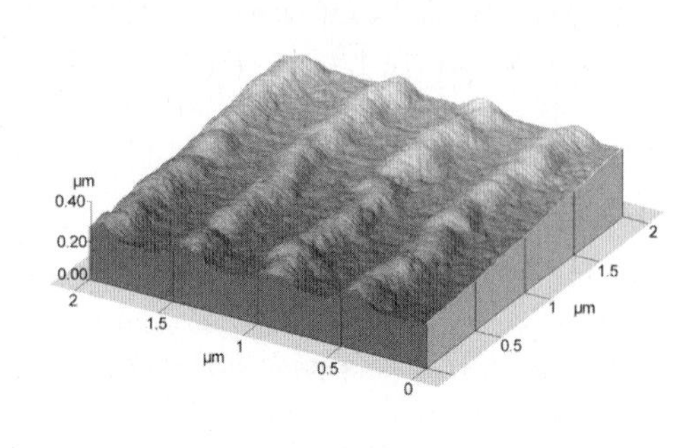

Fig. 1. Scanning electron microscopy (SEM, *left*) and atomic force microscopy (AFM, *right*) images of a laser irradiated Au/PP film (wavelength $\lambda = 800$ nm); height of the ripples determined by AFM ≈ 70 nm

in this report, we focus on plasma polymer films containing gold nanoparticles (Au/PP films). In the following, the laser-induced structural changes in relation to important laser and structural parameters as well as the resulting anisotropic optical properties are described.

2 Experimental

2.1 Film Deposition

Thin polymer films containing metal particles were produced by alternating plasma polymerization of the monomer hexamethyldisilazane (HMDSN, $(CH_3)_3Si{-}NH{-}Si(CH_3)_3$) and thermal evaporation of gold in a vacuum chamber. A detailed description of the fabrication and properties of plasma polymer films as a matrix material for nanoparticles is given in [9]. In order to arrange the particles in one plane in the polymer matrix the films were build up on the substrates in the sequence polymer – metal nanoparticles – polymer. Where the thickness of the individual layers is in the range of a few ten nanometers (see cross section in [8]). The films were deposited on glass substrates and for the optical measurements on quartz plates. Because of a specific arrangement of the substrates in the deposition chamber a gradient in the metal content of the films could be achieved. This metal gradient gives the possibility to investigate the correlation between nanostructure and optical properties at different particle sizes and shapes, starting from well-separated metal nanoparticles up to a nearly closed metal layer in produced film [10].

2.2 Laser Irradiation

A commercial system of a mode-locked Ti-sapphire laser with regenerative amplification (Spectra Physics) was used for irradiation. The films were illuminated at normal incidence with linearly polarized 150 fs pulses at a repetition rate of 1 kHz. Pulse energies in the range of about 1 to 150 μJ per pulse were applied. The beam had a nearly Gaussian intensity profile. The sample was placed in front of the focus and the laser spot size ω on the sample was typically of about 150–250 μm. In order to write lines into the films along the gradient in metal content, the samples were moved through the laser beam using a X-Y translational stage with a velocity of 1 mm/s. This corresponds to a movement of the sample between two pulses of 1 μm. Different wave lengths were applied for irradiation: the fundamental wavelength of the laser (800 nm), the second (400 nm) and third harmonic (266 nm) and an intermediate wavelength of 528 nm prepared by optical parametric frequency conversion and sum frequency with the laser fundamental wavelength.

2.3 Film Characterization

The films were studied by different microscopical techniques to characterize the periodical structure modifications. Systematic investigations of a large number of irradiated samples were carried out by scanning electron microscopy (SEM, Hitatchi S-4500, at 10 keV). Fast fourier transformation of the SEM images was applied to obtain the mean value of the structure period. To determine the modifications of particle size and shape, transmission electron microscopy (TEM, Philips CM 20, at 200 keV) in combination with digital image analysis (Soft Imaging System analySIS) was applied. The topography of the laser-irradiated films was studied by atomic force microscope (AFM, Park Scientific Instruments Autoprobe CP) operating in contact mode. In order to characterize the optical properties of the original films, transmission spectra were recorded with a conventional double beam spectrometer (Analytik Jena specord 210). A X-Y translational stage was used to record the spectra along the gradient in metal content. After laser irradiation transmission spectra were recorded locally on the irradiated lines using a microscope spectrometer (Carl Zeiss Axioplan2 imaging microscope coupled with J&M TIDAS spectrometer) in an area of 100x100 μm^2.

3 Results and Discussion

3.1 Characterization of the Original Samples

As already mentioned, a special spatial arrangement of the samples in the deposition chamber was used to create a lateral gradient in the metal content of the films. Figure 2 shows transmission electron micrographs of selected positions at one film with different amount of gold. The images present different stages of the island growth process (Volmer–Weber growth).

Due to the plane particle distribution and the electron transparency of the plasma polymer matrix the area filling factor f_a could be determined

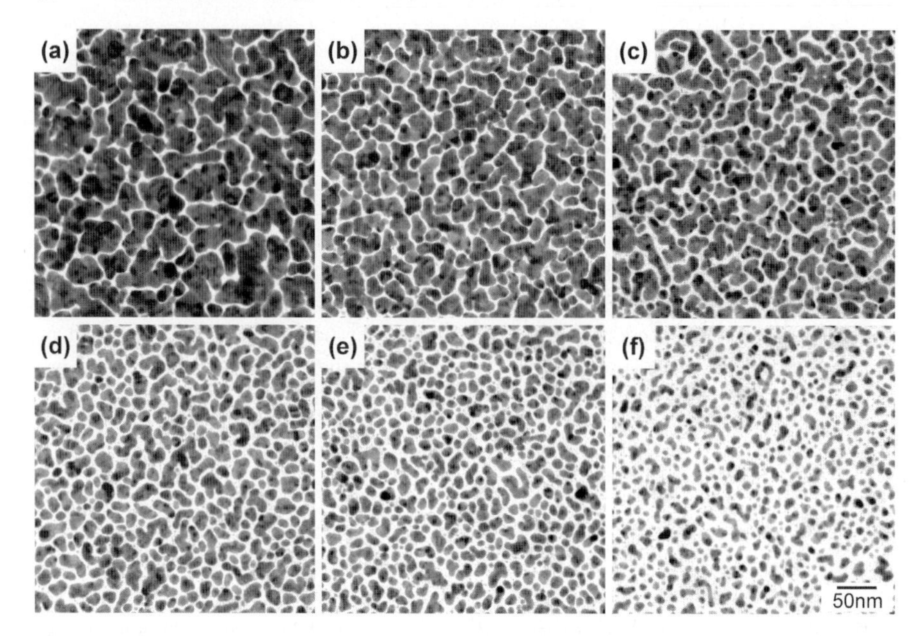

Fig. 2. TEM micrographs of gold nanoparticles embedded in a plasma polymer thin film with decreasing metal content from (a) to (f)

Table 1. Results of digital image analysis of TEM micrographs in Fig. 2 – filling factor f_a, mean values and standard deviation of equivalent circle diameter D and shape factor S

	a	b	c	d	e	f
f_a	0.90	0.85	0.75	0.70	0.60	0.40
D [nm]	-	-	-	22 ± 10	15 ± 6	10 ± 4
S	-	-	-	0.7 ± 0.2	0.8 ± 0.2	0.84 ± 0.16

(Table 1). This factor is given by the relation of the area of all particles to the total film area. It is used for the quantitative characterization of the portion of embedded metal. Computer aided image analysis of the TEM micrographs was not only applied to determine f_a but also to get geometrical information of each single particle of the assembly. For the characterization of the particles the equivalent circle diameter $D = \sqrt{4\ A\pi^{-1}}$ and the shape factor $S = 4\pi A/U^2$ were determined (Table 1). A and U are the area and circumference of the particle, respectively. The values of the particles shape factor S are in the range $0 \leq S \leq 1$, where $S = 1$ corresponds to a circular object.

At higher metal contents the number of junctions between the particles increases and no particle analysis is possible anymore. In Fig. 2a and 2b no

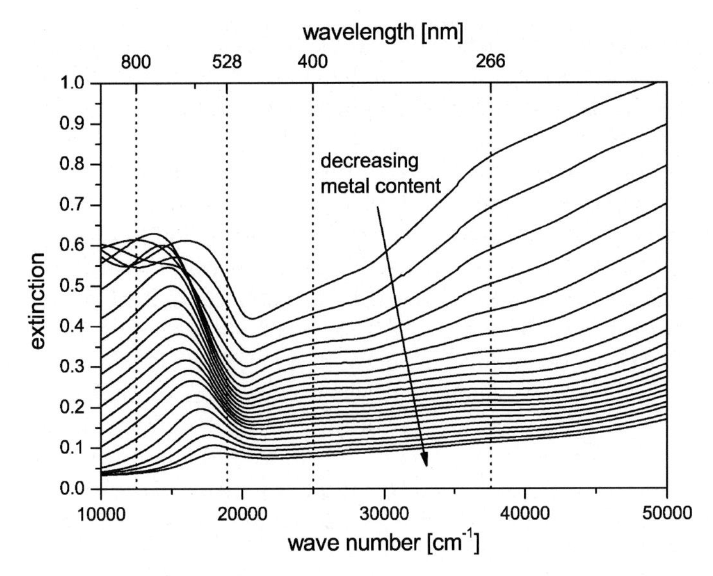

Fig. 3. Optical extinction spectra of a plasma polymer film with embedded gold nanoparticles recorded along gradient of gold content every 2 mm; *dotted vertical lines* indicate position of applied laser wavelengths)

separated particles can be found because the whole presented area consists of a metal network. In the case of Fig. 2c partially interconnected structures with a dimension of about 100 to 300 nm have formed. A decreasing metal content (Fig. 2d to 2f) leads to a reduction of particle size and the size distribution as well as an increase of the shape factor, i.e., there is a tendency to more spherical particles.

Figure 3 presents optical spectra recorded along the gradient in the gold content of a plasma polymer film. The spectra are dominated by surface plasmon resonance, which leads to a characteristic extinction peak. Its amplitude, position and width is determined by the size, shape and material of the nanoparticles and the dielectric host [11]. Due to the broad particle size and shape distribution which is the result of Volmer–Weber growth of the nanoparticles, the spectra show a broad extinction peak as well. With decreasing metal content, i.e., decrease of f_{a} and variation of the particle size and shape distribution as described above, the amplitude of the extinction peak diminishes and the peak position shifts to higher wave numbers. The blue shift results from the smaller and more spherical particles as well as the reduction of dipolar coupling between the particles.

3.2 Effect of Structural Parameters

The irradiation with appropriate laser parameters can result in well developed laser-induced periodic structures in the nanoparticle containing polymer films. They are characterized by clear separated, alternating line-like areas in which the particles show significant size and shape modifications due to coalescence and areas in which the particles remain unchanged in comparison to the original sample (Fig. 1, left). The polarization of the laser beam determines the direction of the periodic structures which is always parallel to the E field vector. Based on the metal content of the Au/PP film prior to laser irradiation, the periodic structure formation was only found in an area filling factor range of $0.59 \leq f_a \leq 0.85$ (Fig. 2). No periodic structures could be observed after irradiation of films with a metal content above or below this region. Instead, samples with $f_a > 0.85$ showed either non-ordered coalescence of the particles or stronger structural changes up to ablation, while in films with $f_a < 0.60$ the changes of the particles are too small to detect a clear periodic structure. In the case of the film presented in Fig. 2, the best developed LIPS were found at $f_a = 0.75$ (Fig. 2c). Going to the upper filling factor boundary (e.g., at $f_a = 0.85$ – Fig. 2b) there is a tendency to unordered coalescence where the line-like areas with typically unchanged particles are affected by the structural modifications as well. Going to the lower boundary (e.g., at $f_a = 0.70$ and 0.60 – Fig. 2d and e) the periodic structure is weak pronounced due to the less changes of the smaller particles. Comparing these observations with the determined geometrical values of the particle assemblies in the former chapter, the best pronounced periodic structures are formed if junctions between several nanoparticles arise. Therefore, the direct coupling between the particles seems to play an important role for the structure formation process.

In addition to the parameters concerning the metal particles, the influence of the plasma polymer layers was investigated. It was found that the layer below as well as the layer above the nanoparticles are necessary for the laser-induced formation of well developed periodic structures. First experiments irradiating films with gold nanoparticles embedded in a sputtered PTFE matrix (polytetrafluoroethylene) have shown similar periodic structures as in the case of a plasma-polymerized HMDSN matrix.

3.3 Effect of Intensity

Besides the size and shape of the particles and their arrangement in the matrix, the laser intensity is another important parameter which determines whether the periodic structure occurs or not (Fig. 4). For the investigations the peak intensity I_{max} of the Gaussian intensity profile was determined and correlated with the particle modifications observed in the center of the laser irradiated lines, i.e., the area with the strongest structural changes. In the case of too low intensity (Fig. 4, left), only slight local particle modifications are found which can already reveal a periodicity similar to that of clear LIPS. If the intensity is too high all particles are modified and the periodicity disappears. This results in mostly spherical particles after the irradiation with relatively high intensities as shown in the right SEM image in Fig. 4. Only in

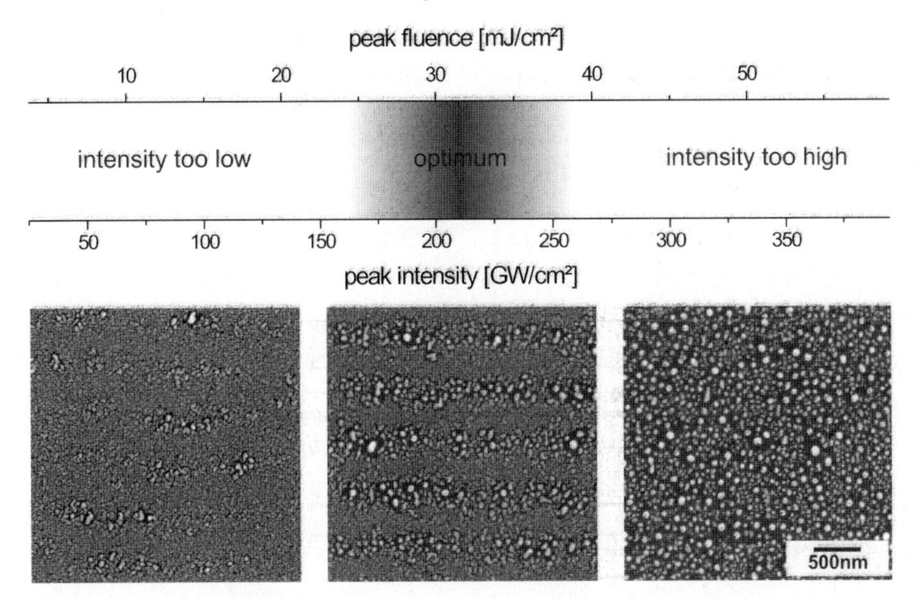

Fig. 4. *Graph* presents intensity and corresponding fluence range for LIPS formation in Au/PP films at $\lambda = 800$ nm. SEM images of Au/PP film at the center of the laser irradiated line (λ = 800nm) in the case of optimum peak intensity (middle – 210 GW/cm^2) and too low and too high peak intensity (*left* and *right* – 168 GW/cm^2 and 1277 GW/cm^2, respectively), in all three cases nearly same filling factor

a relatively narrow intensity range clear developed periodic structures occur as can be seen in the SEM image in the middle of Fig. 4. The values of the required intensities are material specific and depend on the optical absorption of the film at the applied laser wavelength. The optical absorption of the Au/PP films within the filling factor range for LIPS formation is in the same range for $\lambda = 800$ nm, 528 nm and 400 nm. For $\lambda = 266$ nm the value is much higher due to the additional absorption by the plasma polymer. Therefore, less intensity is required to produce periodic structures and a stronger tendency to ablation of the film was observed.

Another important question is the dynamical development of the LIPS. The number of pulses per spot N in the case of moving the sample through the laser beam can be calculated as $N = 2Df/v$ with the spot diameter $D = 2\omega$, the pulse repetition rate f and the velocity of the sample v. Using values for D between 300 and 500 µm, $v = 1$ mm/s and $f = 1$ kHz the number of pulses per spot is about 600 to 1000. Due to the Gaussian intensity profile of the laser beam the energy deposited on a distinct sample position varies from pulse to pulse which complicates an evaluation of the results.

Therefore, experiments without moving the sample and applying a different number of pulses (1–1000) were carried out. After single pulse irradiation

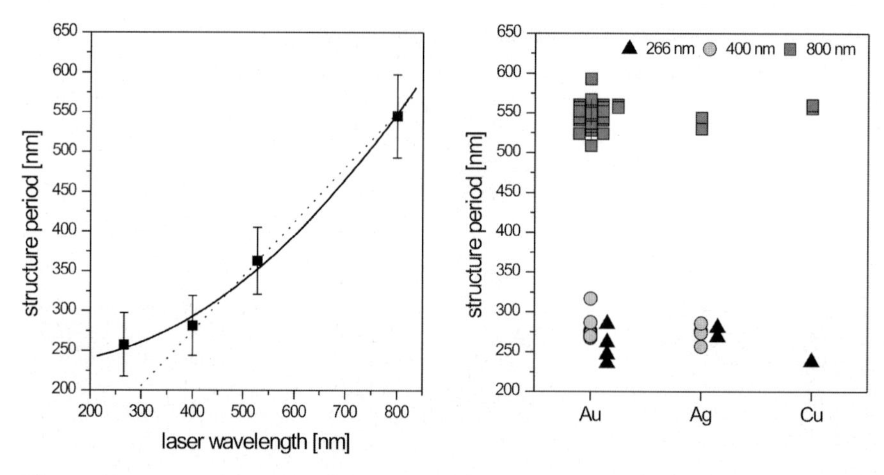

Fig. 5. Laser wavelength used for the film irradiation vs. structure period Λ (left); mean values of the structure period at different wavelengths for various plasma-polymerized films containing different metal particles (*right*)

no periodic structures could be observed whereas an increase of the number of pulses to 400 or more resulted in LIPS formation. Besides the required number of pulses, a modification threshold has to be exceeded which was found to be in the peak intensity range of 35 to 70 GW/cm^2 (corresponding peak fluences 5–10 mJ/cm^2).

3.4 Effect of Wavelength

Besides the fundamental wavelength of the Ti-sapphire laser $\lambda = 800$ nm the second (400 nm) and third harmonic (266nm) and an intermediate wavelength of 528 nm were used for irradiation. In all cases, LIPS formation could be observed although the optical response of the particle system at the four wavelengths differ (Fig. 3). It was found that the applied laser wavelength determines the structure period Λ of the LIPS, i.e., the distance between the areas with modified particles (Fig. 5 left). The period of LIPS obtained in a single experiment can vary within a range of about 50 nm due to slight modulations and branching of the structures. For $\lambda = 400$ to 800 nm a linear dependence of structure period and laser wavelength can be assumed. In the case of $\lambda = 266$ nm larger mean values of Λ, that differ from the linear fit, have been determined. The structure periods of laser-irradiated films containing silver or copper particles are in the same range as values for gold containing films (Fig. 5 right), i.e., the periodicity of the LIPS is not significantly influenced by the particle material.

The dependence of the structure period Λ on the laser wavelength is often described for laser-induced periodic structures [2, 3, 5]. Their formation is described as a result of a self-enhanced interference process between inci-

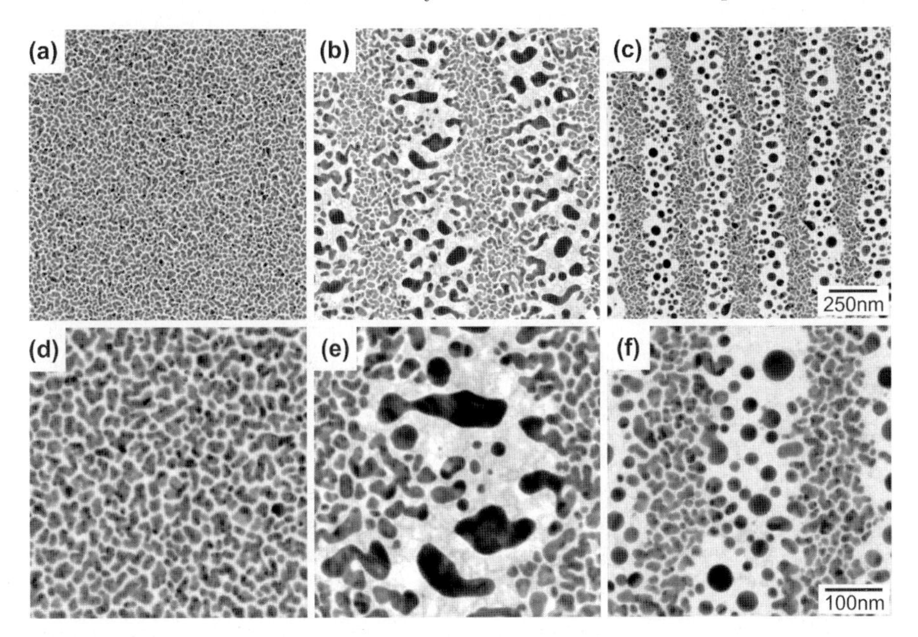

Fig. 6. TEM images of a Au/PP film in the original state (*left*) and after laser irradiation with $\lambda = 800$ nm (middle) and 400 nm (*right*); **(d)**–**(f)** magnification of **(a)**–**(c)**

dent and scattered laser light [3]. Therefor, the equation $\Lambda = \lambda/n(1 \pm sin\theta)$ was derived, where n is the refractive index and θ the angle of incidence, respectively. Considering normal incidence and wavelength $\lambda = 400$to800 nm, n could be determined as $n = 1.46 \pm 0.01$ by a linear fit. The measured refractive index of the pure plasma polymer films is about 1.56. Regarding this difference it has to be mentioned that for the derivation of the interference equation it was assumed that $\epsilon' > 1$ and $\epsilon'' \approx 0$, i.e., the material is non-absorbing. For a better description of the correlation between Λ and λ, the absorption behavior of the Au/PP films should be taken into account by using the wavenumber dependent, effective dielectric function of the composite material. This effective dielectric function can be calculated from the dielectric functions of gold and plasma polymer using a so called effective medium theory [9].

Besides the structure period, the laser wavelength influences the size and shape distribution of the modified particles. Figure 6shows LIPS after irradiation with $\lambda = 800$ nm (left) and 400 nm (right). For 800 nm the particles are larger and have more elongated shapes ($D = (40 \pm 22)$ nm, $S = 0.71 \pm 0.23$) and the particles are smaller and become more spherical ($D = (33 \pm 12)$ nm, $S = 0.88 \pm 0.18$) for 400 nm. A possible explanation for the different laser-induced particle modifications are larger diffusion paths and the higher number of particles "available" for coalescence with increasing wavelength. For

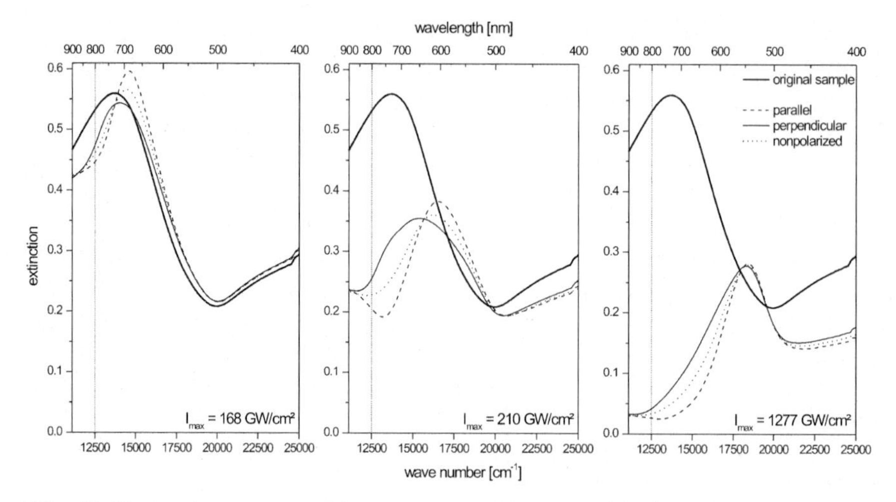

Fig. 7. Extinction spectra of laser irradiated lines in a Au/PP film with different applied peak intensities I_{max}. Each diagram presents the spectra of the original and the laser-irradiated sample. For the irradiated sample, spectra of nonpolarized as well as linearly polarized light parallel and perpendicular to the direction of the laser polarization are given

$\lambda = 266\ nm$ the width of the areas with higher energy input is so small that it is in the range of the particle diameters. This may limit the structure period Λ to a minimum possible value that results in the deviation from the linear dependence of Λ and λ (Fig. 5, left).

3.5 Anisotropic Optical Properties

According to the described modifications of the particle assembly also a modification of the optical properties of the films due to the laser irradiation could be observed. Further, the resulting optical properties are affected by laser irradiation parameters as well. Here, we want to present the influence of the laser intensity. For this purpose several lines were written into the same film along the metal gradient with different laser energies keeping all other laser parameters constant. Afterwards the spectra were recorded from an area of 100x100 μm^2 in the centers of the individual lines at positions with approximately the same filling factor. A spectrum of the not irradiated area between the lines presents the original state of the sample. Figure 7 shows the spectra that correspond to the SEM images in Fig. 4, which display the particle modifications at different laser intensities. The SEM images represent only a small part of the area where the spectra were recorded.

Comparing the spectra of the original and the laser-irradiated film recorded with nonpolarized light (dotted lines in Fig. 7), for all three irradiation intensities a clear shift of the extinction maximum to higher wavelengths and a reduction of the peak width can be observed. With increasing peak intensity a larger shift of the maximum (794, 2535 and 4684 cm^{-1}) and a stronger decrease of the peak width determined as half width at half maximum(-390, -399 and -548 cm^{-1} for $I_{max} = 168, 210$ and1277 GW/cm^2, respectively) is found. This can be attributed to recrystallization and coalescence resulting in more spherical particles and the reduction of dipolar coupling between the particles due to an increase of the particle distances.

Recording optical spectra with linearly polarized light measured parallel and perpendicular to the laser polarization, a distinct difference between the spectra can be seen (Fig. 7). For $I_{max} = 210$ GW/cm^2 (Fig. 7, middle), i.e., in the case of periodic structure formation, the extinction peak measured parallel to the laser polarization is shifted to higher wave numbers and has a smaller width in comparison to the extinction peak measured perpendicular to laser polarization. At less energy ($I_{max} = 168$ GW/cm^2; Fig. 7, left), i.e., only weak periodic particle modifications, the described dichroism is in principle the same, but there is no overall decrease of the extinction as in the former case. At $I_{max} = 1277$ GW/cm^2 (Fig. 7, right) the peak maxima are at the same position but the extinction peak measured parallel to the laser polarization has a smaller width. The equal peak positions correlate with the mostly spherical particle shapes observed after laser irradiation as can be seen in Fig. 4 (left). Possible explanations for the observed dichroism can be a perpendicular orientation of a distinct fraction of elongated particles and a weaker dipolar coupling in the direction of the laser polarization. Both, the short axis excitation of the oriented elongated particles and the higher reduction of dipolar coupling would lead to a blue shift as observed in the spectra recorded parallel to the laser polarization.

Optical modeling using Rayleigh–Gans theory [12, 13] for ellipsoidal particles (half axes a, b and c), which are small compared to the wavelength of the light, was applied to understand the correlation between the laser-induced structural changes and the recorded spectra, especially the observed dichroism. Figure 8 presents calculated spectra of gold spheres ($a = b = c = 10$ nm) and prolate spheroids ($a = 10$ nm; $b = c = 7$ nm) in a polymer matrix with a refractive index of 1.6. Spectra were calculated for incident light which was polarized either parallel or perpendicular to the long axis of the spheroids. The resulting spectra show a difference in the extinction peak position of about 1600 cm^{-1} with the short axis excitation at higher wave number. The double peak structure of the long axis excitation results from the influence of the third axis which corresponds to the height of the particles in the case of the investigated samples. Due to the fact, that there is only one particle layer in the Au/PP films, the influence of this axis is very small and no double peak is found in the recorded spectra. Nevertheless, the distance between the two extinction maxima for light polarization parallel and perpendicular to the long axis is in the same range as in the case of the spectra in Fig. 7. Here the shift is 535 and 1046 cm^{-1} for $I_{max} = 168$ and 210 GW/cm^2, respectively. Therefore, already a relatively small elongation of the particles (ratio major axis/minor axis is about 1.4 – presented in the sketch in Fig. 8) perpendic-

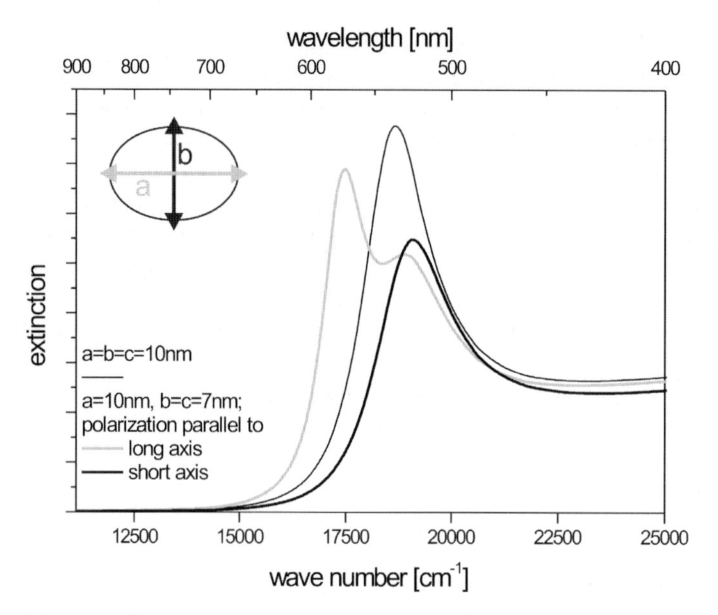

Fig. 8. Calculated extinction spectra of spherical and prolate spheroidal gold nanoparticles in a polymer matrix (n = 1.6) for linear polarized light with polarization direction parallel and perpendicular to the long axis of the spheroids

ular to the laser polarization would be sufficient to produce the observed dichroism with regard to the peak position.

4 Summary and Conclusions

Only a complex interplay of appropriate laser parameters and structural properties of a metal/polymer composite film enables the formation of laser-induced periodic structures. The periodic structuring process occurs at different wavelengths and on various embedded metal nanoparticles. The applied laser wavelength determines the structure period. For an interpretation of the results and an explanation of the mechanism of the periodic structure formation, models based on an interference effect seem to be feasible. Although, the mechanisms and characteristics of LIPS are described in literature, there are still open questions concerning the different feedback mechanisms of various materials. The most important difference between the "common" LIPS observations and the effect we described here is, that the main structural changes do not happen directly at the surface, but in the metal layer located several nanometers below. Furthermore, LIPS formation occurs on a metal particle layer and not on a bulk material. Due to the fact that femtosecond laser pulses can lead to a number of nonlinear effects [14], the observed mechanisms cannot be explained by considering only the initial optical properties of the sample. The behavior of the dielectric function of the composite material under intense excitation has to be taken into account. Closed aggregated nanoparticles interact strongly via dipolar forces, which can affect the modification processes. The relatively high fraction of metal nanoparticles leads to high absorption of the deposited pulse energy in the system so that thermal effects have a substantial contribution to the structural alterations. Thermally induced processes which lead to atomic diffusion through the matrix and along the particle surface, are coalescence, reshapening (recrystallization) and Ostwald ripening. A possible modification of the chemical and physical polymer structure seems to be possible, but could not be detected so far. For a better description, in particular of the kinetic of structural modification, time-resolved laser irradiation experiments will be carried out.

Acknowledgements

The German Research Society (DFG-SFB 418) is gratefully acknowledged for financial support. Many thanks to H. Graener, G. Seifert, M. Dyrba (Department of Physics, Martin Luther University Halle-Wittenberg) for the laser irradiation experiments and for fruitful discussions.

References

[1] M. Birnbaum: J. Appl. Phys. **36**, 3688 (1965)
[2] A. E. Siegman, P. M. Fauchet: J. Quant. Electron. **22** 1384 (1986)
[3] D. Bäuerle: *Laser Processing and Chemistry* (Springer, Berlin 2000)
[4] J. E. Sipe, J. F. Young, J. S. Preston, H. M. van Driel: Phys. Rev. B **27** 1141 (1983)
[5] S. A. Akhmanov, V. I. Emel'yanov, N. I. Koroteev: *Interaction of Strong Laser Irradiation with Solids and Nonlinear Optical Diagnositics of Surfaces*, Teubner-Texte zur Physik, Band 24 (BSB Teubner, Leipzig 1990)
[6] M. Sendova, M. Sendova-Vassileva, J. C. Pivin, H. Hofmeister, K. Coffey, A. Warren: J. Nanosci. Nanotechnol. **6** 748 (2006)
[7] M. Kaempfe, H. Graener, A. Kiesow, A. Heilmann: Appl. Phys. Lett. **79** 1876 (2001)
[8] A. Kiesow, S. Strohkark, K. Loeschner, A. Heilmann: Appl. Phys. Lett. **86** 153111-1 (2005)
[9] A. Heilmann: *Polymer Films with Embedded Metal Nanoparticles* (Springer, Berlin Heidelberg 2003)
[10] A. Kiesow, J. E. Morris, C. Radehaus, A. Heilmann: J. Appl. Phys. **94** 6988 (2003)
[11] U. Kreibig, M. Vollmer: *Optical Properties of Metal Clusters*, Springer Series in Materials Science, Vol. 25 (Springer, Berlin 1995)
[12] A. Heilmann, M. Quinten, J. Werner: Europ. Phys. J. B 3 **4** 455 (1998)
[13] R. Gans: Ann. Physik **38** 881 (1912)

[14] J. C. Diels, W. Rudolph: *Ultrashort Laser Pulse Phenomena: Fundamentals, Techniques and Applications on a Femtosecond Time Scale* (Academic, San Diego 1996)

Proteins and Patients – Magnetic Nanoparticles as Analytic Markers

Meinhard Schilling and Frank Ludwig

Institut für Elektrische Messtechnik und Grundlagen der Elektrotechnik,
Technische Universität Braunschweig,
Hans-Sommer-Strasse 66, 38106 Braunschweig, Germany
m.schilling@tu-bs.de

Abstract. Exciting new applications of low noise magnetic sensors in biotechnology, medical diagnosis and therapy are at hand for magnetic nanoparticles (MNP) , which are already used in the clinical routine. The relaxation of magnetic nanoparticles is a fascinating method which can be employed for the characterization of ferrofluids and is further developed as analytical tool for biochemistry. In this article the basics and methods of the characterization of diluted ferrofluids by magnetorelaxometry (MRX) are described. We start with the properties of the ferrofluid which consists of magnetic nanoparticles. The time schedule for relaxation measurements is discussed. The details of our fluxgate based experimental set-up are presented. To achieve high sensitivity even at low concentrations of magnetic nanoparticles we optimize fluxgate sensors for low intrinsic noise, high sensitivity and large bandwidth.

1 Superparamagnetic Nanoparticles

Magnetic nanoparticles (MNPs) in ferrofluids in general have diameters starting from a few nanometers up to the micrometer range. The magnetic nanoparticles, which can be used in magnetic relaxation measurements, are restricted to diameters of some 10 nm. This is only a factor of 2 to 10 larger than a molecule like a protein. Due to this small size the nanoparticles are not ferromagnetic anymore, even if the material itself in larger sizes behaves ferromagnetically. Instead, due to the high ratio of surface to volume magnetization the nanoparticles show superparamagnetism. For many applications it is advantageous to employ magnetite Fe_3O_4 particles, since they are biocompatible and can be metabolized by many organisms. The chemical stability of these nanoparticles is not as good as necessary and thus organic shells are protecting the magnetic nanoparticles from decomposition. With these shells also functionalization can be achieved, for instance by the binding of antibodies to the organic shell material. The structure of a MNP is depicted in Fig. 1 together with an electron micrograph of dried MNPs, which agglomerated during drying.The magnetic nanoparticles can be rotated by an external homogeneous magnetic field or can be moved in magnetic gradient fields.If diluted ferrofluids with superparamagnetic nanoparticles are investigated, the statistics of the core size distribution and the distribution of the

R. Haug (Ed.): Advances in Solid State Physics,
Adv. in Solid State Phys. **46**, 83–93 (2008)

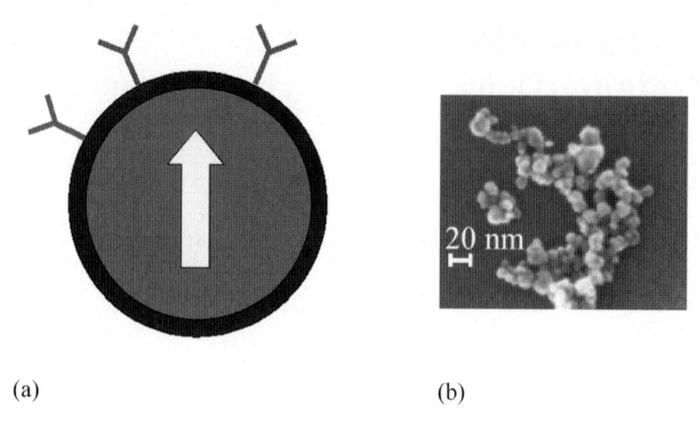

(a) (b)

Fig. 1. (**a**) Magnetic nanoparticle with superparamagnetic core (*light grey*) and organic shell (*dark grey*) with three attached antibodies. (**b**) Electron micrograph of some agglomerated nanoparticles

shell thickness have to be considered. The alignment of the magnetic moment in an external magnetic field depends on the ratio of anisotropy energy and thermal energy, and the net magnetic moment of an ensemble of superparamagnetic nanoparticles is described by the Langevin function.

2 Magnetic Relaxation Measurements

To observe the magnetic relaxation of a diluted ferrofluid an external homogeneous magnetic field is applied, as schematically drawn in Fig. 2. The MNPs start to align with this field, but thermal excitation inhibits complete alignment in the usually employed magnetic flux density of some millitesla. The magnetic signal during magnetization already contains valuable information on the size distribution and the anisotropy constant of the nanoparticle material [1]. The magnetizing field normally aligns the particles for some seconds and is then switched off. The switching time is limited by the inductance of the field coil and the relaxation of eddy currents in the vicinity of the set-up. When the set-up has relaxed only the magnetic signal of the MNPs is measured. Depending on the temperature, viscosity of the medium, the concentration of MNPs and the time constant of relaxation of the magnetization of each particle, the magnetic signal decays with a mean time constant between microseconds and seconds. For the detection of the magnetic signal from the sample under investigation either superconducting quantum interference devices (SQUIDs) [2, 3] or fluxgate sensors [4, 5] are employed.

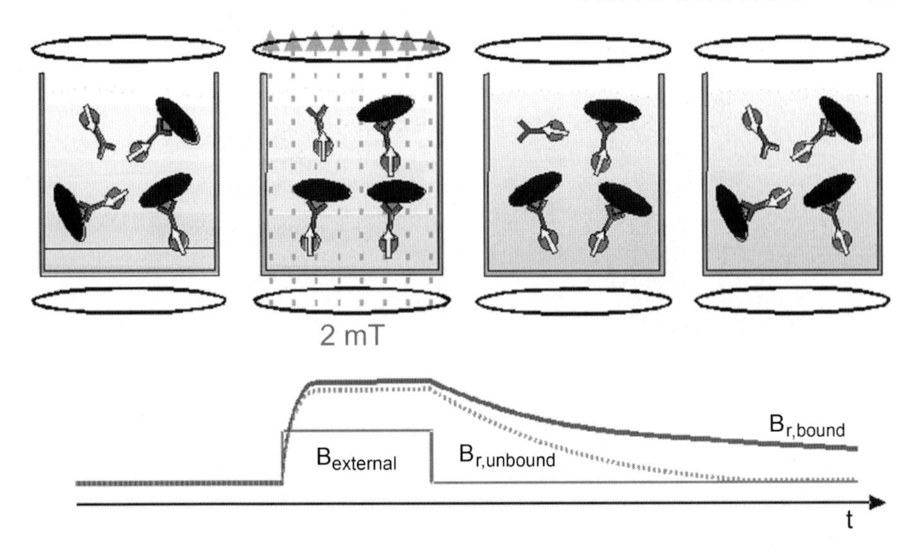

Fig. 2. Time schedule of magnetic relaxation measurements

3 Experimental Set-Up for MRX Measurements

For MRX measurements mostly SQUID magnetometers are employed, since they are the solid-state sensors with the lowest magnetic flux-density noise. In [4] we presented MRX measurements with a single fluxgate and compared this set-up with a SQUID arrangement. The new set-up with two fluxgates in differential arrangement increases the signal-to-noise ratio by a factor of $\sqrt{2}$ [6].This is achieved by measuring the stray field of the sample perpendicularly to the direction of the external magnetic field H_{mag} for magnetization, as depicted in Fig. 3a. Thus, the fluxgates do not measure the magnetizing field directly, but the ferrofluid stray field. This set-up allows one to perform the measurements in unshielded environment with the same resolution and background noise as in magnetically shielded environment, since external disturbing fields from 50 Hz-sources cancel, when the difference of both fluxgate signals is calculated. This can be seen in Fig. 3b, where the signals of both fluxgates and the difference signal are shown.The switching process has been optimized in our set-up to achieve as short as possible switching time and a system relaxation time of below 100 µs has been realized. Currently, only the bandwidth of the fluxgates of 3 kHz limits the onset of data aquisition.

Fluxgate sensors are very low noise magnetic sensors, which use the nonlinear magnetization $M(H)$ dependence of high permeability magnetic core material [7]. For short, a signal generator drives the soft magnetic core periodically into saturation. This drive frequency samples the external field to be measured. The generation of second harmonics of this drive frequency by an slowly varying external field is used to compensate the external field in the core in a control loop. This way, linear, sensitive magnetic field sensors with

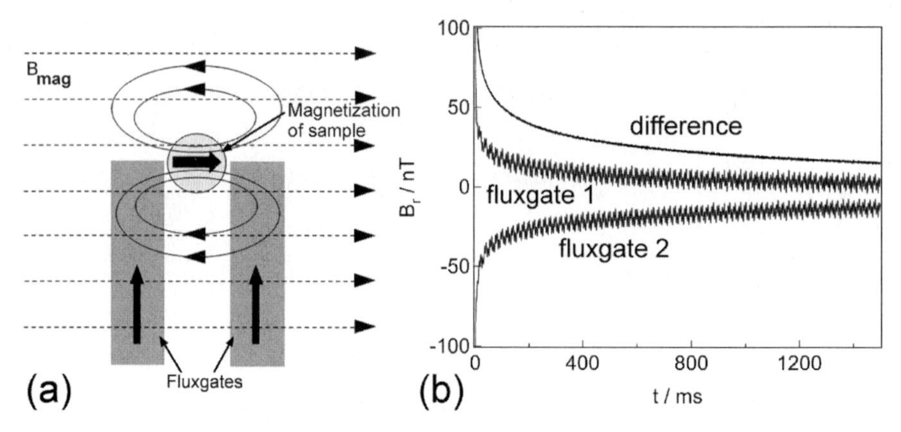

Fig. 3. (**a**) Experimental configuration for magnetic relaxation with differential fluxgate configuration, (**b**) MRX signals of both fluxgates and their difference signal

a very high dynamic range can be realized. Compared to superconducting quantum interference devices (SQUIDs) the noise is still about two orders of magnitude higher, but sensitivity and dynamic range are quite comparable.

4 Analysis and Detection Limits

The relaxation of magnetic nanoparticles can take place either via the Brownian or the Néel mechanism. If a MNP is placed in a liquid and the anisotropy energy is large enough to block the internal magnetization, the particle will rotate as a whole by thermal activation out of the magnetized direction. This process is referred to as Brownian relaxation. The time constant for this process is proportional to the hydrodynamical volume of the particle, i.e, including shell and attached biomolecules etc., proportional to the viscosity of the medium and inversely proportional to temperature. In the Néel relaxation process only the magnetic moment of the particles itself rotates. The Néel mechanism takes place either if the magnetization is not blocked or if the Brownian rotation is inhibited, e.g., by immobilization of the MNP. The time constant of the Néel mechanism depends on the ratio between anisotropy energy KV, where V is the volume of the MNP core and K the anisotropy constant, and thermal energy $k_{\mathrm{B}}T$. The relaxation times for Brownian and Néel relaxation versus hydrodynamic and core diameter, respectively, are depicted in Fig. 4. In a magnetorelaxometry experiment always the mechanism with the shorter time constant determines the behavior of each individual MNP. As can be seen, magnetite MNPs with core diameters below about 15 nm will always relax via the Néel mechanism independently of whether the MNP is mobile or immobilized. A MNP having a core diameter above 20 nm and being freely mobile in a solution relaxes via the Brownian process.

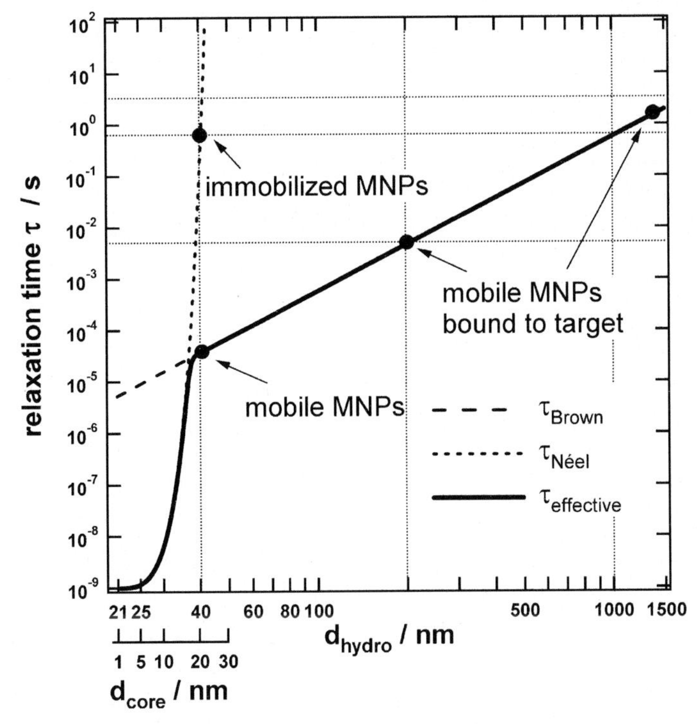

Fig. 4. Dependance of the magnetic relaxation time constants on the core and hydrodynamic diameter of the MNP, respectively

If this MNP is immobilized, e.g., by binding to a substrate, relaxation can take place only via the Néel mechanism.

When the hydrodynamic diameter of the MNP becomes larger, e.g., by binding to biomolecules such as proteins, the relaxation time will increase.In practice, available MNPs with organic shells are not monodisperse but exhibit a relatively wide distribution of core and shell size. The description of the magnetization and relaxation process of ensembles of non-interacting MNPs is generally performed with the magnetic moment superposition model [1, 8]. This model was successfully applied to the characterization of the magnetic core properties of MNPs. To do so, MNPs are immobilized, e.g., by diluting them in a mannite solution and subsequent freeze-drying. Assuming that the MNP cores have circular shape, the relaxation signal is given by

$$m_{\mathrm{r}}(t, t_{\mathrm{mag}}, H) = M_{\mathrm{s}} \int f(d,\mu,\sigma)\frac{\pi d^3}{6}L(d,H,T) \left\{1 - \exp\left[-\frac{t_{\mathrm{mag}}}{\tau_{\mathrm{NH}}(K,d,H)}\right]\right\} \exp\left[-\frac{t}{\tau_{\mathrm{N}}(K,d)}\right]\mathrm{d}d\,, \tag{1}$$

where $f(d, \mu, \sigma)$ is the size distribution function with a particle diameter d, M_s the saturation magnetization, $L(d, H, T)$ the Langevin function and t_{mag} the magnetization time. The term in curly brackets describes the magnetization dynamics of an individual MNP and the last term the decay of the magnetic signal after switching off the magnetizing field. The time constants τ_{NH} for the magnetization and τ_N for the relaxation process can differ; in our data analysis we use Néel's expression [9], which holds for uniaxial symmetry

$$\tau_N = \tau_0 \exp\left(\frac{KV}{k_B T}\right) \text{ and } \tag{2}$$

$$\tau_{NH} = \tau_0 \exp\left[\frac{KV}{k_B T}\left(1 - \frac{2H_{mag}}{H_K}(\cos\Psi + \sin\Psi) + \left(\frac{H_{mag}}{H_K}\right)^2\right)\right], \tag{3}$$

which can be derived by combining Néel's picture with the Stoner-Wohlfarth model [1]. Ψ denotes the angle between the external magnetic field H_{mag} and the magnetic moment m of the individual particle and it is $H_K = 2\mu_0 K/M_s$. For $f(d, \mu, \sigma)$ we assume a lognormal distribution. Note that the first part in (1) can be applied to calculate static $M(H)$ curves. Figure 5 depicts a measured relaxation curve along with the best fit using the model described above. To extract concentrations of bound and unbound MNPs from magnetorelaxation curves, as it has to be done in magnetic relaxation immunoassays (MARIA) [2], much simpler, empirical formulas are applied [3, 8]. The relaxation signal of mobile MNPs can generally be fitted by a stretched exponential function

$$B_r(t) = B_{ub} \exp\left[-\left(\frac{t}{\tau_{ub}}\right)^\beta\right]. \tag{4}$$

The relaxation of immobilized MNPs can be described by

$$B_r(t) = B_b \ln\left(1 + \frac{\tau_b}{t}\right). \tag{5}$$

Measuring the magnetorelaxation curves of dilution series of MNP samples it turned out that the fit parameters τ_{ub}, τ_b and β do not change but are characteristic for the given ferrofluid. On the other hand, B_{ub} and B_b are relative measures for the content of unbound and bound MNPs, respectively. In Fig. 5a the fit parameters deduced from the relaxation curves measured on samples with MNPs diluted in water are shown versus nominal Fe content. For comparison, the results obtained with SQUID MRX on the same samples are depicted [4]. Figure 5b shows the amplitudes B_b obtained for a dilution series of freeze-dried MNP samples, measured with and without magnetic shielding. The detection limit of about 200 nmol Fe achieved at $B_{mag} = \mu_0 H_{mag} \approx 200$ pT is determined by the noise of the fluxgate magnetometers [6]. Further improvements of the detection limit can be achieved by

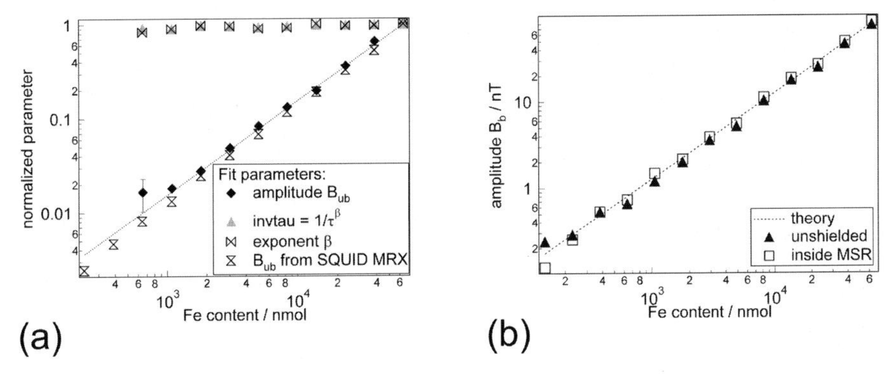

Fig. 5. Dependance of (**a**) normalized fit parameters of MRX curves of mobile MNPs and (**b**) fitted MRX amplitude of immobilized MNPs on the nominal Fe content

1. the use of ferrofluids with narrower size distribution and proper mean core size (the Fe_3O_4 ferrofluids characterized so far exhibit generally mean core sizes around 10 nm and standard deviations of $\sigma \approx 0.4$, consequently only a very small portion of MNPs contributes to the relaxation signal),
2. the reduction of the sample volume (due to the $1/r^3$-dependence of the magnetic field of a magnetic dipole mainly the MNPs close to the sensor contribute to the signal) and
3. the use of fluxgate magnetometers with lower noise and larger bandwidth (the bandwidth of the currently used commercial fluxgates of 3 kHz forces one to start with the data analysis not before about 300 μs after switching off the magnetizing field).

5 Further Characterization of Magnetic Nanoparticles

The size distribution of the MNPs is very important for the correct interpretation of the relaxation data, especially if concentration measurements are intended, where the strength of the magnetic signal during relaxation is directly evaluated. So, for the development of a reliable measurement method other methods are employed to compare the determined size distributions to get a self-consistent picture of the MNP solution. Therefore, we use ac-susceptometry, dc-magnetization measurements with a SQUID together with SEM-, STEM- and AFM-micrographs for characterization.

5.1 Ac-Susceptometry

The response of the magnetic nanoparticles to an external ac-magnetic field contains also valuable information on the distribution of MNP sizes in the

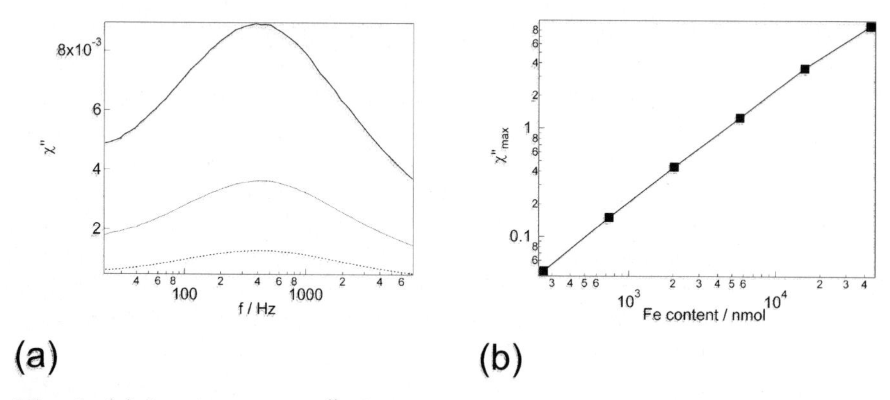

Fig. 6. (**a**) Imaginary part χ'' of the ac-susceptibility versus frequency measured on mobile magnetic nanoparticles of different concentrations and (**b**) dependence of the amplitude at the maximum of the imaginary part of the ac-susceptibility on the concentration of MNPs

sample. Sweeping the ac frequency, the maximum in χ'' is found for a frequency ω equal to the inverse relaxation time τ of MNPs, so that $\omega\tau = 1$. If the MNPs are not monodisperse the maximum in χ'' refers to the mean size of MNPs. In Fig. 6a a typical example for such an ac-susceptibility measurement is depicted. The curves with different strength correspond from top to bottom to samples with decreasing concentration of magnetic nanoparticles. In contrast to magnetorelaxometry measurements, the peak in χ'' provides a direct measure of the mean characteristic time constant τ_{N} and τ_{B}, respectively, and thus of the mean particle size. By analyzing the frequency dependance of the response in more detail, information on the size distribution can be deduced and compared to that obtained from MRX measurements. In Fig. 6b the dependance of the amplitude at the maximum of χ'' on the MNP concentration is evaluated. The detection limit in our ac-susceptibility set-up is about the same as in the MRX measurements.

5.2 Dc-Magnetization by SQUID-Susceptometry

The dependance of the mean magnetization on an external static magnetic field is investigated by dc-magnetization measurements [1]. The $M(H)$ curves are analyzed using the first term of (1) providing independent information on the core size distribution. Here, also consistency can be proven. Note that in contrast to the magnetization and relaxation dynamics the static magnetization $M(H)$ does not depend on the anisotropy constant K and the saturation magnetization M_{s}.

5.3 Comparison of Magnetic Nanoparticle Solutions

With the methods mentioned above various ferrofluids from magnetic nanoparticles with different organic shells have been investigated and compared for their applicability in MRX measurements [10]. For successful application the MNPs have to exhibit both Brownian and Néel relaxation time constants which are accessible within the measurement window between 300 μs and about 10 s. If the mean diameter of the particles is too small and even mobile MNPs show Néel relaxation, immobilization does not change the mechanism and the nanoparticles are not suitable for the realization of a magnetic relaxation immunoassay (MARIA) [10].

6 Fractionation

To decrease the spread of the size distribution found in the investigated ferrofluids, magnetic columns have been used to separate larger from smaller MNPs. Such a commercially available column consists of small plastic-coated iron spheres, which are placed in a container in a static magnetic field of up to 800 mT. Since the magnetic force on a MNP is proportional to its magnetic moment and the magnetic gradient field, for a given field applied to the column, only MNPs with sizes below a certain threshold in diameter will be washed out. Performing successive washing steps for gradually decreasing applied field, fractions of different mean size can be obtained. In Fig. 7 the resulting mean diameters of such fractions after stepwise decreasing the magnetic field from 500 mT to 0 mT as obtained from the analysis of ac-susceptibility measurements are depicted.

7 Applications

7.1 Process Monitoring

The method of magnetic relaxation is investigated primarily for applications in biotechnology and biochemistry. In the Sonderforschungsbereich 578 with the title "From gene to product" research groups from microbiology, biochemistry, and biotechnology cooperate with engineers from mechanical and electrical engineering to establish new routes for the bioengineering production of proteins. For short, the proteins are produced by fermentation of fungi or bacteria in various processes. The proteins, which are partially secreted into the growth medium, have to be monitored for process regulation and to determine the momentary protein yield. As alternative tool to time consuming and more complicated assays, we investigate magnetic nanoparticles as specific markers for the produced proteins.

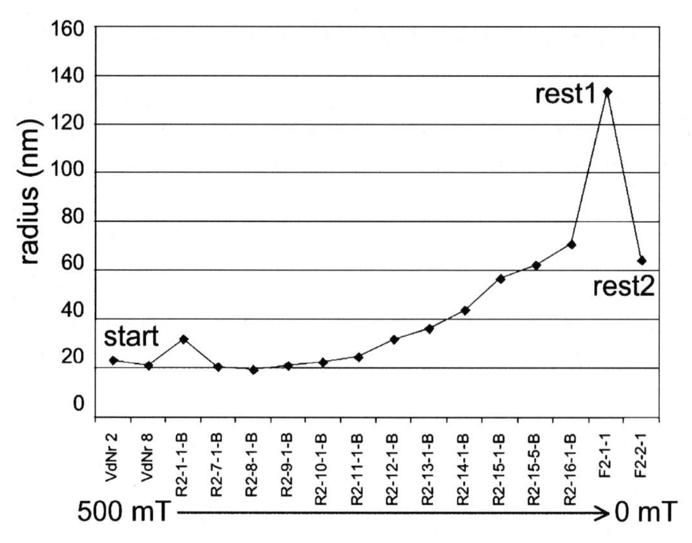

Fig. 7. Fractions of MNP after successive washing steps with decreasing magnetic field. The mean diameters before fractionation (start) and of the rest from the column (rest1, rest2) are depicted

7.2 Drug Delivery

Another important application of magnetic nanoparticles is drug targeting for therapy. Since with external magnetic field gradients the particles with attached therapeutical substances can be moved to the site in the body where the drugs are needed, much lower doses can be applied compared to the unspecific distribution of drugs by the blood circulation alone. This promises cheaper therapies and less side effects as still common, e.g., in cancer chemotherapy. Another important application in drug targeting is the determination of the dissolving behavior of drugs introduced to the body. This can be monitored by MNPs, since during dissolving from the solid matrix of tablets etc. the MNPs are set free and change their relaxation properties. This way a quantitative analysis of the drug release can be monitored [11]. For this purpose we investigate hydrogels, which possess pores large enough for mechanical encapsulation of pharmaceutical substances without chemical bonds. These hydrogels polymerize under UV-radiation and dissolve in aqueous solution depending on the pH-value and on temperature.

8 Summary

In conclusion, we investigate diluted ferrofluids of superparamagnetic nanoparticles for applications in biochemistry as specific, magnetic markers for the determination of protein concentration and their temporal evolution. By this method fast and simple immunoassays without washing steps can be realized. Therefore, the nanoparticle size and magnetic moment distribution have to be well characterized, which can be achieved by a combination of functional measurements like relaxation, ac susceptibility, dc magnetization and structural characterization with electron microscopy and scanning probe microscopy.

Acknowledgements

We would like to acknowledge financial support by the Deutsche Forschungsgemeinschaft by the SFB 578.

References

[1] F. Ludwig, E. Heim, D. Menzel, M. Schilling, J. Appl. Phys. **99**, (2006).
[2] R. Kötitz, L. Trahms, H. Koch, W. Weitschies, in *Biomagnetism: Fundamental Research and Clinical Applications*, edited by C. Baumgartner, L. Deecke, S. J. Williamson (Elsevier Science, Amsterdam, 1995), p. 785.
[3] Y. R. Chemla, H. L. Grossmann, Y. Poon, R. McDermott, R. Stevens, M. D. Alper, J. Clarke, Proc. Natl. Acad. Sci. USA **97**, 14268 (2000).
[4] F. Ludwig, E. Heim, S. Mäuselein, D. Eberbeck, M. Schilling, J. Magn. Magn. Mater. **293**, 690 (2005).
[5] E. Heim, F. Ludwig, M. Schilling, Biomed. Tech. **49**, 420 (2004).
[6] F. Ludwig, S. Mäuselein, E. Heim, M. Schilling, Rev. Sci. Instr. **76**, 106102–1 (2005).
[7] C. Hinnrichs, C. Pels, M. Schilling, J. Appl. Phys. **87**, 7085 (2000).
[8] D. Eberbeck, S. Hartwig, U. Steinhoff, L. Trahms, Magnetohydrodynamics **39**, 77 (2003).
[9] L. Néel, Adv. Phys. **4**, 191 (1955).
[10] E. Heim, F. Ludwig, M. Schilling, Biomed. Tech. **50**, 613 (2005).
[11] E. Heim, S. Harling, F. Ludwig, H. Menzel, M. Schilling, to be published.

Novel Quantum Transport Effects in Single-Molecule Transistors

Felix von Oppen and Jens Koch

Institute for Theoretical Physics, Freie Universität Berlin,
Arnimallee 14, 14195 Berlin, Germany
vonoppen@physik.fu-berlin.de, Jens.Koch@physik.fu-berlin.de

Abstract. Transport through single molecules differs from transport through more conventional nanostructures such as quantum dots by the coupling to few well-defined vibrational modes. A well-known consequence of this coupling is the appearance of vibrational side bands in the current-voltage characteristics. We have recently shown that the coupling to vibrational modes can lead to new quantum transport effects for two reasons. (i) When vibrational equilibration rates are sufficiently slow, the transport current can drive the molecular vibrations far out of thermal equilibrium. In this regime, we predict for strong electron-phonon coupling that electrons pass the molecule in avalanches of large numbers of electrons. These avalanches consist themselves of smaller avalanches, interrupted by long waiting times, and so on. This self-similar avalanche transport is reflected in exceptionally large current (shot) noise, as measured by the Fano factor, as well as a power-law frequency spectrum of the noise. (ii) Due to polaronic energy shifts, the effective charging energy of molecules may be strongly reduced compared to the pure Coulomb charging energy. In fact, for certain molecules the effective charging energy U can become negative, a phenomenon known in chemistry as potential inversion. We predict that transport through such negative-U molecules near charge-degeneracy points, where the Coulomb blockade is lifted, is dominated by tunnelling of electron pairs. We show that the dependence of the corresponding Coulomb-blockade peaks on temperature and bias voltage is characteristic of the reduced phase space for pair tunnelling.

1 Sketch of Molecular Electronics

In recent years, electronic transport through nanostructures has witnessed a shift towards molecular systems [1, 2]. Several ingenious schemes for measuring transport through single molecules have been realized and experimental control over such systems is rapidly improving. The most popular experimental techniques create two closely spaced (gold) electrodes by the use of breakjunctions [3–5] or electromigration [6]. The molecule may or may not be chemically bound (e.g., by thiol groups) to the electrodes. An alternative method first attaches the molecule to gold clusters via thiol groups before maneuvering (using ac-fields) the resulting conglomerate towards two microfabricated electrodes [7]. An intriguing direction is the use of wet molecular junctions, where the molecule is immersed in an electrolytic environment [8]. Another powerful method employs a scanning-tunneling-microscope (STM)

R. Haug (Ed.): Advances in Solid State Physics,
Adv. in Solid State Phys. **46**, 95–105 (2008)

tip as an electrode for measuring the current through a molecule, which rests on a substrate [9].

Experiments on single-molecule junctions observe many effects familiar from conventional nanostructures such as quantum dots, especially in setups that include a gate electrode. Prominent examples are the Coulomb blockade which arises due to the large charging energy of nanostructures, as well as the Kondo effect [10]. Due to the small size of molecules, the Kondo effect persists in these systems to rather high temperatures [11]. It has been argued that Coulomb-blockade physics, via charging fluctuations of nearby traps, is also at the origin of temporal fluctuations in the measured IV characteristics [7].

A prime difference between transport through single molecules as opposed to transport through more conventional nanostructures lies in the coupling of the electronic degrees of freedom responsible for transport to few well-defined vibrational modes (phonons). A prominent effect of this electron-phonon coupling is the appearence of vibrational sidebands in the current-voltage characteristics. These are observed at large bias voltages, comparable to or larger than typical phonon frequencies. While the basic phenomenon has been known for a number of decades [12–14], current interest in the context of single-molecule junctions started with the work of *Park* et al. [15] on a C_{60} molecule located in between two metallic leads.

In the meantime, vibrational features have been observed in transport in a wide variety of molecules, ranging >from hydrogen H_2 [16] to larger conjugated molecules [9]. Depending on the system, vibrations appear in the current-voltage (IV) characteristics as steps or kinks [17]. Typically, molecules more strongly coupled to the leads (as for example in STM experiments where the molecule is lying on the substrate) will exhibit kinks in the IV characteristics [9]. By contrast, steps in the IV characteristic are observed for weakly coupled molecules (as for example in breakjunction experiments without chemical bond between electrode and molecule) [15].

An important parameter in single-molecule junctions is the vibrational relaxation rate. For some systems, relaxation times can be as large as 10 ns (measured for suspended nanotubes) [18]. Thus, these times can be comparable or longer than the time between two consecutive electrons traversing the molecule which are of the order of 100 ps (0.1 ps) for a current of 1 nA (1 μA). Thus, in the intriguing regime of slow vibrational relaxation (large currents), the current drives the molecular vibrations far out of equilibrium [19, 20]. Such nonequilibrium vibrations have so far been observed directly in at least one experiment as *absorption* sattelites of Coulomb blockade peaks [18].

The theoretical description of transport through single molecules rests on two pillars which are at present somewhat disconnected. One approach uses density functional theory to obtain a detailed account of the molecular orbitals including parts of the electrodes [21]. In a second step, the Kohn-Sham potentials of density functional theory are used to set up a (single-particle) transport problem in the spirit of Landauer. While there are impressive successes, differences between experiment and this "ab-initio" theory can be as large as several orders of magnitude [22]. Discrepancies between theory and experiment are particularly pronounced for fully conjugated molecules. It is important to realize that the use of density-functional theory for transport is an uncontrolled (though often useful) approximation as there is no theorem underlying the definition of a scattering problem on the basis of the Kohn-Sham orbitals. There are some efforts under way to improve this situation within the context of time-dependent density functional theory [23, 24]. It is also difficult to account for the coupling to molecular vibrations as well as electronic correlations within density functional theory.

An alternative approach starts from models of electronic transport through nanostructures [19, 25]. This approach finds its justification in the fact that experiments on molecular junctions exhibit several phenomena which are familiar from quantum dots. In addition, such models can in principle be treated systematically in a variety of transport regimes, including the coupling to molecular vibrations and electronic correlation effects. A minimal model assumes that transport through the molecule is dominated by a single, spin-degenerate molecular orbital ε_d [lowest unoccupied molecular orbital (LUMO) or highest occupied molecular orbital (HOMO)], coupled to one vibrational mode of frequency ω. Double occupation of the molecular orbital "costs" a charging energy U. Adding electrons to the molecule shifts the equilibrium position of the nuclei, which provides a coupling of the electronic and vibrational degrees of freedom of the molecule. This physics is sketched in Fig. 1 and encapsulated in the Hamiltonian [13, 14, 19]

$$H = H_{\mathrm{mol}} + H_{\mathrm{vib}} + H_{\mathrm{leads}} + H_{\mathrm{i}} \,, \tag{1}$$

where the separate contributions correspond to the electronic molecular orbital

$$H_{\mathrm{mol}} = \varepsilon_d n_d + U n_{d\uparrow} n_{d\downarrow} \,, \tag{2}$$

to the molecular vibrations

$$H_{\mathrm{vib}} = \hbar\omega b^\dagger b + \lambda\hbar\omega (b^\dagger + b) n_d \,, \tag{3}$$

to the leads

$$H_{\mathrm{leads}} = \sum_{a=L,R} \sum_{\mathbf{p},\sigma} \epsilon_{\mathbf{p}} c^\dagger_{a\mathbf{p}\sigma} c_{a\mathbf{p}\sigma} \,, \tag{4}$$

and to the tunneling between leads and molecule,

$$H_{\mathrm{i}} = \sum_{a=L,R} \sum_{\mathbf{p},\,\sigma} \left(t_a c^\dagger_{a\mathbf{p}\sigma} d_\sigma + \mathrm{h.c.} \right) . \tag{5}$$

Here, the operator d_σ annihilates an electron with spin σ on the molecule, $c_{a\mathbf{p}\sigma}$ annihilates an electron in lead a ($a = L, R$) with momentum $\mathbf{p}$

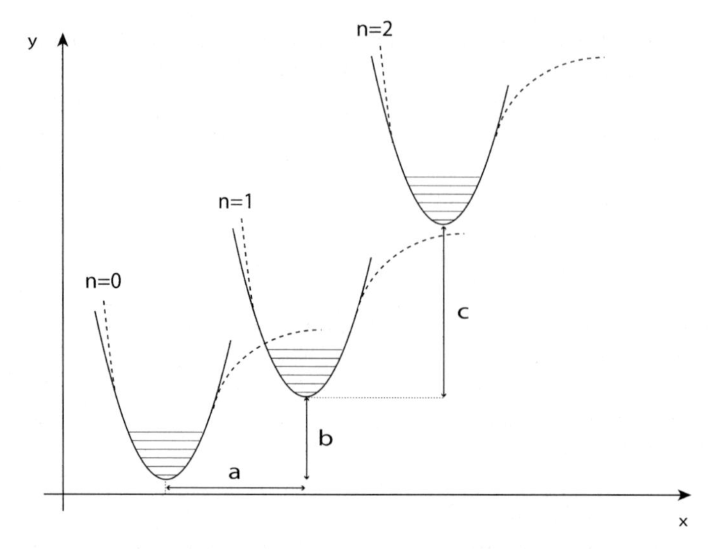

Fig. 1. Potential surfaces corresponding to the model Hamiltonian of a molecule featuring a single, spin-degenerate electronic orbital coupled to a single vibrational mode. The model Hamiltonian approximates the potential surfaces by a harmonic-oscillator potential

and spin σ, vibrational excitations are annihilated by b, and $t_{L,R}$ denotes the tunneling matrix elements. The electron-phonon coupling (with coupling constant λ) can be eliminated by the Lang–Firsov canonical transformation [26], which implies renormalizations of the tunneling Hamiltonian $t_a \to t_a e^{-\lambda(b^\dagger - b)}$, of the orbital energy $\varepsilon_d \to \varepsilon_d - \lambda^2 \hbar\omega$, and of the charging energy $U \to U - 2\lambda^2 \hbar\omega$.

Frequently, the position of the molecular orbital ε_d can be varied experimentally by a gate voltage. Electronic transport proceeds by a variety of processes, depending on the position of ε_d relative to the Fermi energies of the leads. If ε_d lies within the bias voltage window between the Fermi energies of left and right lead, transport occurs by sequential (or resonant) tunneling. On the other hand, electrons will occupy the molecule only virtually while tunneling between left and right lead, if the molecular orbital lies far below or far above this bias-voltage window. The latter process is often referred to as cotunneling in the quantum-dot literature [27].

In the limit of high temperatures (as compared to the molecule-lead coupling), transport through single molecules can be treated systematically within the rate-equation approach [19, 28, 29]. Rate equations allow one to account for the coupling to molecular vibrations, for charging effects, as well as for the spin degrees of freedom. This approach has been successfully employed in the theory of vibrational sidebands [19, 25, 30], in describing transport through magnetic molecules [31], as well as in the description of the regime of vibrational nonequilibrium mentioned above [20].

Beyond the rate-equation limit, no systematic theoretical methods are currently available, when charging effects are relevant. An exception is the theory of Kondo correlations where vibrational sidebands of the Kondo resonance have been predicted [32]. Effects of the charging energy U are less prominent if the molecule is well coupled to at least one of the electrodes, as, e.g., in most STM experiments. In this case the time spent by the electron on the molecule is small compared to the inverse phonon frequency, and one expects that perturbation theory in the electron-phonon coupling is appropriate [33].

While experiments to date, with few exceptions, have focused on measurements of IV characteristics, a number of other quantities have been investigated theoretically. A very interesting quantity is shot noise in the transport current which has been widely investigated in the context of conventional nanostructures, both in theory and experiment [34]. It has been shown that vibrational sidebands also appear in the noise power as function of bias voltage [19, 30], and giant shot noise has been predicted in certain situations of vibrational nonequilibrium (see Sect. 3) [30,35]. Thermoelectric and thermal properties of molecular junctions were studied theoretically in several works [36–38]. A particularly interesting quantity is the thermopower which, unlike the IV characteristic, would give information on whether the current flow proceeds predominantly through LUMO or HOMO. The underlying reason is that a nonzero thermopower requires breaking of particle-hole symmetry about the Fermi energy.

2 Novel Quantum Transport Effects

As mentioned above, transport experiments on single-molecule transistors exhibit effects familiar from transport through quantum dots, most notably the Coulomb blockade as well as the Kondo effect. In view of this, it is natural and important to ask the following question: Can we exploit the enormous variety of molecular structures to observe novel quantum transport effects or to access novel regimes of quantum transport? This question is particularly pertinent to those of us, who prefer to approach molecular electronics as a field of science rather than engineering.

A natural place to look for such novel transport physics in single-molecule transistors involves the coupling of the current-carrying electrons to the molecular vibrations. The reason is that unlike quantum dots, where one would be typically dealing with an entire continuum of phonon modes, molecules typically have few well-defined and localized vibrational modes. Indeed, we believe that there are at least two reasons why this coupling to molecular vibrations can lead to novel quantum transport which goes beyond the vibrational side bands reviewed in the previous section.

The first reason is that the transport current can drive the molecular vibrations far out of thermal equilibrium. We will review in Sect. 3 that this

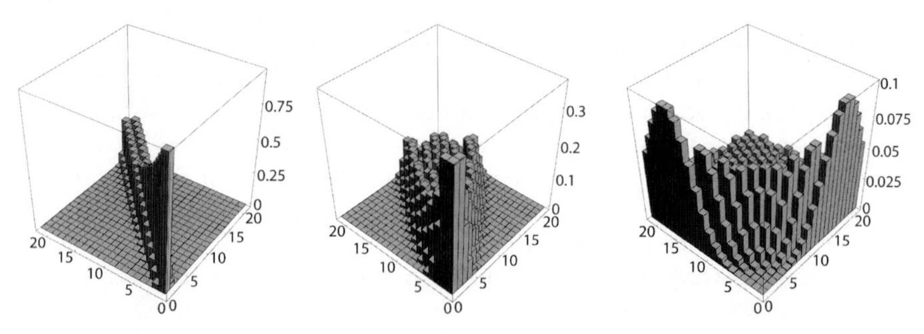

Fig. 2. Franck–Condon matrix elements for weak ($\lambda = 0.2$), intermediate ($\lambda = 1$), and strong ($\lambda = 4$) electron-phonon coupling. The *upper panel* depicts the squared Franck–Condon matrix elements as a function of initial and final phonon states, q_i and q_f. The *lower panel* illustrates the corresponding shifts $\Delta x = \sqrt{2}\lambda\ell_\mathrm{osc}$ of the harmonic oscillator potentials for the two charge states $n = 0$ and $n = 1$

may result in self-similar avalanche transport. The second reason is the polaronic downward renormalization of the charging energy $U \to U - 2\lambda^2\hbar\omega$ mentioned in Sect. 1. The renormalization is particularly intriguing when the effective charging energy becomes negative. This situation is familiar from negative-U centers in semiconductors and referred to as potential inversion in electrochemistry. We show in Sect. 4 that transport in this regime is characterized by pair tunneling of electrons.

3 Franck-Condon Blockade and Self-Similar Avalanche Transport

In the high-temperature limit, the tunneling Hamiltonian H_i may be treated perturbatively, and the formalism of rate equations can be employed for the computation of current and noise [19, 28, 29]. The central input to this formalism consists of the rates for the tunneling-induced transitions, which can be obtained systematically by an expansion of the T-matrix and the use of Fermi's golden rule. Due to the presence of electron-phonon coupling, all electronic transitions are accompanied by transitions in the space of molecular vibrations. In the transition rates, this is reflected by Franck–Condon (FC) matrix elements which are given by the overlap of the initial and final harmonic-oscillator wavefunction. As depicted in Fig. 2, the FC matrix elements crucially depend on the strength of the electron-phonon coupling.

Strong electron-phonon coupling ($\lambda \gg 1$) is characterized by a large displacement between the molecular potential surfaces for different charge states. As a result, the overlap between vibrational states in the vicinity of the ground state acquire an exponential suppression, and the corresponding FC matrix elements are drastically reduced for low-lying phonon states, see Fig. 2. At low bias voltages where only vibrational states close to the ground states are accessible, this reduction of FC matrix elements causes a strong current suppression which we termed Franck–Condon blockade [30]. In contrast to other current suppressions such as Coulomb blockade, the FC blockade cannot be lifted by tuning a gate voltage. This provides a characteristic fingerprint that may be accessed in experiments with gated setups.

The FC blockade is a generic feature for strong electron-phonon coupling, independent of the vibrational relaxation rate. Surprisingly however, the transport mechanisms for strong and weak vibrational relaxation are entirely different. For strong vibrational relaxation, the occupation of vibrational states is fixed to the thermal equilibrium distribution. When applying a bias, electrons are transferred across the molecule one by one, and the current shot noise is sub-Poissonian, as usually expected in the case of transport by fermionic carriers. By contrast, for weak vibrational relaxation it is the tunneling dynamics which determines the phonon distribution. We have shown that the vibrational nonequilibrium causes avalanches of large numbers of electrons intermitted by long waiting times [30, 35], see Fig. 3. The corresponding shot noise is drastically enhanced ($F \sim 10^2$–10^3 for $\lambda = 4$), where the Fano factor reflects the average number of electrons per avalanche.

The origin of these avalanches can be traced back to the increase of the FC matrix elements towards highly excited phonon states, and to the fact that weak vibrational relaxation allows for the accumulation of phonon excitations. As a concrete example, we consider the situation of a low bias voltage that limits the maximal number of phonon excitations to $\Delta q = 1$ per tunneling event. Then, long waiting times always occur whenever the system resides in the vibrational ground state. When the system succeeds in making a transition, the increase of FC matrix elements towards higher phonon numbers will favor the excitation of one phonon. The accessible transition rates from the state $q = 1$ are significantly larger due to the enhanced overlap of harmonic-oscillator wavefunctions. Consequently, the system undergoes a rapid series of tunneling events, proceeding to even higher phonon excitations. This dynamics forms an avalanche, which is only terminated when by chance, the system returns to the vibrational ground state.

Zooming into one such avalanche, we have shown that it consists of smaller avalanches, see Fig. 3. This self-similarity is explained by the fact that the situation in the first excited phonon state $q = 1$ very much resembles the ground state situation: Again, the transition rates for $q = 1$ are suppressed and increase when reaching higher phonon levels, resulting in waiting times and (sub-)avalanches. Experimentally, this self-similarity may be probed by a measurement of the noise frequency spectrum, which exhibits a characteristic power-law decay as a function of frequency, $S \sim f^{-1/2}$. Interestingly, the self-similar nature of the avalanches allows for a completely analytical treatment of the transport, and the derivation of a compact formula for the Fano factor as well as the full counting statistics [35].

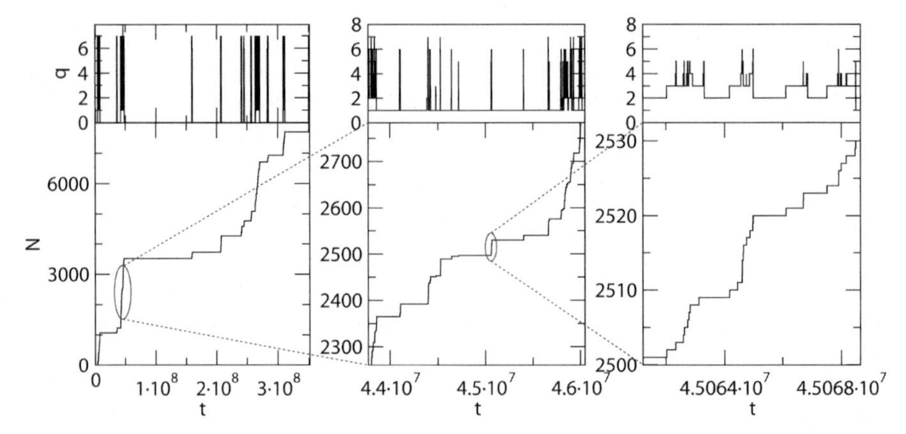

Fig. 3. Self-similar electron avalanches for strong electron-phonon coupling ($\lambda = 4.5$). The lower three panels depict the number of transferred electrons N as a function of time t (in units of $\hbar/\Gamma$). The *middle* and *right hand plot* are magnifications of the respective previous plot, resolving the self-similar structure. The *upper panels* depict the number q of excited phonons, revealing that waiting times of different hierarchy levels correspond to different qs

4 Negative-U Molecules

Apart from the emergence of FC matrix elements in transition rates, the coupling between electronic and vibrational degrees of freedom also leads to a renormalization of the charging energy $U \to U - 2\lambda^2\hbar\omega$, the polaron shift. The most intriguing consequence of this downward shift is the possibility of a negative charging energy, i.e., an effectively attractive electron-electron interaction. In electrochemistry, examples for this scenario are in fact known [41]. Negative U causes the system to favor even occupation numbers, and by tuning the gate voltage the charge states $n = 0$ and $n = 2$ can become degenerate.

At very low temperatures, this scenario leads to the charge Kondo effect [42], and several papers have recently investigated this regime numerically [43, 44]. In the high-temperature regime $T \gg T_K$, Kondo correlations are irrelevant up to logarithmic corrections. We have shown that in this limit transport in the negative-U model is dominated by tunneling of electron pairs [45]. In distinction to sequential tunneling and elastic cotunneling, this two-particle process is associated with a reduced phase space, resulting in a characteristically different behavior of the corresponding rates. While sequential and elastic cotunneling lead to constant rates (whenever the one-particle level is located in the bias window), pair-tunneling rates increase roughly linearly with the detuning $\varepsilon_d + U/2$ of the two-particle level below the lead Fermi energy.

At the degeneracy point $2\varepsilon_d + U = 0$, the tunneling of electron pairs generates a peak in the linear conductance as a function of gate voltage.

As a result of the linear increase ofrates with detuning, the peak width is proportional to temperature while its height is fixed to $G = 24e^2\Gamma_L\Gamma_R/U^2h$, where $\Gamma_a = 2\pi\nu\left|t_a\right|^2$ denotes the partial level width induced by tunneling in junction a. It is important to note that this scaling of the peak with temperature is entirely different from the conventional Coulomb-blockade peak, for which peak height and width are correlated so that its integral value remains constant.

Additional differences between the conventional Coulomb blockade and the negative-U scenario emerge at finite bias. Apart from the cotunneling regime, the nonlinear IV characteristics for negative U show regions dominated by pair tunneling. In contrast to conventional Coulomb diamonds, inside which the current is essentially constant, the pair-tunneling regions are signalled by a phase-space induced current increase with growing bias voltage.

Intriguingly, the realistic scenario of coupling asymmetry, $\Gamma_L \neq \Gamma_R$ causes a negative-U device to develop transistor-like properties. Specifically, we have shown that for $\Gamma_L \ll \Gamma_R$ the current strongly depends on the bias direction, resulting in current rectification. Additionally, the sense of rectification may be altered by the gate voltage, allowing for a gate-controlled switching of the device. We have explained this behavior in terms of the dominant transport mechanism for asymmetric coupling, which involves a transition with an electron pair leaving to (or coming from) opposite leads. It is the gate-voltage dependence of such a process that results in the transistor-like behavior.

5 Summary

We have shown that single-molecule transistors are a promising arena for novel quantum transport physics for at least two reasons. First, the current can drive the molecular vibrations far out of thermal equilibrium. We find that in the regime of strong electron-phonon interaction, this may lead to current flow in the form of a self-similar hierarchy of avalanches of large numbers of electrons. Specifically, this effect occurs in a regime of transport which we term Franck–Condon blockade where the low-bias current is suppressed due to the coupling to vibrational modes, even when the molecular orbital is located in the bias-voltage window between the Fermi energies of the two leads. Experimental signatures of self-similar avalanche transport are giant Fano factors, a power-law shot noise spectrum, as well as strongly non-Gaussian full counting statistics. We also mention in passing that interesting non-equilibrium effects occur for weak electron-phonon interactions where the degree of vibrational nonequilibrium grows as the electron-phonon interaction decreases. Vibrational nonequilibrium may also affect vibrational sidebands for anharmonic or charge-dependent molecular vibrations [39, 40].

Second, the coupling to vibrations renormalizes the molecular charging energy by a polaronic energy shift. As familiar from the theory of superconductivity, the electron-phonon interaction counteracts the Coulomb repulsion and reduces the effective charging energy of the molecule. This downward renormalization of the charging energy becomes particularly interesting when the resulting effective charging energy becomes negative. In this case, the molecular ground states as function of gate voltage always involves an even number of electrons. We have shown that transport through such negative-U molecules is dominated by tunneling of electron pairs near those gate voltages where two molecular charge states become degenerate. The corresponding Coulomb blockade peaks directly reflect the characteristic phase space for pair tunneling, which provides convenient experimental signatures of pair tunneling.

Acknowledgements

The work reviewed in this paper has been supported by Sfb 658 and the Studienstiftung d. dt. Volkes. We have benefitted from discussions with many colleagues. We would like to mention specifically A. Nitzan, Y. Oreg, M. Raikh, E. Sela, as well as M. Semmelhack, with whom we have collaborated on various aspects of molecular electronics.

References

[1] M. A. Reed and J. M. Tour, Sci. Am. **282**, 86 (2000).
[2] C. Joachim, J. K. Gimzewski, and A. Aviram, Nature (London) **408**, 541 (2000).
[3] J. Moreland and J. W. Ekin, J. Appl. Phys. **58**, 3888 (1985).
[4] C. Zhou, C. J. Muller, M. R. Deshpande, J. W. Sleight, and M. A. Reed, Appl. Phys. Lett. **67**, 1160 (1995).
[5] J. M. van Ruitenbeek, A. Alvarez, I. Piñeyro, C. Grahmann, P. Joyez, M. H. Devoret, D. Esteve, and C. Urbina, Rev. Sci. Instrum. **67**, 108 (1996).
[6] H. Park, A. K. L. Lim, A. P. Alivisatos, J. Park, and P. L. McEuen, Appl. Phys. Lett. **75**, 301 (1999).
[7] T. Dadosh, Y. Gordin, R. Krahne, I. Khivrich, D. Mahalu, V. Frydman, J. Sperling, A. Yacoby, and I. Bar-Joseph, Nature (London) **436**, 677 (2005).
[8] L. Grüter, F. Cheng, T. T. Heikkilä, M. T. Gonzalez, F. Diederich, C. Schönenberger, and M. Calame, Nanotechnology **16**, 2143 (2005).
[9] X. H. Qiu, G. V. Nazin, and W. Ho, Phys. Rev. Lett. **92**, 206102 (2004).
[10] J. Park, A. N. Pasupathy, J. I. Goldsmith, C. Chang, Y. Yaish, J. R. Petta, M. Rinkoski, J. P. Sethna, H. D. Abruñas, P. L. McEuen, and D. C. Ralph, Nature (London) **417**, 722 (2002).
[11] L. H. Yu and D. Natelson, Nano Lett. **4**, 79 (2004).
[12] R. C. Jaklevic and J. Lambe, Phys. Rev. Lett. **17**, 1139 (1966).
[13] L. I. Glazman and R. I. Shekhter, Sov. Phys. JETP **67**, 163 (1988).
[14] N. S. Wingreen, K. W. Jacobsen, and J. W. Wilkins, Phys. Rev. Lett. **61**, 1396 (1988).

[15] H. Park, J. Park, A. K. L. Lim, E. H. Anderson, A. P. Alivisatos, and P. L. McEuen, Nature **407**, 57 (2000).
[16] R. H. M. Smit, Y. Noat, C. Untiedt, N. D. Lang, M. C. van Hemert, and J. M. van Ruitenbeek, Nature **419**, 906 (2002).
[17] M. Galperin, M. A. Ratner, and A. Nitzan, J. Chem. Phys. **121**, 11965 (2004).
[18] B. LeRoy, S. Lemay, J. Kong, and C. Dekker, Nature **432**, 371 (2004).
[19] A. Mitra, I. Aleiner, and A. J. Millis, Phys. Rev. B **69**, 245302 (2004).
[20] J. Koch, M. Semmelhack, F. von Oppen, and A. Nitzan, Phys. Rev. B **73**, 155306 (2006).
[21] Y. Xue and M. A. Ratner, Phys. Rev. B **68**, 115407 (2003).
[22] F. Evers, F. Weigend, and M. Koentopp, Phys. Rev. B **69**, 235411 (2004).
[23] K. Burke, R. Car, and R. Gebauer, cond-mat/0410352 (2004).
[24] S. Kurth, G. Stefanucci, C.-O. Almbladh, A. Rubio, and E. K. U. Gross, Phys. Rev. B **72**, 035308 (2005).
[25] S. Braig and K. Flensberg, Phys. Rev. B **68**, 205324 (2003).
[26] I. G. Lang and Y. A. Firsov, Sov. Phys. JETP **16**, 1301 (1963).
[27] H. Grabert and M. H. Devoret, eds., *Single Charge Tunneling in Coulomb Blockade Phenomena in Nanostructures* (Plenum Press, New York and London, 1992).
[28] C. W. J. Beenakker, Phys. Rev. B **44** (1646).
[29] D. V. Averin, A. N. Korotkov, and K. K. Likharev, Phys. Rev. B **44**, 6199 (1991).
[30] J. Koch and F. von Oppen, Phys. Rev. Lett. **94**, 206804 (2005).
[31] F. Elste and C. Timm, Phys. Rev. B **71**, 155403 (2005).
[32] J. Paaske and K. Flensberg, Phys. Rev. Lett. **94**, 176801 (2005).
[33] S. Gao, M. Persson, and B. I. Lundqvist, Phys. Rev. B **55**, 4825 (1997).
[34] Y. M. Blanter and M. Büttiker, Phys. Rep. **336**, 1 (2000).
[35] J. Koch, M. E. Raikh, and F. von Oppen, Phys. Rev. Lett. **95**, 056801 (2005).
[36] D. Segal and A. Nitzan, J. Chem. Phys. **117**, 3915 (2002).
[37] D. Segal, A. Nitzan, and P. Hänggi, J. Chem. Phys. **119**, 6840 (2003).
[38] J. Koch, F. von Oppen, Y. Oreg, and E. Sela, Phys. Rev. B **70**, 195107 (2004).
[39] J. Koch and F. von Oppen, Phys. Rev. B **72**, 113308 (2005).
[40] M. R. Wegewijs and K. C. Nowack, New J. Phys. **7**, 239 (2005).
[41] C. Kraiya and D. H. Evans, J. Electroanal. Chem. **565**, 29 (2004), and references therein.
[42] A. Taraphder and P. Coleman, Phys. Rev. Lett. **66**, 2814 (1991).
[43] P. S. Cornaglia, H. Ness, and D. R. Grempel, Phys. Rev. Lett. **93**, 147201 (2004).
[44] L. Arrachea and M. J. Rozenberg, Phys. Rev. B **72**, 041301(R) (2005).
[45] J. Koch, M. E. Raikh, and F. von Oppen, Phys. Rev. Lett. **96**, 056803 (2006).

Part III

Nanowires and 1D Systems

Growth Evolution and Characterization of PLD Zn(Mg)O Nanowire Arrays

Andreas Rahm[1], Thomas Nobis[1], Michael Lorenz[1], Gregor Zimmermann[1], Nikos Boukos[2], Anastasios Travlos[2], and Marius Grundmann[1]

[1] Institut für Experimentelle Physik II, Universität Leipzig, Linnéstr. 5 04103 Leipzig, Germany
andreas.rahm@physik.uni-leipzig.de
[2] Institute of Materials Science, National Center for Scientific Research "Demokritos", 15310 Ag. Paraskevi Attikis, POB 60228 Athens, Greece

Abstract. ZnO and $Zn_{0.98}Mg_{0.02}O$ nanowires have been grown by high-pressure pulsed laser deposition on sapphire substrates covered with gold colloidal particles as nucleation sites. We present a detailed study of the nanowire size and length distribution and of the growth evolution. We find that the aspect ratio varies linearly with deposition time. The linearity coefficient is independent of the catalytic gold particle size and lateral nanowire density. The superior structural quality of the whiskers is proven by X-ray diffraction and transmission electron microscopy. The defect-free ZnO nanowires exhibit a FWHM(2θ-ω) of the ZnO(0002) reflection of 22 arcsec. We show (0-11) step habit planes on the side faces of the nanowires that are a few atomic steps in height. The microscopic homogeneity of the optical properties is confirmed by temperature-dependent cathodoluminescence.

1 Introduction

Zinc oxide as a II-VI semiconductor has been rediscovered and attracted increasing attention in recent years [1]. ZnO exhibits interesting properties like a high exciton binding energy of 61 meV, radiation hardness, biocompatibility and piezoelectricity [2]. In particular quasi one-dimensional ZnO nanostructures are being considered a key material for electronic and optical nanodevices including light emitters for the UV range. In addition to high device-related expectations, ZnO whiskers can be utilized for the understanding of fundamentally interesting phenomena [3]. There are reports on the synthesis of quasi-1D and 2D nanostructures [4] and combinations thereof [5], of tetrapod zinc oxide structures [6], of tapered tip morphologies [7] and of flowerlike ZnO [8], etc. A considerable amount of work has been done on gold catalyst driven processes. Pulsed laser deposition (PLD) has been proven to be a versatile technique for ZnO and MgZnO nanostructure growth [9]. Alloying ZnO with Mg enables the tuning of the band gap energy. However, detailed studies on growth evolution and the formation mechanisms, as well as on the lateral homogeneity of samples, are rare in the literature [10]. Here

R. Haug (Ed.): Advances in Solid State Physics,
Adv. in Solid State Phys. **46**, 109–120 (2008)

we report on a morphology evolution investigation of Zn(Mg)O nanowhiskers grown from gold colloidal particles by high-pressure PLD.

2 Experimental Details

The whiskers were synthesized on 10×10 mm^2 *a*-plane sapphire substrates by a special high-pressure PLD process for nanostructure growth [9] at a temperature of about 850°C. The target-to-substrate distance was 25 mm and 500-16,000 pulses of a KrF excimer laser were used to ablate a target with a laser energy density of about 2 J/cm^2. The pulse repetition rate was 10 Hz. In this work, growth from ZnO and $Mg_{0.26}Zn_{0.74}O$ rotating targets made from pressed and sintered 5N powders was investigated [11]. Ar was used as a carrier gas at a background pressure of 100 mbar and a constant flow rate of 100 sccm. As a growth catalyst we used 2 sizes of colloidal gold particles (Ted Pella, Inc.) suspended in water: 10 nm and 50 nm (coefficient of diameter variation $< 10\%$). The substrates were immersed into Poly-L-Lysine (0.1%) for two minutes, rinsed with distilled water and dried with nitrogen. The surface becomes adhesive to gold colloidal particles when coated with Poly-L-Lysine [12]. The nanoparticle solution was 1:10 diluted with distilled water before it was applied to the substrates for 120 s. (60 s. for the 50 nm colloid solution) and again rinsed with water and dried with N_2. The topography of the substrates covered with gold particles was characterized by a Vecco Metrology Dimension 3100 /Nanoscope IV atomic force microscope (AFM) with standard Si cantilevers. We used the tapping mode at a frequency of 300 kHz.

Field emission scanning electron microscopy (SEM) images of the nanowhisker samples were acquired using a FEI Nanolab 200. The crystalographic relations were revealed by high resolution X-ray diffraction (HRXRD) at a Philips Xpert with a Bonse–Hart collimator using Cu-$K_{\alpha 1}$ radiation. Transmission electron microscopy (TEM) investigations were done at a Philips CM20 operated at 200kV and equipped with a Gatan GIF200 imaging filter. The ZnO nanorods were separated from the substrate and subsequently dispersed on the holey carbon support film of a Cu grid. The optical properties of PLD-grown ZnO nanostructures have been investigated by temperature-dependent cathodoluminescence spectroscopy (CL). CL experiments were performed using a scanning electron microscope CamScan CS44 with 10 keV electron energy and a beam current of 200 pA. For the spectral analysis a Jobin Yvon HR 320 monochromator was used with a grating of 2400 lines per mm and an entrance slit width of 50 μm. The luminescence of the nanostructures was detected by a Spectrum One back-illuminated liquid-nitrogen cooled CCD array with 1024×256 pixels. The spectral resolution of such setup amounts to about 0.9 meV. The temperature T of the investigated samples was chosen in a range of $10\ \mathrm{K} \leq T \leq 150\ \mathrm{K}$ using a heatable helium-flow cryostat. Further details about the used CL equipment can be found in [13,14].

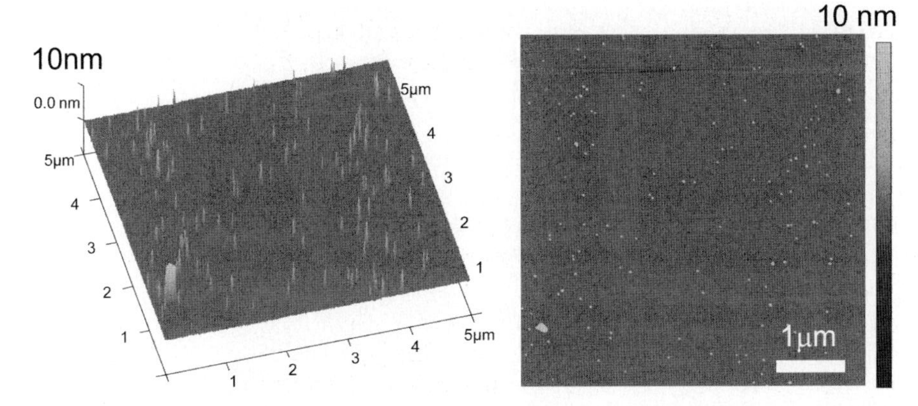

Fig. 1. AFM images of 10 nm gold colloidal particles on an a-plane sapphire surface attached using Poly-L-Lysine as an adhesive

3 Detailed Growth Evolution Studies

3.1 ZnO

Figure 1 depicts a typical substrate with 10 nm gold colloids prior to growth. The particles exhibit a narrow size distribution and are randomly arranged over the substrate surface. In order to observe the nanowhisker growth at different stages, a series of 5 samples with different growth times or, correspondingly, a different number of laser pulses (1,000-16,000) with a constant repetition frequency of 10 Hz has been produced. Note that there is always a factor of two between the growth times of successive samples. All other PLD parameters remained unchanged. Typical as-grown ZnO nanowire structures are shown in the SEM images in Fig. 2. For the purpose of accurate determination of the nanostructure dimensions we have cut the substrate along the axis of symmetry (cf. PLD chamber geometry [9]) by scratching the sapphire with a diamond needle tip. This facilitates an open view down to the nanowire base and the substrate cleaving edge. For better comparability, all SEM pictures shown in this paper were taken exactly at the center of the samples. Fig. 2 a–c) represent tilted views for the samples grown from 10 nm Au dots with 1,000, 4,000 and 16,000 PLD pulses, respectively. Fig. 2d) is the corresponding top view of the sample in Fig. 2c. Nanowires with hexagonal cross section are clearly resolved mimicking the wurzite crystal structure of ZnO. The prism side faces are oriented parallel to each other indicating that there is a very good in-plane epitaxial relationship. This behavior has been reported earlier for similar ZnO nanocrystals [15]. It is apparent from the pictures that the length of the whiskers increases upon increase of deposition time. The growth evolution is shown in more detail in Fig. 3a and b. In the first growth steps the aspect ratio (length/diameter) increases almost linearly with deposition time (Fig. 3b). Note that the x-scale in the figure is loga-

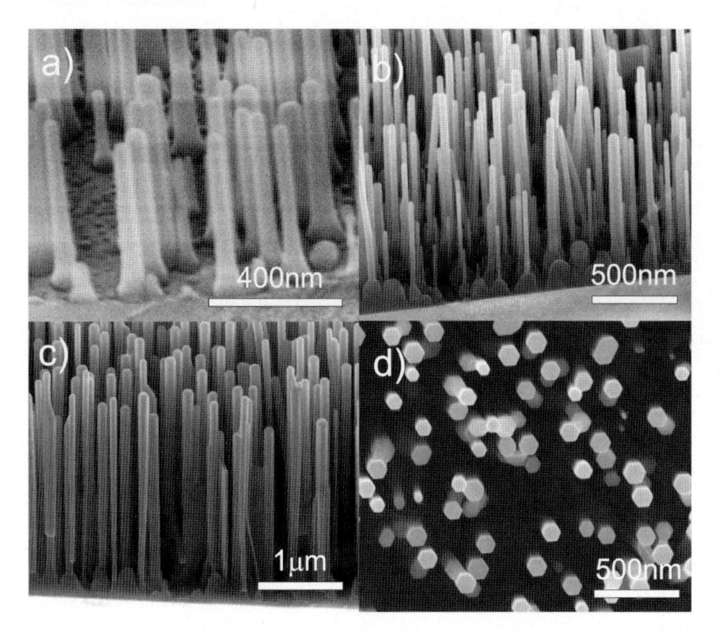

Fig. 2. Typical SEM images of PLD grown ZnO nanowires taken at different growth stages after application of: **(a)** 1,000, **(b)** 4,000 and **(c)** 16,000 laser pulses. **(d)** represents the *top view* of the sample shown in **(c)**

rithmic for better visualization. Most of the whiskers are between 20 nm and 70 nm in diameter (Fig. 3c and d). Figure 3c shows a typical corresponding histogram (sample with 8,000 pulses). The sample with 16,000 pulses plays an exceptional role. Most of its nanowires are thicker than 70 nm while the diameter distribution functions (Fig. 3d) are similar for all other samples. In this last growth step the length and the diameter increase by similar relative amounts, resulting in a nearly constant aspect ratio. The lateral nanowire density of the first 4 samples scattered between 56 and 80 wires per μm^2 and for the last sample (16,000 pulses) it was as low as 23 μm^{-2}. It is evident from Fig. 2c and d) that at later growth stages the whiskers tend to grow together thus forming structures with larger diameter but reduced lateral density.

For comparison we have synthesized a sample based on 50 nm Au colloids with 8,000 pulses. The substrate in the PLD chamber is oriented in such a way that one side faces the PLD target (see [9]), i.e., the surface is parallel to the symmetry axis of the plasma plume. In Fig. 4 the mean whisker diameter is plotted against the relative position on the sample. The inset of Fig. 4 shows schematically 5 different positions on a substrate where position 1 corresponds to the side with the smallest distance to the target and position 5 is 10 mm further apart from it. The diameter decreases systematically with increasing distance from the target. An analogous finding was already presented earlier for different substrate positions [9]. The narrowest diameter

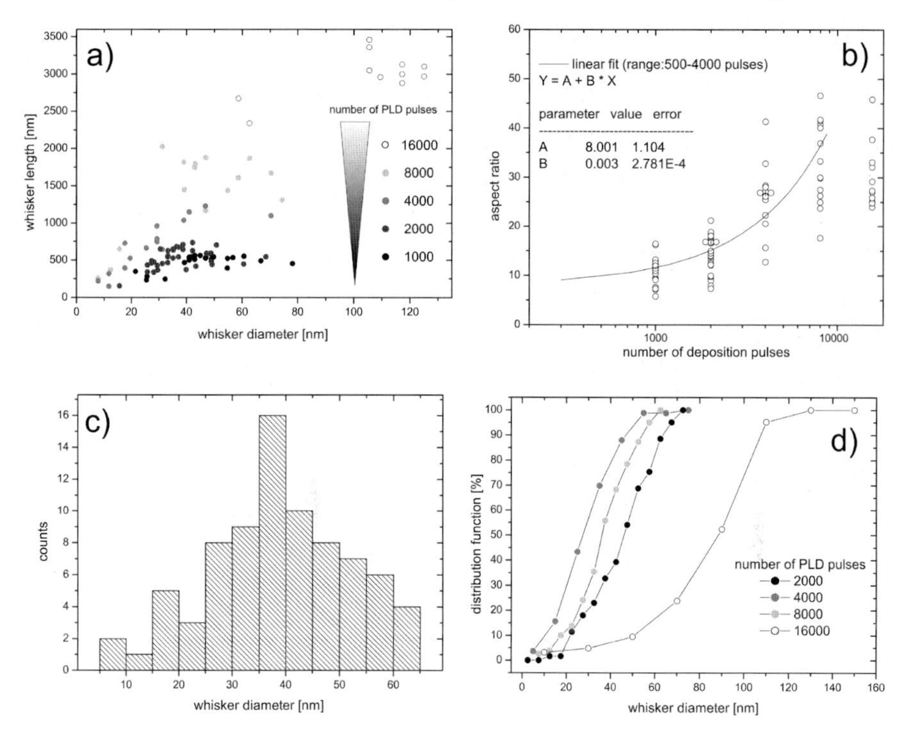

Fig. 3. Detailed growth evolution study of ZnO nanowires. **(a)**, **(b)** Diameter, length and the aspect ratio in dependence of the number of PLD pulses. The linear fit in **(b)** includes the data points up to 8,000 laser pulses. **(c)** Histogram of the whisker diameter of the sample with 8,000 pulses **(d)** distribution functions of the diameter for different samples

variation (represented by the bars in Fig. 4) can be found in the middle of the sample, at position 3. Also the height of the nanowires vary over the substrate surface (Fig. 5). This difference of growth morphology on one and the same sample arises from the expansion dynamics of the plasma plume. The energy and density of the ablated species depend strongly on the distance from the target. The availability of ZnO whiskers with different and defined sizes on one and the same sample is of great use for basic research, as for example for optical cavity experiments [3].

3.2 MgZnO

A second set of samples was prepared from a target containing 13 at-% Mg. Again, the growth time was varied while all other parameters remained unchanged. The magnesium content in the wires was determined from the excitonic CL peak energy to be approximately 2 percent (see [9]). The non-stoichiometric transfer of Mg from the target to the nanowires is not unexpected [9,11,16,17]. Figure 6 visualizes the different growth stages of MgZnO nanowires using 50 nm gold colloids with 500 pulses (Fig. 6a) up to 8,000

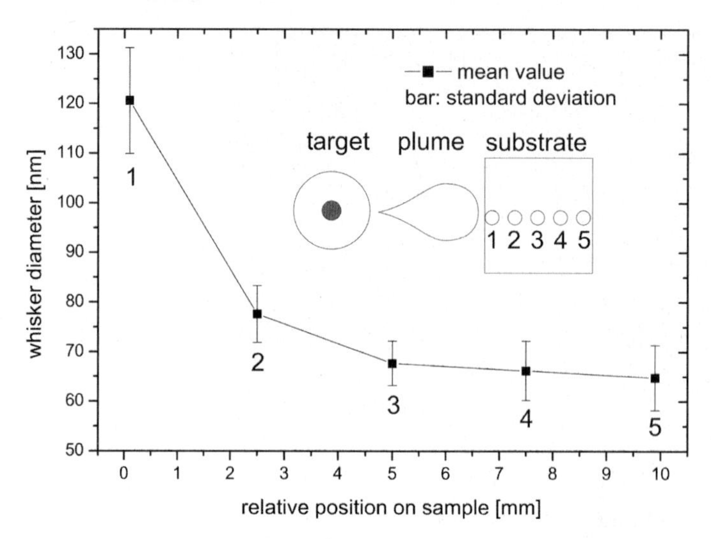

Fig. 4. Lateral homogeneity investigation over a sample. 1–5 represent different spots on the substrate, from front edge (near plasma, spot 1) to the back edge (spot 5, see *inset*). The *bar* indicates the standard deviation from the mean ZnO whisker diameter

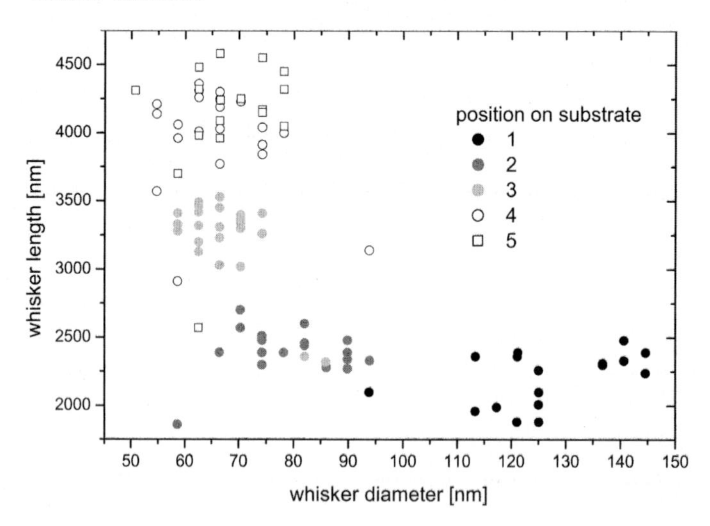

Fig. 5. ZnO whisker length vs. diameter for different positions on one and the same substrate. The positions 1 to 5 correspond to the inset of Fig. 4

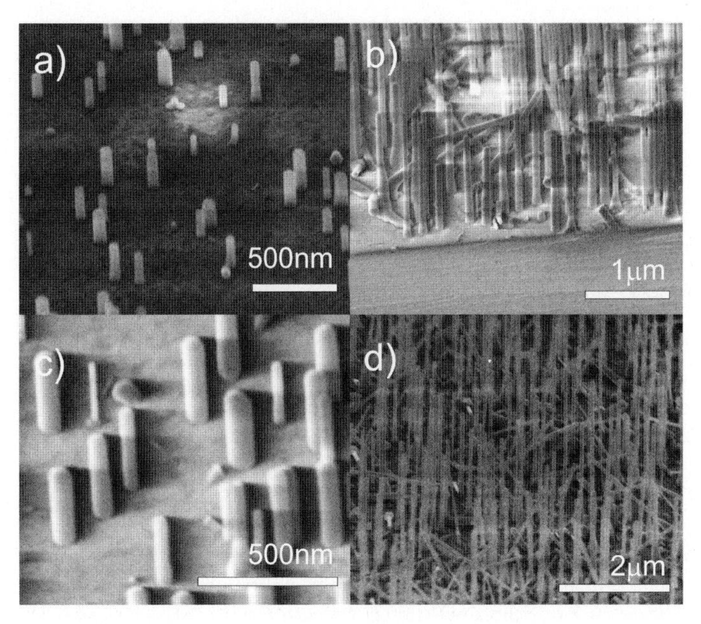

Fig. 6. SEM images of a series of MgZnO nanowhiskers with about 2% Mg content. The growth time was varied: **(a)** 500, **(b)** 1,000, **(c)** 4,000 and **(d)** 8,000 laser pulses with 10 Hz repetition frequency

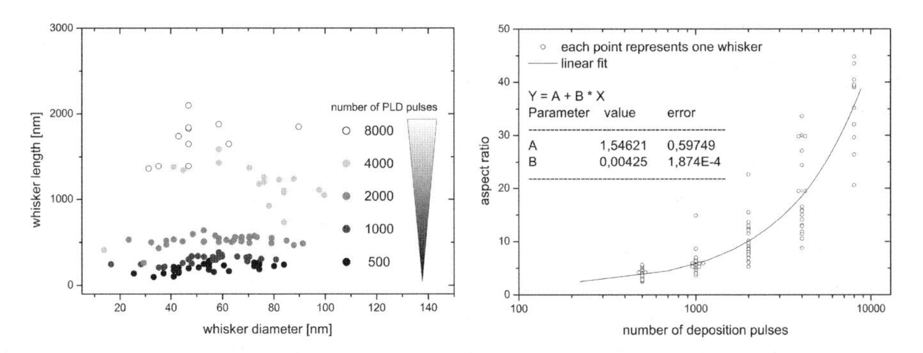

Fig. 7. MgZnO nanowire size (*left*) and aspect ratio (*right*) analysis in dependence on growth time

pulses (Fig. 6d). The diameter for all of those samples varies between 20 nm and 80 nm (Fig. 7 left) and the density was between 7 and 9 whiskers per μm^2, i.e., about one order of magnitude less dense than for the first set of samples (ZnO). The aspect ratio follows a linear dependence on growth time (Fig. 7). Interestingly, the slope of the curves in Fig. 7 right and Fig. 3b are approximately equal (B = 0.0043 ± (2e-4) compared to B = 0.003 ± (3e-4), respectively). Unlike as for the first set of samples no join of the whiskers

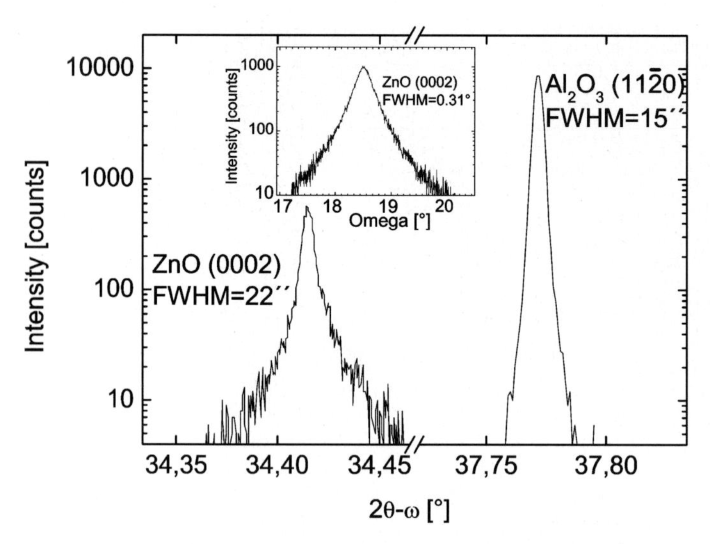

Fig. 8. XRD diffractogram of a ZnO nanowire sample. The peak widths of the ZnO and the substrate reflection are in the range of the instrumental line broadening (12 arcsec). The Rocking curve of the ZnO (0002) reflection with a FWHM of 0.31° is shown in the *inset*

at later growth stages is observed presumably due to the lesser whisker density. We can conclude that the overall morphology and growth behavior of MgZnO whiskers with low Mg content ($< 2\%$) is not different from pure ZnO whiskers. This opens the possibility for ZnO/MgZnO heterostructure growth which could form the basis of future nanodevices. Since the diameter distribution (Figs. 3 and 7) for the two sets of samples is independent of the gold colloidal particle size we conclude that the growth in not only governed by the vapor-liquid-solid (VLS) mechanism [18]. Although the catalyst is needed for the formation of seed particles [9, 10], it seems to play no role at later stages. We can derive from the analogous growth law for different whisker densities but identical whisker size and length distributions, that our specific PLD conditions provide a supersaturated environment, i.e., growth rate is not determined by the supply of new particles from the plasma plume.

4 Structural Properties

XRD measurements show that the crystallographic c-axis of the Zn(Mg)O nanowires is oriented perpendicular to the surface of the a-plane sapphire substrate. Figure 8 shows a HRXRD diffractogram of the ZnO(0002) and the $Al_2O_3(11\bar{2}0)$ substrate reflection of the ZnO nanowire sample treated in Figs. 4 and 5. The corresponding peak widths (FWHMs) are 22 arcsec and 15 arcsec, respectively. Compared to the instrumental line broadening of

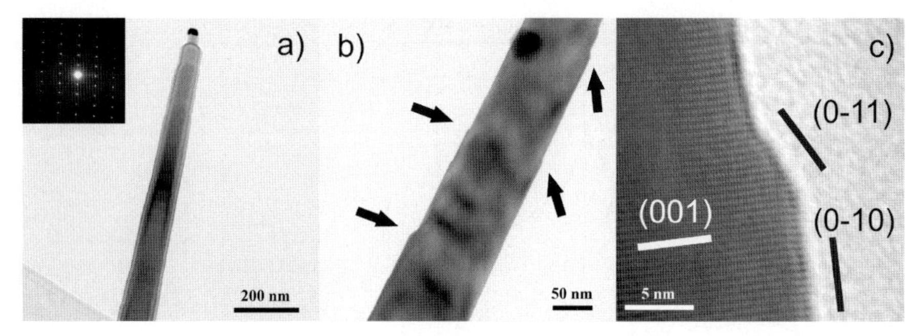

Fig. 9. TEM images of a single ZnO nanowire. The inset in **(a)** shows the corresponding SAD pattern of the whisker. In **(b)** the arrows indicate the positions of steps on the side faces which are shown in more detail in **(c)**. **(b)** and **(c)** were studied in the [100] zone axis

12 arcsec this underlines the superior crystalline quality of the whiskers. The corresponding rocking curve of the ZnO (0002) reflection has as FWHM of 0.31° (inset of Fig. 8) which is comparable to the best values reported for ZnO nanowires and PLD thin films [11, 16, 17]. Hence, as it can be seen already in the SEM pictures, there is very less tilt between the whiskers. The c-axis lattice constant for this sample was 5.207 Å which is exactly the value for bulk ZnO. For the MgZnO nanowire samples a slightly lower lattice constant of 5.204 Å was determined.

The typical rod-like growth morphology is studied in more detail by TEM (Fig. 9). A single crystalline wurzite ZnO whisker with a little gold droplet on top is revealed. The inset of Fig. 9a shows the selected area diffraction pattern (SAD) of ZnO. As already deduced from the XRD measurements, the growth direction is the c-axis of ZnO and in accordance with the very low FWHMs of the XRD ZnO(0002) reflection, dislocations in the nanowires are very rare. Figure 9b and c depict steps on the side surfaces of the whiskers which are not periodic along the growth direction and not symmetric on both wire sides. They exhibit various heights usually between 1 and 15 atomic steps. In Fig. 9c the important ZnO planes as well as the step habit plane are indicated. We do not observe the angled {11-23} faces reported in [20].

5 Optical Properties

Figure 10 shows typical CL spectra of arrays of ZnO nanowires within the spectral region of donor-bound exciton (DBX) emission for $T = 10$ K. The size of the excited nanowire array has been increased from a few $(1-3)$ wires up to $100 - 300$ wires. All CL spectra are dominated by a strong emission line at 3.360 eV indicating the recombination of excitons bound to a neutral aluminium donor (I_6, [21]). The FWHM of this line varies from 1.4 meV to 2.0 meV and is thus comparable to values for high-quality PLD-grown thin

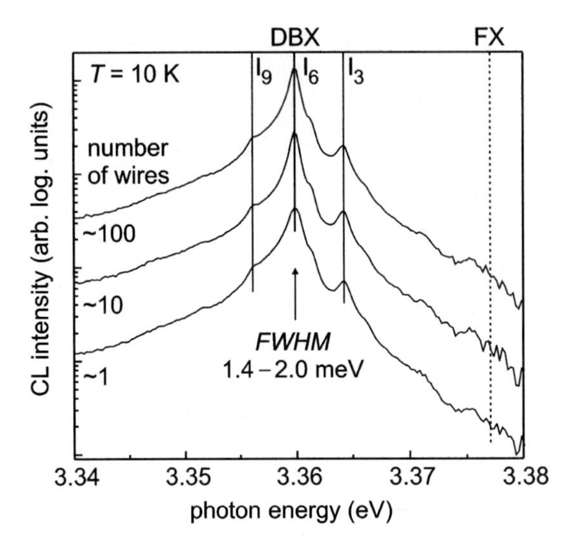

Fig. 10. CL spectra (shifted vertically for clarity) of arrays of ZnO nanowires within the spectral region of donor-bound exciton (DBX) emission. The excited lateral area of the nanowire sample has been increased to include 1 – 3 nanowires (*lower curve*), 10 – 30 nanowires (*middle curve*) and 100 – 300 nanowires (*upper curve*). *Vertical lines* refer to dominating DBX transitions. The *dashed vertical line* markes the expected position of free exciton (FX) emission

films and single crystals [13, 19]. A transition at 3.364 eV possibly belongs to an ionized-donor bound exciton (I_3, [21]). The luminescence intensity of further DBXs (3.356 eV, Indium, I_9, [21]) is weaker to about one order of magnitude. Free-exiton (FX) emission cannot be resolved. The absence of significant differences between the CL spectra originating from different array sizes proves again the lateral homogeneity of the nanowire sample.

With increasing temperature (Fig. 11) all transitions broaden and shift to lower energies. DBX emission vanishes (dashed line in Fig. 11a), while FX emission gains in intensity (solid line). Figure 11b focusses on photon energies < 3.34 eV. Phonon replicas and two-electron satellites (TES) of the DBX-lines can be observed [21]. Both types of transitions vanish with increasing temperature (dashed lines) due to their physical correlation to donor bound excitons. Additional peaks that can be detected even at elevated temperatures indicate phonon replicas of the FX transition (solid line).

6 Summary

Summarizing, we have investigated 2 sets of samples based on different gold colloidal particle sizes, namely ZnO nanowire arrays grown from 10 nm Au droplets with a lateral density of about 70 wires per μm^2 (set 1) and MgZnO nanowire arrays with about 2% Mg content and a density of ~ 8 wires

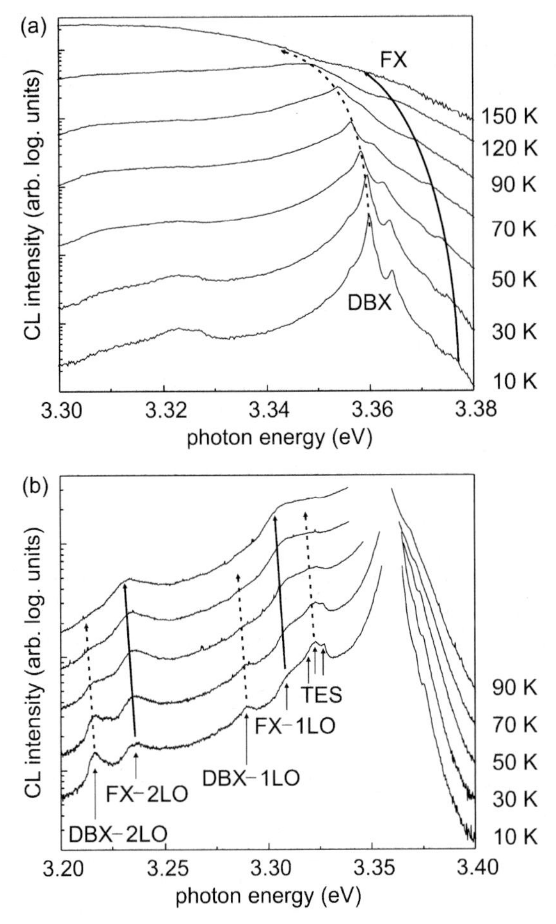

Fig. 11. Temperature dependence of CL spectra of an array of about 10 nanowires. Results are given for the spectral region of DBX emission **(a)** and for the spectral region of phonon replicas **(b)**, with DBX peaks faded out

per μm^2. In both cases, we found the same linear increase of the aspect ratio on growth time and the ZnO nanowire diameter distribution for sets 1 and 2 were comparable. HRXRD and TEM investigations revealed the superior crystalline quality of the whiskers. Furthermore, HRTEM images showed (0-11) steps at the whisker side faces that were a few atomic steps in height. The lateral homogeneity of selected areas of nanowire arrays, including 1, 10 and 100 whiskers was proven by CL measurements.

Acknowledgements

The authors thank G. Wagner, University of Leipzig for helpful discussions on the growth mechanism. This work was supported by the DFG FOR 522 (Gr 1011/11-1, -2) and by the EU STREP NANDOS (FP6-016924).

References

[1] C. Klingshirn, M. Grundmann, A. Hoffmann, B. Meyer, A. Waag, Physik-Journal **5(1)**, 33 (2006)
[2] S.C. Minne, S.R. Manalis, C.F. Quate, Appl. Phys. Lett. **67**, 3918 (1995)
[3] Th. Nobis, E. M. Kaidashev, A. Rahm, M. Lorenz, M. Grundmann, Phys. Rev. Lett. **93**, 103903 (2004)
[4] Z. L. Wang, J.Phys.: Condens. Matter **16**, R829 (2004)
[5] P. A. Hu, Y. Q. Liu, L. Fu, X. B. Wang, D. B. Zhu, Appl. Phys. A **80**, 35 (2005)
[6] H. Yan, R. He, J. Pham, P. Yang, Adv. Mater. **15**, 402 (2003)
[7] W. I. Park, G. Yi, M. Kim, S. J. Pennycook, Adv. Mater. **14**, 1841 (2002)
[8] X. Gao, X. Li, W. Yu, J. Phys. Chem. B **109**, 1155 (2005)
[9] M. Lorenz, E. M. Kaidashev, A. Rahm, Th. Nobis, J. Lenzner, G. Wagner, D. Spemann, H. Hochmuth, M. Grundmann, Appl. Phys. Lett. **86**, 143113 (2005)
[10] H. J. Fan, W. Lee, R. Hausschild, M. Alexe, G. L. Rhun, R. Scholz, A. Dadgar, K. Nielsch, H. Kalt, A. Krost, M. Zacharias, U. Gösele, Small **2**, 561 (2006)
[11] M. Lorenz, E. M. Kaidashev, H. v. Wenckstern, V. Riede, C. Bundesmann, D. Spemann, G. Benndorf, H. Hochmuth, A. Rahm, H.-C. Semmelhack, M. Grundmann, Solid State Electronics **47**, 2205 (2003)
[12] M. A. Wood, M. Riehle, C. D. W. Wilkinson, Nanotechnology **13**, 605 (2002)
[13] M. Lorenz, J. Lenzner, E. M. Kaidashev, H. Hochmuth, M. Grundmann, Ann. Phys. (Leipzig) **13**, 39 (2004)
[14] Th. Nobis, E. M. Kaidashev, A. Rahm, M. Lorenz, J. Lenzner, M. Grundmann, Nano Lett. **4**, 797 (2004)
[15] A. Rahm, M. Lorenz, Th. Nobis, G. Zimmermann, M. Grundmann, B. Fuhrmann, F. Syrowatka, Appl. Phys. A, *in press*
[16] A. Ohtomo, M. Kawasaki, T. Koida, K. Masubuchi, H. Koinuma, Y. Sakurai, Y. Yoshida, T. Yasuda, and Y. Segawa, Appl. Phys. Lett. **72,** 2466 (1998)
[17] S. Choopun, R. D. Vispute, W. Yang, R. P. Sharma, T. Venkatesan, H. Shen, Appl. Phys. Lett. **80,** 1529 (2002)
[18] A. P. Levitt (Ed.), *Whisker Technology,* (Wiley-Interscience, New York, 1970)
[19] E. M. Kaidashev, M. Lorenz, H. v. Wenckstern, A. Rahm, H.-C. Semmelhack, K.-H. Han, G. Benndorf, C. Bundesmann, H. Hochmuth, M. Grundmann, Appl. Phys. Lett. **82**, 3901 (2003)
[20] J.B. Baxter, F. Wu, E.S. Aydil, Appl. Phys. Lett. **83**, 3797 (2003)
[21] B. K. Meyer, H. Alves, D.M. Hofmann, W. Kriegseis, D. Forster, F. Bertram, J. Christen, A. Hoffmann, M. Straßburg, M. Dworzak, U. Haboeck, A.V. Rodina, Phys. Stat. Sol. (B) **241**, 231 (2004)

Spin-Transfer Torques in Single-Crystalline Nanopillars

D. E. Bürgler[1], H. Dassow[1], R. Lehndorff[1], C. M. Schneider[1], and A. van der Hart[2]

[1] Institute of Solid State Research – Electronic Properties (IFF6) and cni – Center of Nanoelectronic Systems for Information Technology, the parts of your address Forschungszentrum Jülich GmbH
52425 Jülich, Germany
d.buergler@fz-juelich.de

[2] Institute of Thin Films and Interfaces – Process Technology (ISG-PT) and cni – Center of Nanoelectronic Systems for Information Technology, Forschungszentrum Jülich GmbH
52425 Jülich, Germany

Abstract. Current-induced magnetization switching (CIMS) due to spin-transfer torques is an advanced switching concept for magnetic nanostructures in spintronic devices. Most previous studies employed sputtered, polycrystalline samples. Here, we report on the first measurements of CIMS in single-crystalline nanopillars. Fe(14 nm)/Cr(0.9 nm)/Fe(10 nm)/Ag(6 nm)/Fe(2 nm) multilayers are deposited by molecular beam epitaxy. The central Fe layer is coupled to the 14 nm-thick Fe layer by interlayer exchange coupling over Cr, and the topmost Fe layer is decoupled (free layer). The maximum observed giant magnetoresistance with current perpendicular to the layers (CPP-GMR) is 2.6% at room temperature and up to 5.6% at 4 K. Nanopillars with a diameter of 150 nm are prepared by optical and e-beam lithography. The opposite scattering spin asymmetries of Fe/Cr and Fe/Ag interfaces enable us to observe CIMS at small magnetic fields and opposite current polarity in a single device. The critical current density for switching is $j_c \approx 10^8$ A/cm^2. At high magnetic fields, step-like resistance changes are measured at positive currents and are attributed to current-driven magnetic excitations.

1 Introduction

In a magnetic multilayer containing two ferromagnetic layers and a nonmagnetic spacer (FM/NM/FM), an electric current flowing perpendicularly to the layers (CPP) gets spin-polarized by the FM layers, leading to a giant magnetoresistance (GMR) [1, 2]. Thus, spin currents can sense the magnetization state of the magnetic system. *Slonczewski* [3] and *Berger* [4] first predicted that spin currents of appropriate strength can also directly influence the magnetizations without applying an external magnetic field. This phenomenon termed *current-induced magnetization switching (CIMS)* is shown in Fig. 1. We consider two FM layers separated by a non-FM spacer with a thickness below its spin diffusion length. The FM layers are different in such

R. Haug (Ed.): Advances in Solid State Physics,
Adv. in Solid State Phys. **46**, 121–133 (2008)

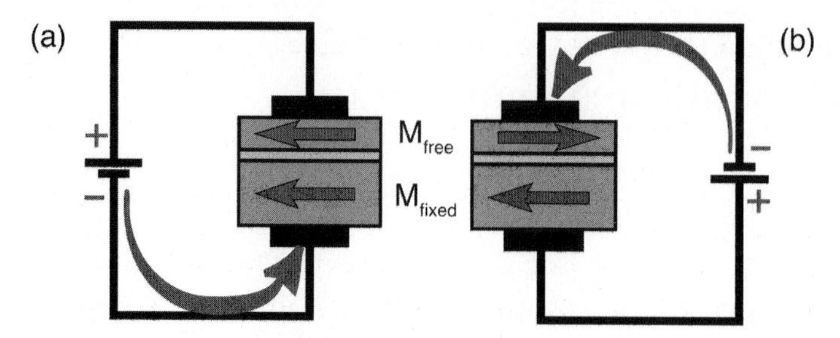

Fig. 1. Phenomenology of current-induced magnetization switching: The stable alignment of the magnetizations depends on the polarity, *i.e.*, the direction of the current flowing perpendicularly through the trilayer

a way, that one of them – the free layer – can be remagnetized more easily than the other, the so-called fixed layer. The pinning of the magnetization of the fixed layer M_{fixed} can be achieved by different ferromagnetic layer thicknesses [5–7], by the exchange bias effect [8], or by making use of interlayer exchange coupling, as in our case, where we use Fe/Cr/Fe(001) antiferromagnetically exchange coupled structures. When electrons flow from the fixed to the free layer [9], the magnetization of the free layer M_{free} aligns parallel to M_{fixed} and this alignment is stabilized (Fig. 1a). When the current direction is reversed, however, the antiparallel alignment is more stable and adopted (Fig. 1b) as will be explained in Sect. 2. Therefore, M_{free} can be switched from the parallel to the antiparallel configuration with respect to M_{fixed} back and forth by repeatedly reversing the current polarity. The observation of CIMS requires laterally confined samples with diameters below about 1 μm for two reasons: 1. high current densities of the order of $10^7 - 10^8$ A/cm^2 are needed [3, 4] and 2. the competing Oersted field generated by this current must be minimized. The fabrication of suitable nanostructures, so-called *nanopillars*, will be described in Sect. 3.

In order to achieve large spin-torque effects a high spin polarization P of the current is needed. The present work with Fe/Ag/Fe(001) nanopillars is thus motivated by two publications of *Stiles* and *Penn* [10] and *Stiles* and *Zangwill* [11], in which the authors predict high spin polarization for single-crystalline Fe/Ag interfaces. Single crystalline layered structures can also serve as model systems for comparison with theory and micromagnetic simulations due to the well known structure, the small amount of defects, and the homogeneous magnetic properties when prepared by molecular beam epitaxy (MBE). In particular, the single-crystalline Fe(001) layers employed here show well-defined 4-fold in-plane magnetocrystalline anisotropy.

2 Current-Induced Magnetization Switching

2.1 Physical Picture of Current-Induced Magnetization Switching

For developing a physical picture of CIMS, it is sufficient to realize that an unpolarized current gets spin-polarized when it enters a ferromagnetic material. Similarly, a polarized current gets repolarized when its initial polarization direction deviates from the quantization axis in the ferromagnet. The basic reason is the exchange splitting of the bandstructure, which gives rise to different density of states for majority and minority electron states. A detailed quantum mechanical description of the processes for polarization (or repolarization) has been given by *Stiles* and *Zangwill* [11]. The basic result is that – to a good approximation – the incident spin current component transverse to the magnetization direction of the ferromagnet is absorbed in the interface region. It acts as a torque – the so-called *spin-transfer torque* – on the magnetization of the ferromagnet. The longitudinal component parallel to the magnetization is transmitted and constitutes the (re-)polarized spin current.

A schematic FM/NM/FM trilayer structure is shown in Fig. 2. We assume that the magnetization directions of the two ferromagnets are slightly tilted by the angle θ. This is possible, *e.g.*, due to thermal fluctuations when the two magnetic layers are separated by a non-magnetic spacer layer. In Fig. 2a the left, fixed layer polarizes the unpolarized current entering from non-magnetic leads as describes above. The polarization is *not* modified at the interface to the *non-magnetic* spacer layer because its density of states is not spin-split. The only requirement is that the spacer layer thickness is below its spin diffusion length to prevent significant depolarization by spin-flip scattering.

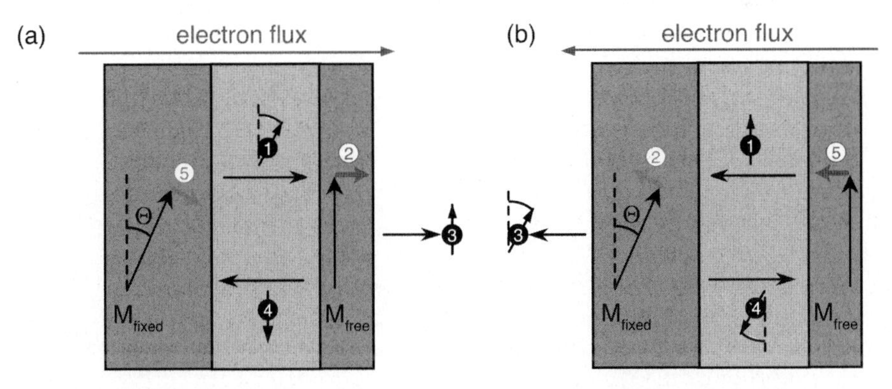

Fig. 2. Physical picture of CIMS: Due to the assumed magnetic asymmetry M_{fixed} does not respond to the torque (*short gray arrows*) acting on it, whereas M_{free} can follow the torque (*short red* and *green arrows*). The numbers refer to the sequence of the description. **(a)** and **(b)** show the situation for opposite electron flux directions, which result in stable parallel or stable antiparallel alignment, respectively

The polarized current label (1) in Fig. 2a now hits the free layer and transfers its transversal component as a torque to the free layer (2). The current is partly transmitted (3) and another part is reflected (4). This reflected current must now be considered as a polarized current impinging on the fixed layer. Again, the transversal component will be absorbed and acts as a torque on the fixed layer (5). However, due to the assumed magnetic asymmetry the fixed layer will resist to the torque, and only M_{free} starts to rotate in order to reach the stable alignment parallel to M_{fixed}. For the opposite direction of the electron flux in Fig. 2b, we obtain a similar situation but the torques point in the opposite directions. Therefore, the stable state corresponds to the antiparallel alignment of M_{free} and M_{fixed}. Note, that in this case the torque on M_{free} arises from the current which first has been reflected from the fixed layer [(4) in Fig. 2b]. Obviously, the magnetic asymmetry (fixed $\leftrightarrow$ free) plays an important role, which is very reasonable, because "left" and "right" cannot be distinguished for the symmetric case.

2.2 Normal and Inverse Current-Induced Magnetization Switching

In Fig. 2, we have tacitly assumed that electrons with their spin moment parallel to the magnetization direction pass through the NM/FM interface more easily than those with opposite spin. This is not necessarily the case and only holds for NM/FM material combinations with positive scattering spin asymmetry

$$\gamma = \frac{r^{\downarrow} - r^{\uparrow}}{r^{\downarrow} + r^{\uparrow}} > 0\,, \tag{1}$$

where $r^{\uparrow(\downarrow)}$ is the interface resistivity for the majority (minority) spin channel [12]. In general, there are two contributions to the spin-dependent resis-

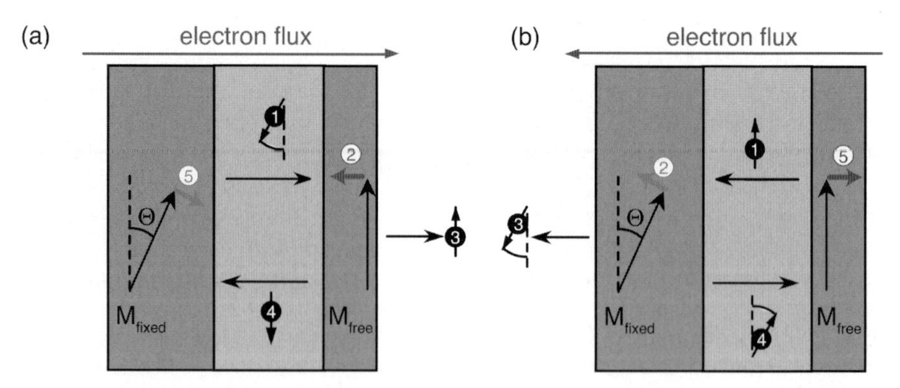

Fig. 3. Inverse CIMS: We assume a negative γ for the fixed layer and a positive γ for the free layer. The direction of the spin-transfer torque (*red* and *green arrows*) in the free layers are reversed compared to the situation in Fig. 2

tance, one arising >from the bulk of the FM layers and the other from the NM/FM interfaces [13]. In order to keep the notation simple, we use here the parameter γ for the interface scattering spin asymmetry. In Sect. 6 we will justify why we can restrict ourselves to the interface contribution in the present case. Scattering spin asymmetry parameters (for bulk and interfaces) are well known from the GMR effect, for which the signs of the scattering spin asymmetries of the left and right side of the spacer, $\gamma_{\mathrm{L,R}}$, determine whether the GMR is normal ($\gamma_{\mathrm{L}}\gamma_{\mathrm{R}} > 0$) or inverse ($\gamma_{\mathrm{L}}\gamma_{\mathrm{R}} < 0$). Normal (inverse) GMR is characterized by higher resistivity for antiparallel (parallel) magnetization alignment. For CIMS, the sign of the scattering spin asymmetry of the fixed (polarizing) layer determines the sign (direction) of the spin-transfer torques. For normal CIMS as shown in Fig. 2 and our sign convention (positive current denotes an electron flow from free to fixed layer) the antiparallel (parallel) alignment is stabilized by a positive (negative) current. For inverse CIMS this relation is inverted as can bee seen in Fig. 3. Note, that we have used a positive γ for the free layer in Fig. 3. A negative γ would only reverse the polarization directions of the currents transmitted through [(3) in Fig. 3a and (1) in Fig. 3b] and reflected from [(4) in Fig. 3a] the interface between the spacer and the free layer. The signs of the spin-transfer torques exerted on the free layer would not be affected.

3 Sample Preparation and Magnetic Properties

In order to achieve single-crystalline growth the magnetic multilayers are deposited in a standard MBE system. The native oxygen layer of the GaAs(001) substrates ($10 \times 10\,\mathrm{mm}^2$) is desorbed by annealing for 60 min at 580°C under UHV conditions. We deposit 1 nm Fe and 150 nm Ag at 100°C to get a flat buffer system after annealing at 300°C for 1 h [14]. The Ag buffer also acts as a bottom electrode for the transport measurements

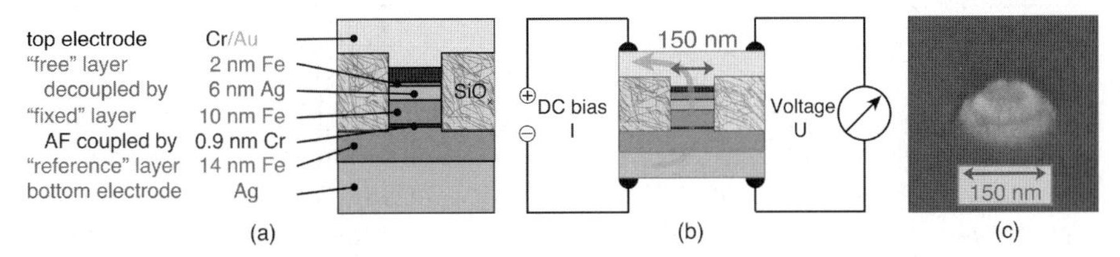

Fig. 4. **(a)** Sequence, thicknesses, and functions of the layers within the multilayer stack. **(b)** Scheme of the junction geometry and the contacts for transport measurements. The DC current is confined to a diameter of $d \approx 150$ nm by the nanopillar. The voltage drop is measured across the pillar in 4-point geometry. **(c)** SEM micrograph of a nanopillar after ion-beam etching showing typical redeposition effects at the edges

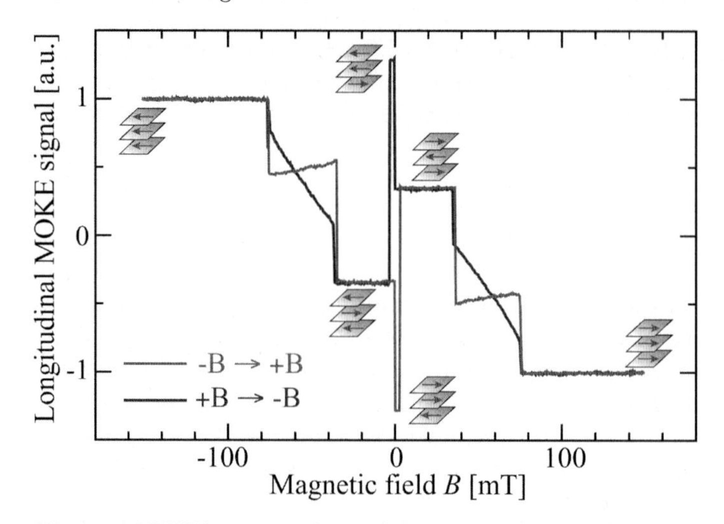

Fig. 5. MOKE hysteresis loop of the extended layered system measured with the external magnetic field parallel to one of the easy [100]-axes of Fe. The interlayer exchange coupling stabilizes the fully antiferromagnetic state between ±3 and ±35 mT

(Fig. 4a and b). The following layers are then deposited at room temperature: Fe(14 nm)/Cr(0.9 nm)/Fe(10 nm)/Ag(6 nm)/Fe(2 nm).

We check the crystalline surface structure after each deposited layer by low-energy electron diffraction (LEED). The spots characteristic of (001) surfaces slightly broaden with increasing total thickness, but still indicate high crystalline quality, even for the final 50 nm Au(001) capping layer. Thicknesses are controlled by quartz crystal monitors. The bottom and central FM layers [Fe(14 nm) and Fe(10 nm)] are antiferromagnetically (AF) coupled by interlayer exchange coupling over the Cr interlayer. Therefore, the central Fe(10 nm) layer is magnetically harder with respect to the top Fe(2 nm) layer.

The longitudinal magneto-optical Kerr effect (MOKE) is used to measure the magnetic properties of the samples. In Fig. 5 we present the hysteresis loop with the magnetic field parallel to one of the easy [100]-axes of the Fe layers in the film plane. The saturation field of the system is $|B_S| = 76$ mT. For smaller magnetic fields the central Fe(10 nm) layer remagnetizes via a canted state to the fully antiferromagnetic configuration of the trilayer stack below the switching field of ±35 mT. After reversing the field direction we measure another jump in the signal, which corresponds to the reversal of the topmost 2 nm-thick Fe layer at ±0.3 mT. At ±3 mT the two coupled Fe layers reverse simultaneously due to their unequal thickness.

By fitting the MOKE measurements and additional Brillouin light scattering measurements (BLS) [17] we extract the magnetic properties of each layer as compiled in Table 1. The saturation magnetization M_S and the crystalline anisotropy constant K_1 of the Fe(14 nm) and Fe(10 nm) layers have bulk values [18, 19] and indicate the high layer quality. The thin Fe(2 nm) layer has reduced M_S and K_1, which can be understood by the lower thickness and by reduced growth quality. The negligible coupling constants J_1 and J_2 for the Ag spacer layer show that the Fe(2 nm) layer is decoupled. K_S denotes the interface anisotropy.

4 Fabrication of Nanopillars

In order to measure the spin-transfer effects in the CPP-geometry we have developed a combined process of optical and e-beam lithography. First, we define the leads and contact pads of the bottom electrode by using AZ5206 photoresist and ion beam etching (IBE) (Fig. 6b). We then employ HSQ (hydrogen silsesquioxane) as negative e-beam sensitive resist [20] and a Leica EBPG 5HR e-beam writer to define small nanopillars. The resist structures are circular and transferred into the magnetic layers by IBE (Fig. 6c). The timed etching process is stopped inside the magnetic multilayer. Typical dimensions of the developed resist structures are 100–150 nm (measured with an atomic force microscope). Due to redeposition of etched material during IBE [21], the nanopillars broaden to 150–200 nm. An SEM micrograph of the free-standing nanopillars in stage (c) of Fig. 6 is shown in Fig. 4c. The pillars are planarized by spin-coating HSQ (Fig. 6d). Subsequent e-beam exposure turns HSQ into SiO_x, which electrically insulates the pillars [20]. In order to improve the insulation, especially at the side walls of the bottom electrodes, a 50 nm-thick Si_3N_4 layer is deposited by plasma enhanced chemical vapor deposition (PECVD). We open the top of the nanopillars by IBE and use an optical lift-off process of 300 nm Au for the preparation of the top electrode for the 4-point resistance measurements.

Table 1. Saturation magnetization M_S, in-plane magnetocrystalline anisotropy constant K_1, and interface anisotropy constant K_S extracted from fits to MOKE and BLS measurements for each Fe layer of the extended trilayer stack [15]. J_1 and J_2 denote the bilinear and biquadratic coupling strength of the interlayer exchange coupling across Cr and Ag [16]

Property	Fe(14 nm)/Fe(10 nm)	Fe(2 nm)	Cr	Ag
M_S [10^6 A/m]	1.75	1.6		
K_1 [kJ/m^3]	56	33		
K_S [mJ/m^2]	0.5	0.3		
J_1 [mJ/m^2]			−0.97	≈ 0
J_2 [mJ/m^2]			−0.01	≈ 0

5 Results

5.1 CPP-GMR

The DC voltage drop of a constant current I applied to the junction is measured (Fig. 4b), and by dividing by I we calculate the absolute resistance R. The differential resistance $\mathrm{d}U/\mathrm{d}I$ is recorded with lock-in technique by mixing a constant current with a small modulated voltage ($\approx 300\,\mu\mathrm{V}$ and $\approx 12\,\mathrm{kHz}$). Typical junction resistances lie in the range between 1 and $3\,\Omega$. The temperature can be controlled with a He flow cryostat between 4 and 300 K.

The magnetoresistance loop of a junction without applying a DC bias current is shown in Fig. 7. The solid blue (dashed green) line represents the data with magnetic field along the easy (hard) axis of Fe(001). The curves show a completely different behavior for the two field directions, but are the same along the second pair of easy and hard axes. Thus, the structure is still single-crystalline and exhibits 4-fold magnetocrystalline anisotropy. The saturation field of the structured sample is 190 mT, which is more than twice the saturation field of the extended layers (see Fig. 5). Another difference becomes obvious in the minor loop (inset of Fig. 7), where the absolute resistance is measured with a small DC current of 1 mA. Coming from large positive magnetic field, the resistance drops to a smaller value at small reversed fields between 1 and 3 mT and jumps back to the high resistance state at larger negative fields. On the way back, the resistance stays at the maximum value. The drop in the first half of the cycle does not occur in every measurement. Thus, the patterning has modified the magnetic configuration,

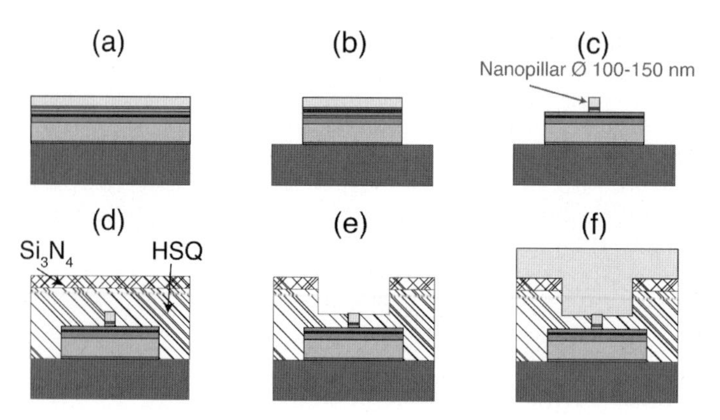

Fig. 6. Lithographic process: **(a)** Extended epitaxial multilayer grown by MBE, **(b)** definition of $10\,\mu\mathrm{m}$-wide bottom electrodes by optical lithography and IBE, **(c)** definition of nanopillars of 150 nm diameter by e-beam lithography and IBE, **(d)** planarization by HSQ and additional insulation by Si_3N_4; e-beam exposure converts HSQ into insulating SiO_x, **(e)** opening of a $8{\times}6\,\mu\mathrm{m}^2$ window to the top of the nanopillar by IBE, and **(f)** definition of the top electrodes by optical lift-off. The colors of different materials correspond to those of Fig. 4

and the structured Fe(2 nm) nanomagnet is presumably coupled to the rest of the system by dipolar stray fields at the edges or by domain wall coupling. This is a common feature in these devices also seen in Co nanopillars [5]. Due to this effect, we cannot separate the contributions of the Fe/Ag/Fe and Fe/Cr/Fe subsystems to the total GMR, and therefore cannot gauge the resistance jumps measured under the influence of a large DC current in Fig. 8. The overall GMR ratio defined as $(R_{AP} - R_P)/R_P$, where R_{AP} is the highest resistance value in the antiferromagnetic configuration and R_P denotes the smallest resistance in the saturated state, amounts to 2.6% at RT and 5.6% at 4 K. The dramatic increase in the saturation field can be explained by the competition between the interlayer exchange coupling, external, and dipolar fields.

5.2 CIMS

A DC current influences the resistance R and, at some critical values, the magnetization state of the junction (Fig. 8). Positive current corresponds to an electron flow from the "free" Fe(2 nm) to the "fixed" Fe(10 nm) layer. We observe a parabolic background, which has been measured previously [5–8] and is usually explained by Joule heating of the junction. On top of that, we measure field dependent resistance changes, which can be attributed to spin-transfer torque effects. For instance at -20 mT (Fig. 8b), the resistance

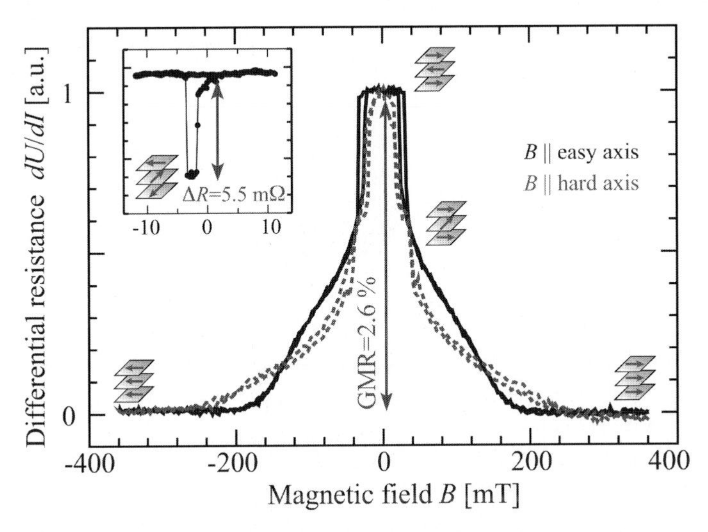

Fig. 7. GMR hysteresis loop with magnetic field parallel to an easy (*solid blue*) and hard (*dashed green*) axis. *Inset*: Minor GMR loop. Only in the first half of the loop ($+B \rightarrow -B$) the resistance drops to a smaller value corresponding to a canted magnetization state. These drops do not occur in every cycle. In the second half of the loop ($-B \rightarrow +B$) the resistance stays at the maximum value

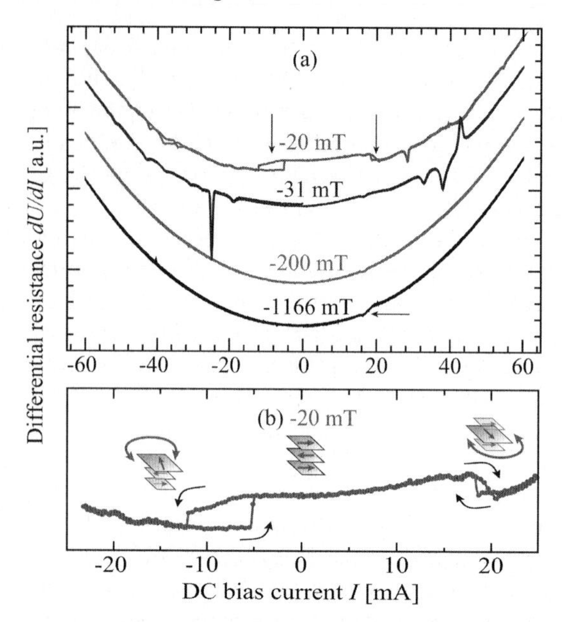

Fig. 8. **(a)** DC current loops with magnetic field parallel to an easy axis. Curves measured at different external fields as indicated are vertically offset for clarity. **(b)** Expanded view of the curve for $-20\,\mathrm{mT}$ with the interpretation of the switching processes occurring at positive and negative DC bias current

drops at $I_C^+ = +18.2\,\mathrm{mA}$ from the high-resistive to an intermediate state. After reducing the current again, the resistance jumps back to the large value. But also at negative bias the resistance changes at $I_C^- = -12.1\,\mathrm{mA}$ from large to small. With an estimated junction diameter of $d = 150\,\mathrm{nm}$ the corresponding critical current densities are $j_c^+ = 1 \cdot 10^8\,\mathrm{A/cm^2}$ and $j_c^- = -0,7 \cdot 10^8\,\mathrm{A/cm^2}$. As already mentioned in the discussion about the GMR data of Fig. 7 we cannot directly relate the resistance jumps to changes between specific magnetization states.

If we start the measurement in the intermediate resistance state at a field of $-31\,\mathrm{mT}$, the canted alignment is the initial state. But at large positive and negative currents ($+31.4$ and $-24.8\,\mathrm{mA}$) we observe strong deviations from the parabolic background that may indicate current-driven high-frequency excitations of the magnetization.

At large magnetic fields exceeding the saturation field (*e.g.*, -1166 mT in Fig. 8a), the two thick layers [Fe(14 nm) and Fe(10 nm)] are stronger stabilized by the Zeeman energy than the Fe(2 nm) layer, and therefore only one step-like resistance change due to magnetic excitations of Fe(2 nm) at $I > 0$ is observed under these conditions.

6 Discussion

The occurrence of jumps at both polarities of the current at small fields is at first glance surprising, but can be explained by taking into account that both Fe/Cr and Fe/Ag interfaces contribute and have scattering spin asymmetries with opposite signs [10, 22], namely $\gamma_{\mathrm{Fe/Ag}} > 0$ and $\gamma_{\mathrm{Fe/Cr}} < 0$. This leads for the Fe/Cr subsystem to inverse current-induced magnetization switching, very similar to inverse GMR [23, 24]. Thus, the spin-transfer torques for the two subsystems are inverted. For instance, a negative current stabilizes the parallel state for Fe/Ag/Fe and the antiparallel state for Fe/Cr/Fe. At low fields, the central Fe(10 nm) layer points opposite to the external magnetic field (Fig. 5). At positive currents, the spin-transfer torque generated in the Fe/Cr subsystem destabilizes this direction and switches the Fe(10 nm) layer (Fig. 8b). At negative currents, the Fe(2 nm) layer gets unstable by the torque created from the Fe/Ag subsystem, while the Fe/Cr subsystem is even stronger stabilized in the antiparallel state. Our layer sequence suggests that the opposite scattering spin asymmetries arise from *interface* scattering because all involved FM layers consists of iron grown by MBE at room temperature. Therefore, we do not expect different *bulk* scattering spin asymmetries in the present case. This justifies the restriction to interface scattering spin asymmetries in Sect. 2.2. In the general case, however, both interface and bulk scattering spin asymmetries may contribute. *AlHajDarwish et al.*, for instance, have induced an inversion of the bulk scattering spin asymmetry by diluting Fe with 5% of Cr [22].

7 Summary

We have prepared single-crystalline nanopillars by molecular beam epitaxy and a combined process of optical and e-beam lithography. The extended multilayers are characterized by MOKE and compared to CPP-GMR data of the nanopillars, which clearly show the 4-fold magnetocrystalline anisotropy of Fe and, thus, reveal the single-crystalline structure of the nanopillars. The large GMR ratio of up to 5.6% at 4 K reflects the high spin polarization for Ag/Fe interfaces predicted in [10, 11]. After the patterning process the magnetic properties change so that the 2 nm-thick "free" Fe layer is now coupled to the rest of the system by dipolar stray fields. Under the influence of a DC current we are able to measure distinct resistance changes, which give clear evidence of spin-transfer torque effects at current densities of about $10^8\,\mathrm{A/cm^2}$. This value is mostly determined by the sizable dipolar coupling in the nanopillar. At high magnetic fields, step-like resistance changes are measured at positive currents and are attributed to current-driven magnetic excitations. The opposite scattering spin asymmetries of Fe/Cr and Fe/Ag interfaces enable us to observe CIMS at small magnetic fields for both subsystems in a single device, for one at positive and for the other at negative DC currents. The switching at opposite current polarity provides opportunities for optimizing the CIMS behavior and realizing further magnetic excitation modes, *e.g.* by exciting one subsystem at higher current density while simultaneously suppressing excitations of the fixed layer with the torque exerted by the second subsystem.

References

[1] M. N. Baibich, J. M. Broto, A. Fert, F. Nguyen Van Dau, F. Petroff, P. Etienne, G. Creuzet, A. Friedrich, J. Chazelas: Giant magnetoresistance of (001)Fe/(001)Cr magnetic superlattices, Phys. Rev. Lett. **61**, 2472–2475 (1988)
[2] G. Binasch, P. Grünberg, F. Saurenbach, W. Zinn: Enhanced magnetoresistance in layered magnetic structures with antiferromagnetic interlayer exchange, Phys. Rev. B **39**, 4828–4830 (1989)
[3] J. C. Slonczewski: Current-driven excitation of magnetic multilayers, J. Magn. Magn. Mater. **159**, L1–L7 (1996)
[4] L. Berger: Emission of spin waves by a magnetic multilayer traversed by a current, Phys. Rev. B **54**, 9353–9358 (1996)
[5] F. J. Albert, J. A. Katine, R. A. Buhrman, D. C. Ralph: Spin-polarized current switching of a Co thin film nanomagnet, Appl. Phys. Lett. **77**, 3809–3811 (2000)
[6] J. Grollier, V. Cros, A. Hamzic, J. M. George, H. Jaffrès, A. Fert, G. Faini, J. B. Youssef, H. Legall: Spin-polarized current induced switching in Co/Cu/Co pillars, Appl. Phys. Lett. **78**, 3663–3665 (2001)
[7] S. Urazhdin, N. O. Birge, W. P. Pratt Jr., J. Bass: Current-driven magnetic excitations in permalloy-based multilayer nanopillars, Phys. Rev. Lett. **91**, 146803 (2003)
[8] I. N. Krivorotov, N. C. Emley, J. C. Sankey, S. I. Kiselev, D. C. Ralph, R. A. Buhrman: Time-domain measurements of nanomagnet dynamics driven by spin-transfer torques, Science **307**, 228–231 (2005)
[9] Whenever we refer to the direction of a current, we mean the direction of the electron flux rather than the (opposite) technical current direction.
[10] M. D. Stiles, D. R. Penn: Calculation of spin-dependent interface resistance, Phys. Rev. B **61**, 3200–3202 (2000)
[11] M. D. Stiles, A. Zangwill: Anatomy of spin-transfer torque, Phys. Rev. B **66**, 014407 (2002)
[12] T. Valet, A. Fert: Theory of the perpendicular magnetoresistance in magnetic multilayers, Phys. Rev. B **48**, 7099–7113 (1993)
[13] A. Barthélémy, A. Fert, F. Petroff: *Handbook of Magnetic Materials*, vol. 12 (Elsevier, Amsterdam 1999) Chap. Giant Magnetoresistance in Magnetic Multilayers
[14] D. E. Bürgler, C. M. Schmidt, J. A. Wolf, T. M. Schaub, H.-J. Güntherodt: Ag films on Fe/GaAs(001): From clean surfaces to atomic Ga structures, Surf. Sci. **366**, 295–305 (1996)
[15] M. Buchmeier: unpublished work

[16] D. E. Bürgler, P. Grünberg, S. O. Demokritov, M. T. Johnson: *Handbook of Magnetic Materials*, vol. 13 (Elsevier, Amsterdam 2001) Chap. Interlayer Exchange Coupling in Layered Magnetic Structures, pp. 1–85
[17] M. Buchmeier, B. K. Kuanr, R. R. Gareev, D. E. Bürgler, P. Grünberg: Spinwaves in magnetic double layers with strong antiferromagnetic interlayer exchange coupling: Theory and experiment, Phys. Rev. B **67**, 184404 (2003)
[18] C. Kittel: *Einführung in die Festkörperphysik* (R. Oldenbourg Verlag, München, Wien 1996)
[19] E. P. Wohlfarth (Ed.): *Ferromagnetic Materials* (North-Holland Publishing Company, Amsterdam, New York, Oxford 1980)
[20] H. Namatsu, T. Yamaguchi, M. Nagase, K. Yamazaki, K. Kurihara: Nanopatterning of a hydrogen silsesquioxane Resist with reduced linewidth fluctuations, Microelectron. Eng. **41/42**, 331–334 (1998)
[21] P. G. Glöersen: Ion-beam etching, J. Vac. Sci. & Tech. **12**, 28–35 (1975)
[22] M. AlHajDarwish, H. Kurt, S. Urazhdin, A. Fert, R. Loloee, W. P. Pratt, Jr., J. Bass: Controlled Normal and Inverse Current-Induced Magnetization Switching and Magnetoresistance in Magnetic Nanopillars, Phys. Rev. Lett. **93**, 157203 (2004)
[23] J. M. George, L. G. Pereira, A. Barthélémy, F. Petroff, L. Steren, J. L. Duvail, A. Fert, R. Loloee, P. Holody, P. A. Schroeder: Inverse spin-valve-type magnetoresistance in spin engineered multilayered structures, Phys. Rev. Lett. **72**, 408–411 (1994)
[24] M. Buchmeier, R. Schreiber, D. E. Bürgler, P. Grünberg: Inverse giant magnetoresistance due to spin-dependent interface scattering in Fe/Cr/Au/Co, Europhys. Lett. **63**, 874–880 (2003)

Andreev Reflection in Nb-InAs Structures: Phase Coherence, Ballistic Transport and Edge Channels

Jonathan Eroms[1,2] and Dieter Weiss[2]

[1] Kavli Institute of Nanoscience, Delft University of Technology, Lorentzweg 1, 2628 CJ Delft, The Netherlands
eroms@qt.tn.tudelft.nl

[2] Institut für Experimentelle und Angewandte Physik, Universität Regensburg, Universitätsstraße 31, 93053 Regensburg, Germany
dieter.weiss@physik.uni-regensburg.de

Abstract. We present our experimental work on transport in superconductor-semiconductor structures. Using high-quality contacts, Andreev reflection dominates the transport properties in a range of experimental parameters, including high magnetic fields. We investigated periodic arrays of Nb filled stripes or antidots in an InAs-based 2DEG. Depending on the geometry and magnetic field, Andreev reflection modifies transport in different ways. At magnetic fields up to a few flux quanta per unit cell, we observe phase-coherent behavior, such as flux-periodic oscillations. At slightly higher fields, the Andreev reflection probability is determined by induced superconductivity in the 2DEG, which is gradually suppressed by an increasing magnetic field. The impact of Andreev reflection on the ballistic motion in antidot lattices is particularly intriguing: the commensurability peaks commonly found in the magnetotransport in those lattices are strongly suppressed. At fields of several Tesla we enter the regime of the quantum Hall effect in the 2DEG, and we find a pronounced increase of the amplitude of $1/B$-periodic magnetoresistance oscillations. The latter can be traced to an enhanced backscattering of Andreev-reflected edge channels, which contain both electrons and holes.

1 Introduction

When different material classes are combined in solid state physics, new and unexpected behavior is frequently observed. In this article we describe experiments on superconductor-semiconductor structures fabricated from the Nb-InAs material system. With semiconductors we typically associate properties such as tunable electron density and high carrier mobility, and effects such as the quantum Hall effect or ballistic transport [1, 2], to name a few examples. Superconductors, on the other hand, are known for their perfect conductivity, phase coherence or the Josephson effect. If both materials are brought into contact, many new phenomena arise, which are based on *Andreev* reflection [3]. This process takes places at the boundary between a superconductor and a normal conductor (which can be a metal or a semiconductor). When an electron from the Fermi edge of a normal conductor impinges on

R. Haug (Ed.): Advances in Solid State Physics,
Adv. in Solid State Phys. **46**, 135–146 (2008)

the superconductor, the energy gap in the superconductor prevents it from entering, unless a Cooper pair is immediately formed in the superconductor. Since a Cooper pair is composed of two electrons with opposite spin and momentum, a suitable extra electron has to be taken from the Fermi sea of the normal conductor, creating a hole in the normal conductor. Note that the hole is still located in the conduction band of the normal conductor and is not to be confused with a hole in the valence band of a p-type semiconductor. The incoming electron is thus converted into a hole, whose properties can be calculated by solving the Bogoliubov-de Gennes equations [4] or by considering conservation laws. Most importantly, the hole is retroreflected, i.e., it exactly retraces the trajectory of the incoming electron, provided that no magnetic field is applied. It was shown that a ballistic billiard with Andreev-reflecting boundaries is always regular, regardless of its shape, and chaotic motion is only obtained in a non-zero magnetic field [5]. With Andreev reflection the hole wavefunction also acquires an extra phase factor containing the phase of the superconducting condensate. In this way, quasiparticle interferometers can be constructed, and Andreev reflection provides a unified treatment for supercurrents in Josephson junctions and SNS junctions in the short or long limit [6]. Naturally, Andreev reflection can also convert a hole back into an electron, thereby removing a Cooper pair from the condensate.

A vast number of experiments have been carried out, demonstrating phase-coherent transport [7] or tunable supercurrents in hybrid decives [8, 9]. Furthermore, Andreev reflection could be the basis for generating spin entangled electrons in a semiconductor [10].

In our experiments, we focussed on ballistic transport of electrons in a high-mobility two-dimensional electron gas (2DEG) in contact to a superconductor. Although a number of theoretical articles have treated the Andreev-reflecting billiard [5, 11, 12], experimental studies were scarce. For the antidot lattice [1, 13] – a periodic array of etched holes in a 2DEG – it is well known that chaotic and regular motion determine its magnetotransport properties [14]. Filling the etched regions with superconductor created an ideal model system to study the Andreev billiard in experiment [15]. We also found that Andreev reflection persisted up to magnetic fields of several Tesla and we observed striking effects on transport in the quantum Hall regime [16].

2 Sample Preparation and Characterization

Many experiments on superconductor-semiconductor physics were performed with niobium and indium arsenide [7, 17, 18]. InAs does not form a Schottky barrier with a metal and is therefore well suited for high-quality contacts. Nb has a rather high critical temperature and magnetic field while still being reasonably easy to deposit. To obtain high electron mobilities, the InAs was embedded into an MBE-grown AlGaSb-InAs heterostructure [19]. Mesas and contacts were prepared with optical lithography, wet etching and Cr/Au

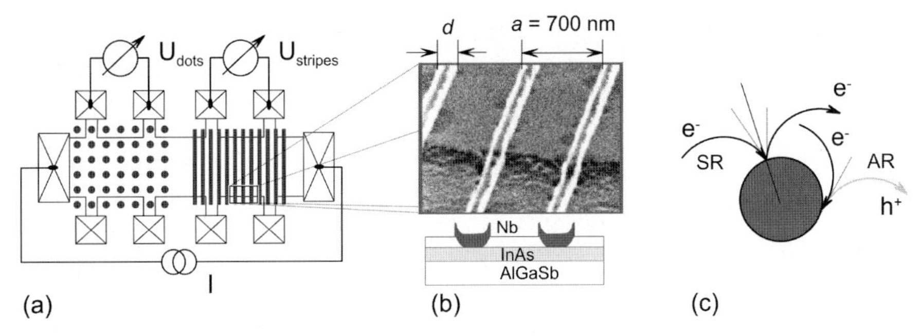

Fig. 1. (**a**): A sketch of the Hall bar geometry used for the magnetotransport measurements. Arrays of etched and filled stripes or dots are placed on an InAs-based heterostructure. (**b**) A SEM-image of a stripe array and a schematic cross section of the geometry: Only the top AlGaSb-layer is etched, while the InAs channel continues underneath the niobium. (**c**) Specular (SR) or Andreev (AR) reflection of an electron trajectory at a non-zero magnetic field

evaporation. The Nb filled structures were defined with electron beam lithography, selective wet etching or reactive ion etching and Nb sputter deposition. An in situ sputter cleaning step of the InAs surface proved to be essential for high transparency contacts. More fabrication details can be found in [15]

Measuring the differential conductivity with respect to the bias voltage on single SNS-junctions and stripe arrays, we could determine the Z-parameter of the OTBK-model [20] to be $Z = 0.4 \ldots 0.6$. In short junctions we also observed multiple Andreev reflections both due to the superconducting gap of niobium and due to an induced gap [21, 22] in the InAs channel. The latter ranged between 300 and 500 μeV depending on the geometry. In our experiments we also assume that Andreev reflection occurs between the InAs 2DEG and the induced superconducting region in the InAs underneath the Nb film.

3 Phase-Coherent Experiments

At low magnetic fields and low temperatures, the phase acquired on an Andreev reflection event can lead to coherent coupling between the niobium structures and a supercurrent can flow through the stripe or dot arrays. The critical current oscillates as a function of the magnetic field with the fundamental period given by a flux quantum through the unit cell of the lattice. The oscillations are more easily measured at slightly elevated temperatures when a finite resistance appears. The shape of the oscillations is related to the shape of the basis element of the lattice. This can be seen in Fig. 2 where we plot the magnetoresistance oscillations for three different arrays of cross-shaped dots. While the crosses with the thinnest lines show more than 30 minima in the magnetoresistance, the wider crosses display much less oscillations, but more

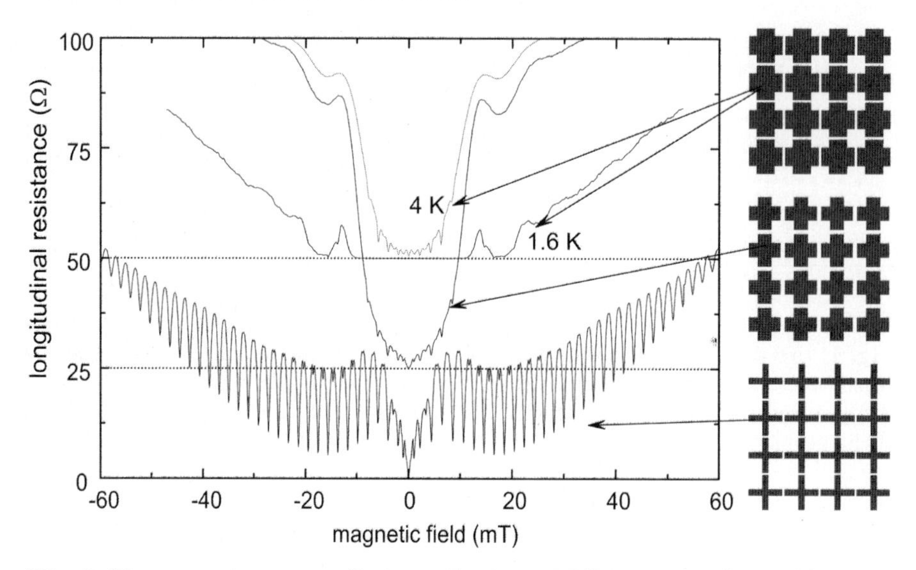

Fig. 2. Magnetoresistance oscillations of lattices of different cross-shaped Nb structures. The lattice period was $a = 1\,\mu\text{m}$ in all cases. The curves are offset by $25\,\Omega$ for clarity

substructure per oscillation period. The situation is reminiscent of an optical diffraction grating where the shape of the slits determines the envelope of the interference pattern. The fractional minima stem from coherent coupling in larger groups of unit cells [23]. In dot arrays, we observed coherent effects in up to 3×3 unit cells. In stripe arrays we observed coherent coupling between neighboring stripes and next-to-nearest neighbors, implying that parts of the supercurrent flow through the InAs underneath the niobium stripes. More recent measurements of the temperature dependence of the critical current in Nb/InAs stripe arrays allowed to determine the induced superconducting gap in the InAs channel, which amounted to $300\,\mu\text{eV}$ [24]. The same experiment also revealed the transparency $\tau = 0.63$ of the interface between the unetched InAs 2DEG and the Nb-covered InAs underneath the stripes. It is smaller than unity due to different electron densities in both regions.

4 Andreev Reflection and Chaotic Motion

In an antidot lattice, the interplay of regular and chaotic motion leads to peculiar effects in the magnetotransport. This was studied in detail in [14]. Most prominently, a peak appears in the magnetoresistance when the diameter of the cyclotron orbit $2R_c$ matches the lattice period a. Depending on the dot diameter and the mean free path, more features can be observed at lower (corresponding to larger orbits encircling more antidots) or higher fields (when the orbits bounce off the antidot walls several times [25]). If the

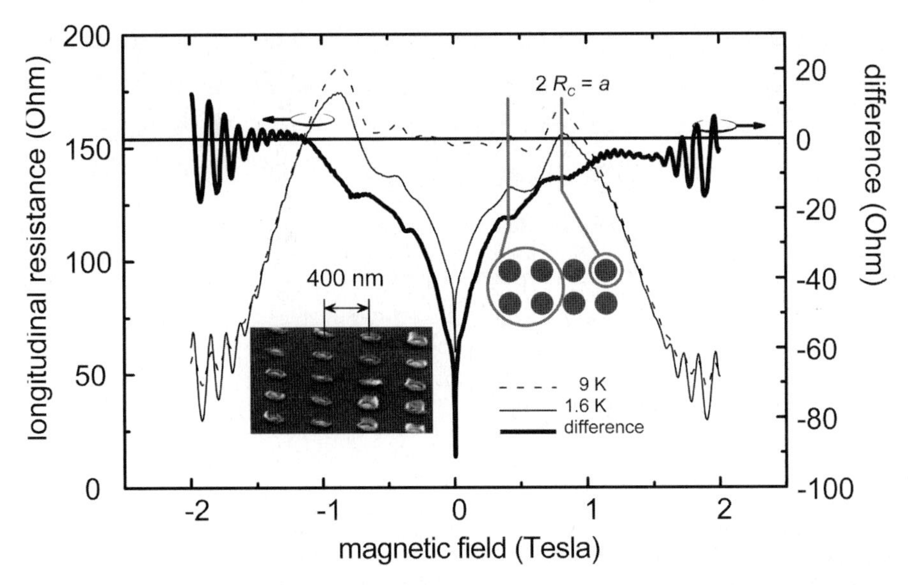

Fig. 3. Magnetoresistance traces above and below T_c for an array of niobium-filled antidots. The magnetic field positions of the orbits encircling one and four antidots are given by *vertical lines*. The difference of both traces shows minima at those positions, which points to a suppression of the antidot peaks by Andreev reflection

specular reflection at the antidot boundary is replaced by Andreev reflection, the character of the orbits should change and different features should be observed.

A typical measurement is reproduced in Fig. 3, where we plot the magnetoresistance of an array of niobium-filled antidots above and below T_c. The measurements above T_c clearly show two antidot peaks corresponding to orbits around one and four antidots. Below T_c, the resistance around $B = 0$ is markedly reduced and recovers gradually when the field is increased. Subtracting both traces, another striking effect is observed: The difference of both graphs shows dips at the positions of the antidot peaks. In a conventional antidot lattice, the opposite would be observed as the ballistic features are better resolved when going to lower temperatures. Evidently, Andreev reflection acts to remove the antidot peaks even at magnetic fields of up to one Tesla.

To clarify this point, we calculated the magnetotransport properties using the Kubo formula:

$$\sigma_{ij} = \frac{m_{\text{eff}}}{\pi\hbar^2} \int_0^\infty \mathrm{d}t\, e^{-t/\tau} \langle q(t)v_i(t)\, q(0)v_j(0)\rangle \,. \tag{1}$$

In this approach, a large number of classical trajectories are calculated and the velocity correlations yield the coefficients σ_{ij} of the conductivity tensor. The finite mean free path is included with the exponential damping term. To

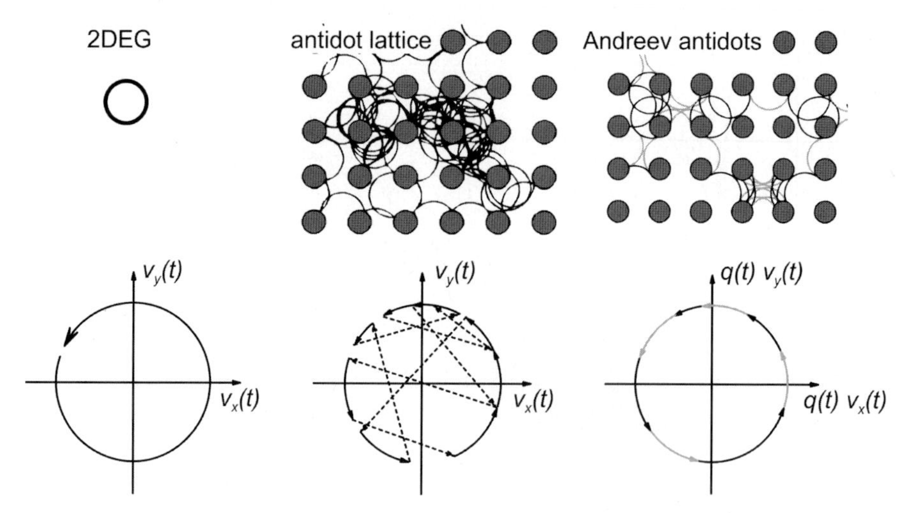

Fig. 4. Trajectories in real space and velocity space of an unpatterned 2DEG, an antidot array and an Andreev antidot array. While the real space trajectories are chaotic for both antidot arrays, the velocity correlations are only chaotic for the specularly reflecting antidot lattice. *Black* and *gray lines* correspond to electrons and holes

account for Andreev reflection we incorporated a time-dependent charge $q(t)$, which is positive for holes and negative for electrons. Both specular and Andreev reflection were allowed with an Andreev reflection probability p_{AR}. The results are shown in Fig. 5. If Andreev reflection is excluded ($p_{\mathrm{AR}} = 0$), the magnetoresistance of a conventional antidot lattice is obtained. For perfect Andreev reflection ($p_{\mathrm{AR}} = 1$), all features disappear, and the magnetoresistance looks flat, just as for an unpatterned 2DEG.

To gain more insight into the mechanisms at work in an Andreev reflecting antidot lattice, we show typical trajectories in an unpatterned 2DEG, an antidot lattices with specularly reflecting walls and an Andreev antidot lattice (see Fig. 4). The magnetic field corresponds to the main commensurability peak. While in both cases the trajectories in real space are chaotic, the velocity correlations are quite different for the two types of antidot lattices. In the specularly reflecting case, the trajectories are also chaotic in velocity space and their correlations give rise to the features in the magnetoresistance. If only Andreev reflection takes place, the velocity is exactly reversed (i.e., multiplied by -1), but because the charge is also reversed, the product of both is unchanged. Since the Kubo formula (1) only evaluates the velocity correlations (including the charge), there is no difference between a cyclotron orbit in an unpatterned 2DEG and the orbit composed of circular pieces of alternating charge, which are found in the Andreev antidot lattice.

To reproduce the experimental data, we assumed a magnetic field dependence of p_{AR}, which we took empirically to be exponentially decreasing.

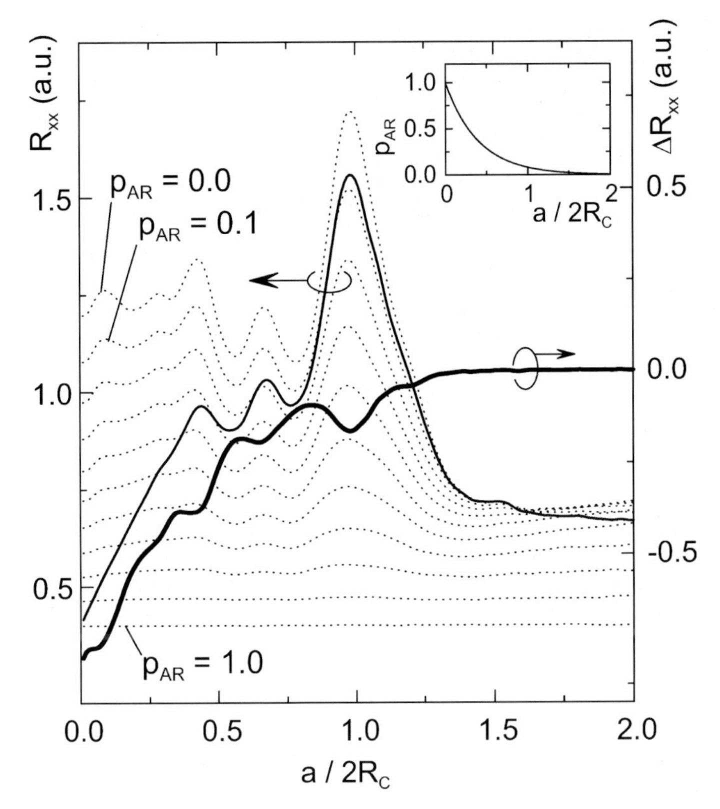

Fig. 5. Calculated magnetoresistance traces for different constant Andreev reflection probabilities (*dotted lines*). For perfect Andreev reflection $p_{\mathrm{AR}} = 1$, no peaks are left. To match the experimental data, an exponentially decreasing $p_{\mathrm{AR}}(B)$ was assumed (*inset*), yielding the *thin solid line*. *Thick solid line*: difference of the *thin solid line* and the uppermost *dotted line*

This allowed us to reproduce the experimental curve, and also the difference graph with the characteristic dips. The magnetic field dependence of Andreev reflection was treated in a recent theoretical paper [26]. It was shown that the induced gap in the semiconductor depends on screening currents in the superconductor and that Andreev reflection indeed depends on the magnetic field. While exact expression for $p_{\mathrm{AR}}(B)$ is not a simple exponential as we assumed, our conclusions are still qualitatively correct.

5 Magnetotransport in Andreev-Reflected Edge Channels

In high magnetic fields, transport in edge channels comes into play. In a bulk 2DEG, highly degenerate Landau levels are formed, with a Landau level split-

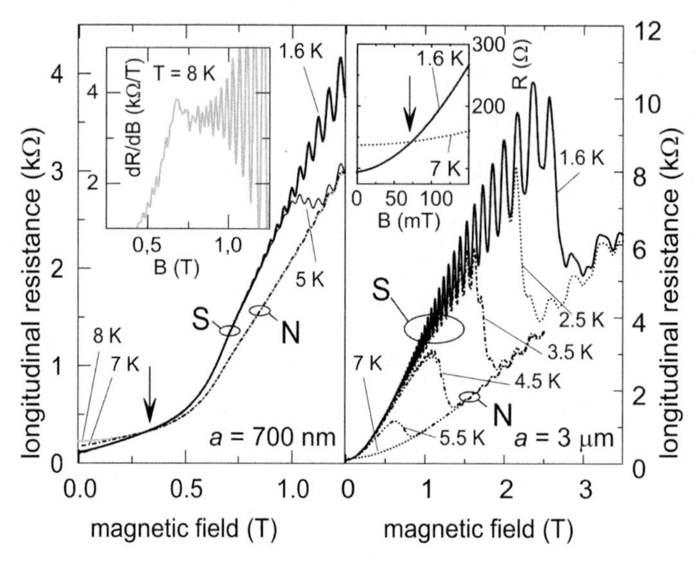

Fig. 6. T-dependent magnetotransport curves for two different samples. *Arrows*: crossing points of the graphs above T_c and below T_c. *Left inset*: The peak in dR/dB corresponds to a change in slope of the resistance trace. *Right inset*: Enlarged view of the crossing point for the sample with $a = 3\,\mu$m. The critical temperatures were 7.4 K (*left*) and 6.9 K (*right*). Letters 'N' and 'S' denote the normal and superconducting branch, respectively

ting of $\hbar\omega_c = \hbar eB/m_{\text{eff}}$, where m_{eff} is the effective mass. At the sample edge, the confinement potential bends the levels upwards, and all Landau levels below E_F have to cross the Fermi level. At the crossing points, edge channels are formed, which are one-dimensional conduction channels contributing one conductance quantum (of $h/2e^2$ for spin-degenerate electrons) each [2]. Since the strong magnetic field forces the electrons to move in one direction only, backscattering is suppressed, and a zero resistance state can form in the Shubnikov-de Haas minima, where the filling factor is integer. At non-integer filling factors, the innermost edge channel can be backscattered, reducing the conductance and giving rise to a finite resistance.

How does the situation change if Andreev reflection is involved? This was treated in a theory article by *Hoppe*, *Zülicke* and *Schön* [27]. When electrons impinge on a superconductor-2DEG boundary, they are Andreev reflected, but in the strong magnetic field the holes move in the same direction as the electrons. They hit the boundary again to be converted to electrons and so forth. A quantum mechanical treatment showed that the counterpart of edge channels in the quantum Hall regime are Andreev edge channels, which are composed of electron- and hole-like quasiparticles. The proportions of both constituents are governed by the interface transparency and the filling factor. Importantly, the proportion of electrons and holes also determines the con-

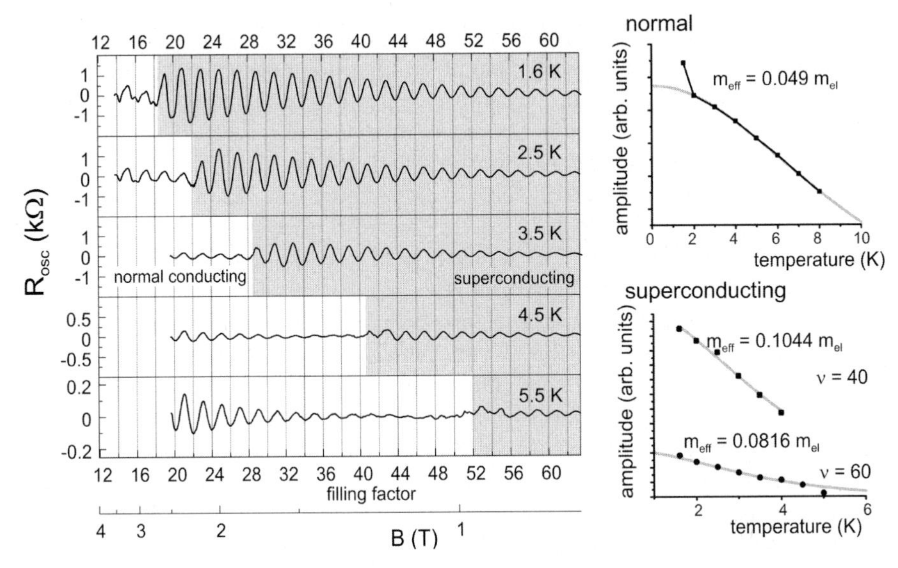

Fig. 7. *Left*: Same data as in Fig. 6, right, after subtracting the slowly varying background and plotted against the filling factor. For ease of comparison, the value of B is also given. *Shaded regions*: Nb stripes are superconducting, as extracted from Fig. 6. Note the strong increase of the amplitude at the superconducting transition. *Right*: Amplitude of the $1/B$-periodic oscillations at $B > B_c$ (*upper part*) and $B < B_c$ (*lower part*). A fit to (2) and the corresponding effective masses is also shown

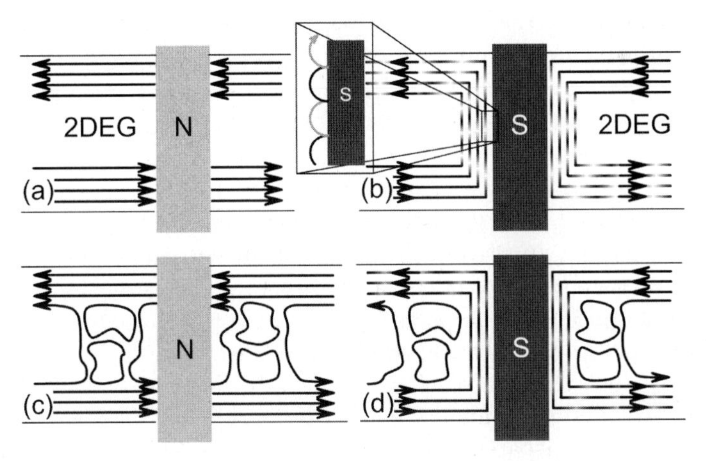

Fig. 8. Edge channels in a 2DEG hosting a normal (*left*) or superconducting (*right*) electrode. (**a**), (**b**): Integer filling factor (i.e., resistance minimum). (**c**), (**d**): In between integer filling factors. With a normal electrode, only the innermost channel is backscattered due to impurities in the 2DEG. In the superconducting case, edge channels hitting the electrode are Andreev reflected (see *inset*) and contain electrons and holes (*gray*). The amount of current which is backscattered depends on the hole probability, which oscillates in a magnetic field

ductance of such an edge channel. If electrons and holes have equal occupancy, the edge channel cannot carry current at all since both quasiparticle types move in the same direction, but due to their opposite charge the currents exactly cancel. This situation occurs at any magnetic field if the interface is perfectly transparent. If an interface barrier is present, the electron/hole ratio oscillates with the filling factor, i.e., periodic in $1/B$.

We observed those effects in Nb-InAs stripe arrays with lattice periods between $0.7\,\mu$m and $3\,\mu$m. The critical field of the stripes can be as high as 2.5 T, permitting us to enter the edge channel regime. Still higher fields can be obtained using NbN as a superconductor, but at the expense of more complicated fabrication [28, 29]. Magnetoresistance traces of two representative samples are shown in Fig. 6. Once the stripes are only coupled via edge channels, the longitudinal resistance in the superconducting state is higher than in the normal state. More importantly, the amplitude of the magnetoresistance oscillations is enhanced in the superconducting state. This can be seen more clearly in Fig. 7 where the non-oscillatory background is subtracted and the data is plotted on a $1/B$-scale. Comparing the amplitudes in the vicinity of B_c, the oscillations are enhanced by a factor of up to 6. This cannot be entirely due to the higher background resistance, because even when normalizing to the background resistance (as a worst case scenario), the enhancement is still around 2. From the temperature dependence of the oscillations we can also conclude that the oscillations in the superconducting state are not the usual Shubnikov-de Haas oscillations. In that case, the T-dependence is determined by thermal activation across the Landau gap $\hbar\omega_c = \hbar eB/m_{\mathrm{eff}}$. This is routinely utilized [30] to determine the effective mass in a 2DEG with the formula

$$\frac{A(T_1)}{A(T_2)} = \frac{T_1 \sinh(2\pi^2 k_{\mathrm{B}} T_2 m_{\mathrm{eff}}/\hbar eB)}{T_2 \sinh(2\pi^2 k_{\mathrm{B}} T_1 m_{\mathrm{eff}}/\hbar eB)} \tag{2}$$

for the amplitudes A at different temperatures T_1 and T_2. In the normal conducting case, this yields $m_{\mathrm{eff}} \approx 0.4 \ldots 0.5\ m_0$, which is comparable to what is found in high-density InAs 2DEGs [31]. In the superconducting case, the effective masses are clearly unreasonable and also seem to depend on the filling factor. This rules out a simple explanation in terms of SdH-oscillations. Instead, we can explain our results in the model of Andreev edge channels, if we take our geometry into account (see Fig. 8). On each Nb stripe in contact with the 2DEG, Andreev edge channels are formed. Since in our samples the interface is not perfectly transparent, the conductance of those channels oscillates strongly with $1/B$. This leads to oscillating backscattering across the Hall bar which in turn results in strong magnetoresistance oscillations. Contrary to ordinary SdH-oscillations, where only the innermost edge channel can be backscattered, all Andreev edge channels are involved in this process. This explains why the oscillation amplitude is much higher in the superconducting case.

The non-oscillatory part of the resistance is also higher in the superconducting state at high fields. This can also be explained in the edge channel picture. At high fields, neighboring stripes are only connected via edge channels. The edge channels emitted from a superconducting electrode always contain both electrons and holes, which means that their conductance is diminished compared to usual electron-only edge channels. Therefore, the resistance background is also increased.

6 Summary

We have performed a number of experiments in periodic Nb-InAs structures in different transport regimes. At very low magnetic fields we could observe supercurrents and phase-coherent oscillations. The ballistic motion of electrons in an antidot lattice with superconducting boundaries is strongly affected by Andreev reflection. Ideally, Andreev reflection should suppress all signatures of chaos in magnetotransport, but due to the decreasing Andreev reflection probability at higher fields this is only partly realized in experiment. We also clearly observed edge channels containing Andreev reflected electrons and holes. This should open up the way for more advanced experiments with ballistic transport and edge channels.

Acknowledgements

This work was based on collaborations with a number of people. Most importantly, the MBE-grown InAs/AlGaSb material was kindly provided by J. De Boeck and G. Borghs from IMEC in Belgium. We thank G. Bayreuther for permission to use the sputtering equipment in his group. The calculations based on the Kubo formula were done by M. Tolkiehn and U. Rössler at the University of Regensburg. For the Andreev reflected edge channels we collaborated with U. Zülicke at Massey University, New Zealand. We also enjoyed fruitful discussions with J. Keller, R. Kümmel, K. Richter, F. Rohlfing, C. Strunk, and G. Tkachov. This work was supported by the Deutsche Forschungsgemeinschaft (GRK 638).

References

[1] T. Ando *et al.* (Ed.): *Mesoscopic Physics and Electronics*, (Springer, Berlin 1998).
[2] S. Datta: *Electronic Transport in Mesoscopic Systems*, (Cambridge University Press, Cambridge 1995).
[3] A. F. Andreev: Sov. Phys. JETP **19**, 1228 (1964) [Zh. Eksp. Teor. Fiz. **46**, 1823 (1964)].
[4] P. G. de Gennes: *Superconductivity of Metals and Alloys*, (W. A. Benjamin, New York, 1966).

[5] I. Kosztin, D. L. Maslov, and P. M. Goldbart: Phys. Rev. Lett. **75**, 1735 (1995).
[6] T. Schäpers: *Superconductor/Semiconductor Junctions*, (Springer, Berlin, Heidelberg 2001).
[7] A. F. Morpurgo, S. Holl, B. J. van Wees, T. M. Klapwijk, and G. Borghs: Phys. Rev. Lett. **78**, 2636 (1997).
[8] H. Takayanagi, T. Kawakami: Phys. Rev. Lett. **54**, 2449 (1985).
[9] Y.-J. Doh, J. A. van Dam, A. L. Roest, E. P. A. M. Bakkers, L. P. Kouwenhoven, and S. De Franceschi: Science **309**, 272 (2005).
[10] P. Recher, E. V. Sukhorukhov, D. Loss: Phys. Rev. B **63**, 165314 (2001).
[11] J. A. Melsen, P. W. Brouwer, K. M. Frahm, and C. W. J Beenakker: Europhys. Lett. **35**, 7 (1996).
[12] W. Ihra, M. Leadbeater, J. L. Vega, and K. Richter: Eur. Phys. J. B **21**, 425 (2001).
[13] D. Weiss, M. L. Roukes, A. Menschig, P. Grambow, K. von Klitzing, and G. Weimann: Phys. Rev. Lett. **66**, 2790 (1991).
[14] R. Fleischmann, T. Geisel, and R. Ketzmerick: Phys. Rev. Lett. **68**, 1367 (1992), R. Onderka, M. Suhrke, and U. Rössler: Phys. Rev. **B 62**, 10918 (2000).
[15] J. Eroms, D. Weiss, J. de Boeck, G. Borghs: Physica C **352**, 131 (2001); J. Eroms, D. Weiss, M. Tolkiehn, U. Rössler, J. De Boeck, G. Borghs: Physica E **12**, 918 (2002); Europhys. Lett. **58**, 569 (2002); in H. Takayanagi and J. Nitta (Eds.): *Towards the Controllable Quantum States*, (World Scientific, Singapore, 2003).
[16] J. Eroms, D. Weiss, J. De Boeck, G. Borghs, and U. Zülicke: Phys. Rev. Lett. **95**, 107001 (2005).
[17] T. Kawakami, and H. Takayanagi: Appl. Phys. Lett. **46**, 92 (1985).
[18] C. Nguyen, H. Kroemer, and E. L. Hu: Phys. Rev. Lett. **69**, 2847 (1992).
[19] M. Behet, S. Nemeth, J. De Boeck, G. Borghs, J. Tümmler, J. Woitok, J. Geurts: Semicond. Sci. Technol. **13**, 428 (1998).
[20] M. Octavio, M. Tinkham, G. E. Blonder, T. M. Klapwijk: Phys. Rev. B **27**, 6739 (1983); K. Flensberg, J. Bindslev Hansen, M. Octavio: Phys. Rev. B **38**, 8707 (1988).
[21] A. Chrestin, T. Matsuyama, U. Merkt: Phys. Rev. B **55**, 8457 (1997).
[22] A. F. Volkov, P. H. C. Magnée, B. J. van Wees, T. M. Klapwijk: Physica C **242**, 261 (1995).
[23] R. F. Voss and R. A. Webb: Phys. Rev. **B 25**, 3446 (1982); R. A. Webb R. F. Voss, G. Grinstein, and P. M. Horn: Phys. Rev. Lett. **51**, 690 (1983).
[24] F. Rohlfing, C. Strunk, personal communication.
[25] J. Eroms, M. Zitzlsperger, D. Weiss, J. Smet, C. Albrecht, R. Fleischmann, M. Behet, J. De Boeck, and G. Borghs: Phys. Rev. **B 59**, 7829 (1999).
[26] G. Tkachov, and K. Richter: Phys. Rev. B **71**, 094517 (2005).
[27] H. Hoppe, U. Zülicke, G. Schön: Phys. Rev. Lett. **84**, 1804 (2000), U. Zülicke, H. Hoppe, G. Schön: Physica B **298**, 453 (2001).
[28] H. Takayanagi, T. Akazaki, Physica B **249–251**, 462 (1998).
[29] I. E. Batov, Th. Schäpers, A. A. Golubov, A. V. Ustinov: J. Appl. Phys. **96**, 3366 (2004).
[30] T. Ando, A. B. Fowler, F. Stern: Rev. Mod. Phys. **54**, 437, (1982).
[31] C. Gauer, J. Scriba, A. Wixforth, J. P. Kotthaus, C. R. Bolognesi, C. Nguyen, B. Brar, H. Kroemer: Semicond. Sci. Technol. **9**, 1580 (1994).

Selective Edge Excitation – Inter-Edge Magnetoplasmon Mode and Inter-Edge Spin Diode

Frank Hohls[1,2], Gennadiy Sukhodub[2], and Rolf J. Haug[2]

[1] Cavendish Laboratory, University of Cambridge, J J Thomson Avenue, Cambridge CB3 0HE, UK
[2] Institut für Festkörperphysik, Universität Hannover, Appelstr. 2, 30167 Hannover, Germany

Abstract. The selective excitation and detection of quantum Hall edge states has opened access to several interesting phenomenons, two of which are addressed in this paper: Firstly we conducted time-resolved current measurements of edge magnetoplasmons. At filling factors close to $\nu = 3$ we observe two decoupled modes of edge excitations, one of which is related to the innermost compressible strip and is identified as an inter-edge magnetoplasmon mode. From the analysis of the propagation velocities of each mode the internal spatial parameters of the edge structure are derived. Secondly we have studied the tunnelling between spin polarised edge states at high magnetic fields up to 28 T. Measurements of the inter-edge I–V characteristic in tilted magnetic fields B allow to determine the effective g-factor $g^*(B)$. We also observe a dynamical nuclear.

1 Introduction

Soon following the discovery of the quantum Hall (QH) effect [1] *Halperin* suggested the crucial role of extended one-dimensional edge states at the perimeter of the sample [2]. These so called edge states proofed their value in the interpretation of a number of transport experiments [3]. The edge state picture was further refined considering the self-consistent electrostatic potential which increases step like towards the edge due to the formation of alternating strips of incompressible and compressible electron liquid [4]. An example of this distribution is shown in Fig. 1a. We study the consequences of the edge reconstruction using the powerful tool of edge state selection due to partial depletion of the electron system [5].

In the first part of the paper we focus on time resolved measurements of collective excitations at the edge of a two-dimensional electron system (2DES) in a magnetic field, the so called edge magnetoplasmons (EMPs) [6]. Our experiments on EMPs in quantum Hall systems reveal a new mode, the inter-edge magneto plasmon, an EMP travelling at the boundary between regions of different electron density [7], which previously was only observed for electrons on helium [8, 9]. In the QH system it can occur within an inner edge channel as depicted in Fig. 1b.

R. Haug (Ed.): Advances in Solid State Physics,
Adv. in Solid State Phys. **46**, 147–159 (2008)

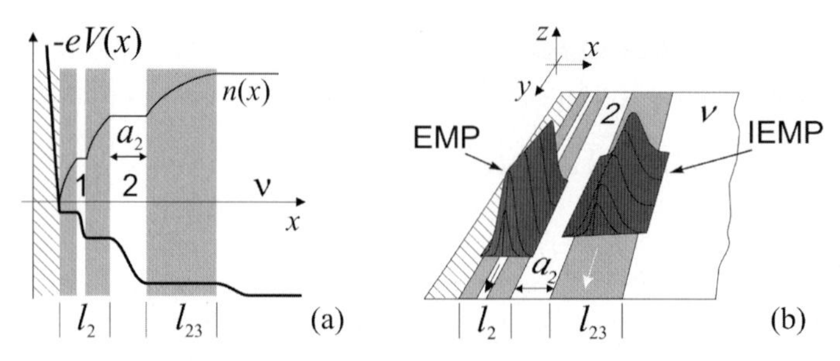

Fig. 1. (a) Potential (*thick line*) and electron density (*thin line*) at the edge for a bulk filling factor of $\nu = 3$. Areas of constant potential (*grey*) form compressible strips while areas of constant density (*white*) are incompressible. **(b)** Due to the wide $\nu = 2$ incompressible strip the edge supports two decoupled modes of edge-magnetoplasmons as indicated by the wave packets

The second part of the paper addresses the spin splitting within the edge channels. For a sufficient large magnetic field the two spin species of one Landau band form two edge channels separated by a small incompressible strip as shown in Fig. 1a. Using independent contacts to the different edge channels [10,11] we measure the tunnelling current between the channels. The current-voltage characteristic is highly directive as spin conservation allows tunnelling only for one bias direction, earning this device its name inter-edge spin diode. This device allows us to extract the Zeeman energy in a wide range of magnetic field [12]. In addition we observe a partial nuclear spin polarisation [13].

1.1 Compressible and Incompressible Strips

The self consistent potential and density profile as discussed first by *Chklovskii*, *Shklovskii* and *Glazmann* [4] are vital for the understanding of our experiments. We therefore discuss them briefly for the case of a filling factor $\nu = 3$ as depicted in Fig. 1a. Without Coulomb interaction between the electrons the potential $-eV(x)$ would rise smoothly towards the edge due to e.g. surface depletion at a mesa edge or the applied potential of a surface gate. Simultaneously the carrier density would drop smoothly until the potential exceed the Fermi energy, being zero thereafter (hatched area in Fig. 1a).

Taking into account Coulomb energy this is not the confirmation lowest in energy. Instead electrons redistribute such that stripes of changing carrier density n_e are interleaved with stripes of constant n_e in which the filling factor ν equals integer values. Within the latter stripes, marked by the white area in Fig. 1a, the Fermi energy lies in the Landau gap and the electrons are localised and therefore cannot redistribute to screen the potential. Due to this property the strips are called incompressible strips. The increase of the potential occurs within these stripes. Within the regions of non-integer filling factor electrons are not localised and can redistribute to screen the potential, which thereby stays constant. They are called compressible strips and are marked grey. These are the edge channels.

The width of the incompressible strips is determined by the energy gap between the quantum Hall states below and above the Fermi energy. For $\nu = 2$ in the inner incompressible strip this is the Landau gap between the $n = 0$ and the $n = 1$ Landau level. This energy is large and a wide strip is formed which allows us to observe an inter-edge magneto plasmon. The $\nu = 1$ strip is formed due to the Zeeman splitting of the lowest Landau level. The Zeeman gap is much smaller and therefore leads to a narrow strip. The size of the gap will be addressed in our tunnelling experiment in the inter-edge spin diode.

2 Inter-Edge Mangetoplasmon

Edge magnetoplasmons in semiconductors were first observed by *Allen* et al. [14] and since then have been studied extensively both theoretically [6, 15, 16] and experimentally [6, 17–20]. EMPs travel along the edge of the electron system, i.e., where the carrier density changes from a finite value within the sample to zero. A second class of magnetoplasmons travels along the boundary between areas of different but nonzero and constant electron densities and is called inter-edge magnetoplasmon (IEMP) mode [21, 22]. These modes have been observed for electrons on a helium surface [8, 9].

For high magnetic fields in the quantum Hall regime the propagation of EMP's is further constrained to the edge due to the localisation in the bulk of the sample. The influence of the edge channel structure is further enhanced in the presence of a metallic Schottky gate above the electron system which screens the long range Coulomb interaction. Ultimately this should lead to decoupled modes in the isolated edge channels, but previous studies of transport in such systems both in the frequency domain [17, 23] and in the time domain [18, 24] could not isolate this effect. Our experiment proofs that there is indeed a separate mode, an inter-edge magnetoplasmon, travelling in an inner edge channel as depicted in Fig. 1b.

A schematic of the experimental setup is shown in Fig. 2a. Our device consists of a T-shaped mesa etched from a 2DES in a GaAs/AlGaAs heterostructure with AuGeNi ohmic contacts at each end (electron density $n_s = 1.8 \times 10^{15}\,\mathrm{m}^{-2}$, mobility $\mu = 70\,\mathrm{m}^2/\mathrm{Vs}$). A thin metal top gate marked G on the figure allows to selectively deplete the area underneath. The whole mesa is covered by a grounded metallic top gate to screen long range Coulomb interaction. The length of signal path along the edge from source (S) to the drain contacts (D1, D2) is chosen to be equal with $l = 1.56\,\mathrm{mm}$. We apply a pulse to the source contact and detect the signal at the drains using impedance matched amplifiers with 1 GHz bandwidth and record the transient current with a fast oscilloscope (DPO). The amplitude of the source

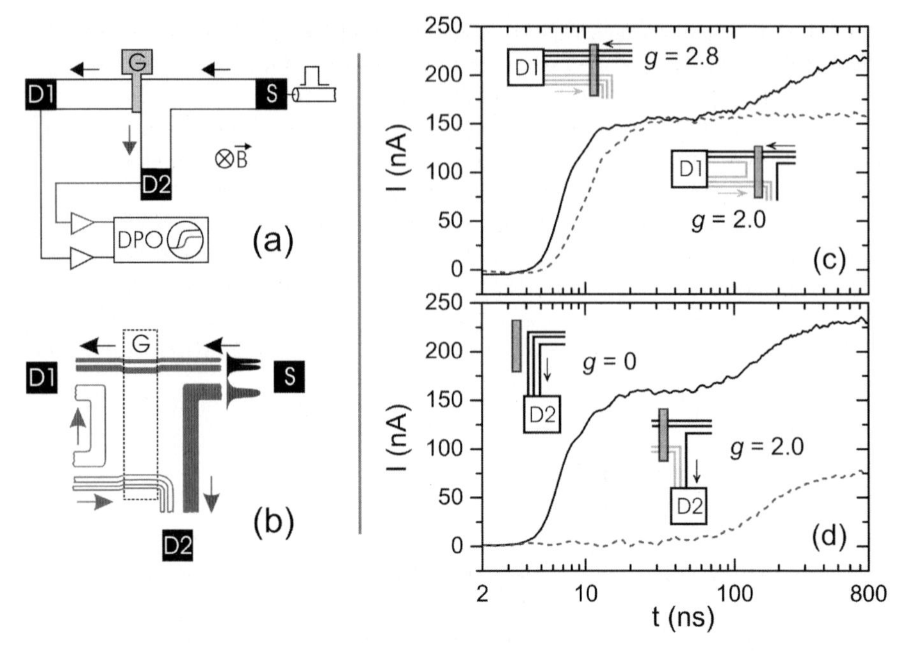

Fig. 2. (**a**) Experimental setup. For clarity we omitted the metallic top gate covering the complete mesa which is grounded throughout all measurements. (**b**) Scheme of the selective detection of edge states. The local filling factor underneath the gate is $g = 2$, the bulk one is $\nu \approx 3$. (**c**,**d**) Transient currents measured at contact D1 (c) and D2 (d) for different local filling factors g underneath the gate ($B = 2.70$ T, $\nu = 2.8$). The *solid lines* are measured with all edge channels transmitted (c) resp. deflected (d) into the detection contact. The *dashed lines* are measured for $g = 2$. A pulse with amplitude $V_{\text{in}} = 2\,\text{mV}$ is applied at $t = 0$ to the source contact.

pulse is $V_{\text{in}} = 2.0\,\text{mV}$ throughout this paper and it is applied at $t = 0$. The device is mounted into a dilution refrigerator in the centre of a superconducting magnet [20].

Throughout the experiment we use a bulk filling factor ν around 3 to achieve the widest possible incompressible strip within the edge structure; for larger filling – lower B – the Landau gap becomes smaller and the strip therefore narrower. The resulting edge structure as depicted in Fig. 1a has three edge channels, two of which are only separated by the rather narrow $\nu = 1$ strip and therefore are strongly Coulomb coupled. We expect therefore to excite two independent modes. The scheme of our selective detection of these modes is displayed in Fig. 2b which shows a blow up of the region near the gate with the course of the edge channels depicted for a filling factor $g = 2$ underneath the gates. For this gate setting the two coupled outer edge channels are transmitted to contact D1 while the inner one is deflected to D2.

The main result of this experiment is summarised in Fig. 2c and d. Figure 2c displays the transient current measured at contact D1 for full transmission of all edge channels through the gate area (solid line) and only selective transmission of the outer two (dashed line). Correspondingly Fig. 2d shows the current measured at D2 for full deflection of all channels into this contact (solid line) and detection of only the innermost state (dashed line). We first focus on the signal due to the complete edge structure, i.e., the solid lines in both graphs. Both curves are very similar and display two main features: We observe a first fast rise of the current starting at only about $t_0 = 4.6$ s with saturation to $I_1 = (2.02 \pm 0.03)(e^2/h)V_{\text{in}}$, $V_{\text{in}} = 2.0\,\text{mV}$ the pulse amplitude, after about $t = 30\,\text{ns}$. A second slower rise in current sets on at $t > 70\,\text{ns}$ and saturates to $I = I_1 + I_2$ with $I_2 = (0.92 \pm 0.04)(e^2/h)V_{\text{in}}$.

If we now switch to selective detection of edge channels, we observe a very nice decomposition of this two-step signal: The dashed curve in Fig. 2c which measures the contribution of the outer two strongly coupled edge channels displays a fast onset of current and saturates for $t < 30\,\text{ns}$ to $2(e^2/h)V_{\text{in}}$ and stays constant. Rise time and amplitude are very similar to the first current step observed when detecting all edge channels together (solid line) but no second riser occurs.

In contrast the signal of the inner edge channel, shown as dashed line in Fig. 2d, stays zero for $t < 60\,\text{ns}$, a range where all other curves already have risen to $2(e^2/h)V_{\text{in}}$. The current then rises slowly within several hundred nanoseconds to a value of only about $(e^2/h)V_{\text{in}}$, revealing a time scale and amplitude very similar to the second step in the solid lines for detection of all edge channels.

Thus we can conclude that the fast signal solely arises from a mode travelling in the outer two edge channels. This is the normal EMP mode. Its amplitude of $I = 2(e^2/h)V_{\text{in}}$ matches the expected conductance $G = 2(e^2/h)$ of two edge states. In contrast, the slow rise corresponds to a second mode exclusively travelling in the innermost edge channel with a current $I = (e^2/h)V_{\text{in}}$ matching the conductance $G = (e^2/h)$ of a single channel. As this compressible region is bounded on both sides by incompressible strips with constant nonzero carrier concentration we can unambiguously identify this mode with an inter-edge mangetoplasmon.

We will now use the opportunity to analyze the velocities of the individual modes which allows us to extract valuable information on the self-consistent edge reconstruction. But before doing so we have to comment on the slight discrepancy of the onset times of the signals detected at contact D1 (Fig. 2c) for full and selective transmission: For selective detection (dashed line) we observe a slight delay with respect to the full transmission case. The reason is a slower propagation underneath the gate for a local filling factor of $g = 2$ due to the enlarged width of the outer two compressible strips. To extract the correct velocities we therefore analyze the double step signal for full transmission of the gate.

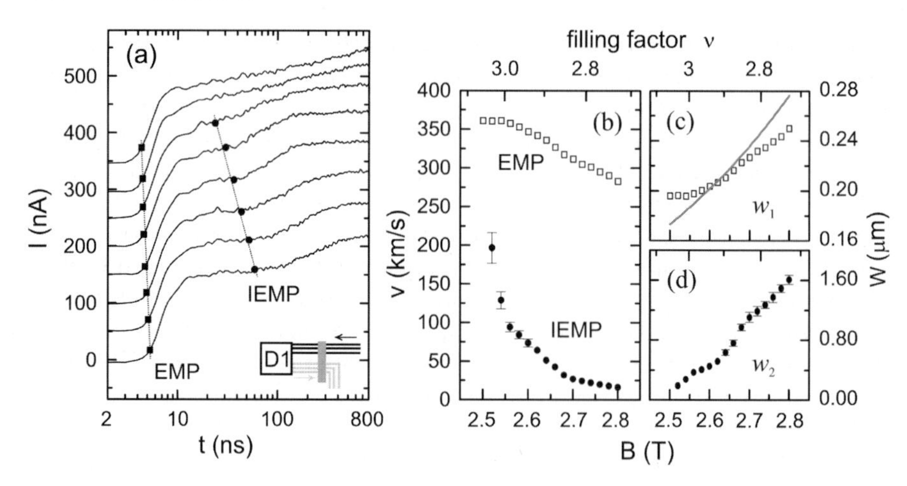

Fig. 3. (**a**) Transient current measurements for different magnetic fields resp. filling factors near $\nu = 3$ (*bottom* to *top*: $B = 2.7\,\mathrm{T}$ to $B = 2.4\,\mathrm{T}$, curves offset by 50 nA for clarity). Both the EMP and IEMP modes are detected at contact D1 (gate G grounded, $g = \nu$). The travel time t_0 of each mode is determined from the onset of the current rise (*filled squares* and *circles*). (**b**) Velocities $v = t/l$ of the EMP (*open square*) and the IEMP (*filled circle*). The travel time t_0 was determined from data as shown in (**a**). (**c**), (**d**) Width w of compressible region to which the EMP (c), w_1) resp./ the IEMP (d), w_2) is confined. The width is calculated from the velocity in (b) using (1). The *line* in (c) shows the width calculated following [4] with a depletion length of 300 nm

Figure 3a displays the transient current of the full edge structure for different magnetic fields resp. filling factors in the vicinity of $\nu = 3$. The onsets t_0 of the two modes are determined by a fit of a two step exponential rise (not shown, see [20]) and are marked by the symbols. The extracted velocities $v = l/t_0$ of the EMP and IEMP mode are displayed in Fig. 3b. For both modes the velocity decreases with rising magnetic field. This reveals a change of the edge structure even while the bulk quantum Hall state stays at $\nu = 3$.

The velocities can be interpreted using a local capacitance approximation developed in [24]. The group velocity is given by

$$v_g = \frac{\partial \omega}{\partial k} = \frac{e^2 \triangle \nu}{h} \frac{d}{\epsilon_0 \epsilon_r w}, \quad d \ll w, \tag{1}$$

with d the distance between the 2DES and the gate, w the width of the strip in which the magnetoplasmon travels, and $\triangle \nu$ the filling factor difference between the two bordering incompressible regions. The EMP mode travels within the outer two edge channels and is therefore confined by the fully depleted edge and the wide $\nu = 2$ incompressible strip, thus $\triangle \nu = 2$ for this mode. For the IEMP it is the difference between the bulk filling factor and again the $\nu = 2$ strip, thus $\triangle \nu = \nu - 2$. We now can use (1) to extract the width w of the compressible regions in which the (I)EMP travels as function of the magnetic field. The result is shown in Fig. 3c for the EMP and in Fig. 3d for the IEMP mode.

The width w_1 of the EMP mode is essentially given by the position of the $\nu = 2$ incompressible strip which we can compare to the prediction of [4] shown by the solid grey line in Fig. 3c. We would like to stress that this comparison was not possible in previous measurements in which the signal was spread over the complete edge structure. We find an acceptable quantitative agreement although the slope of our data is smaller than the predicted one.

We do not have a theory to compare to the extracted width w_2 of the innermost compressible strip in which the IEMP propagates. The initial width for $\nu = 3$ in the centre of the Hall plateau ($w_2 \sim 0.4\,\mu m$) is of the same order of magnitude as $w_1 \sim 0.2\,\mu m$. But with increasing field and decreasing filling factor the width w_2 increases much stronger than w_1 to values of order 1–2 μm, thereby slowing down the IEMP mode considerably. The overall width $w_1 + w_2$ of the edge structure compares well with previous estimates derived from, e.g., capacitance measurements [25] but these measurements could not reveal the inner structure.

3 Inter-Edge Spin Diode

In this second part we will focus on DC-measurements of tunnelling between the spin split edge states within the lowest Landau band for a filling factor arround $\nu = 2$. We apply an extension of the selective detection scheme used in the previous section: Adding another gate we can populate and detect edge channels individually and we also can establish a non-equilibrium between edge channels within a controlled scattering region [10, 11]. The device is sketched in Fig. 4: The two gates are both set to a local gate filling factor of $g = 1$, transmitting only the outer (spin up) edge channel. The chemical potential of the outer edge state going into the scattering region (dashed box) is set by the source contact S, the chemical potential of the inner channel is controlled by the drain contact D. We use additional contacts 1 and 2 to measure the voltage difference between the edge channels without any error due to contact resistances. The active scattering region between the gates with a length of about 1 μm is shown in the SEM micrograph. The device characteristic is the same as in the previous section. For our measurements we placed the sample into a dilution fridge with rotation mechanism. A rotation angle of $\varphi = 0$ denotes a magnetic field perpendicular to the 2DES. We applied total fields of up to 28 T with the rotation angle chosen such that the perpendicular component lies within the range of the $\nu = 2$ quantum Hall plateau.

In a typical measurement we apply a current between contacts S and D and measure the voltage between the inner and outer edge channel between contacts 1 and 2. For each combination of rotation angle φ and magnetic

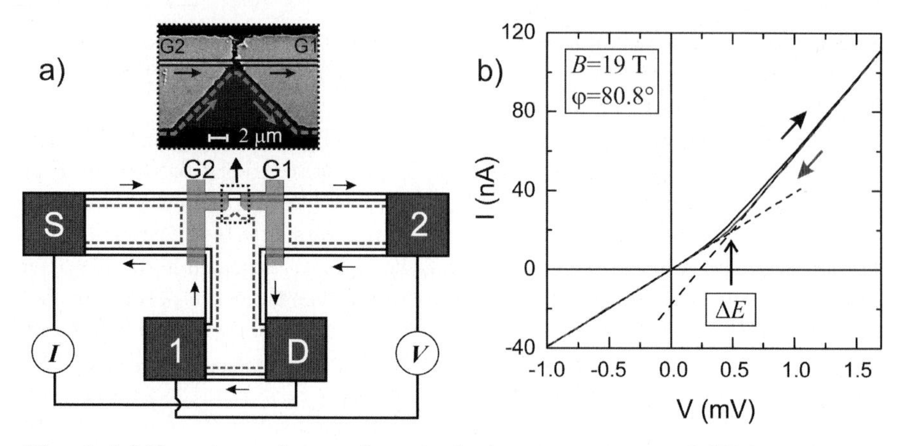

Fig. 4. (**a**) Experimental setup for spin-diode measurement and SEM micrograph of the scattering region. The bulk filling factor is $\nu = 2$, the gates are set to a local filling factor of $g = 1$. *Lines* show the outer (spin up, *solid*) and inner (spin down, *dashed*) edge channels together with the direction of current flow (*arrows*) for positive polarity. A current is biased via contacts S and D and a voltage is measured between 2 and 1. (**b**) Typical measured current-voltage characteristic in a tilted field with total $B = 19\,\mathrm{T}$ and perpendicular component such that $\nu \approx 2$. The current is swept first from negative to positive polarity and then back again. *Dashed lines* extrapolate the linear section of the down sweep. There intersection defines the energy gap $\triangle E$

field B the current is swept from negative to positive polarity and back. For each φ we measure for a sequence of B within the $\nu = 2$ plateau before changing the angle. A typical measurement is shown in Fig. 4b.

To analyze the four terminal conductance measurement we apply the Landauer–Büttiker multichannel formula [26]. For full equilibration between the two edge channels within the scattering region we obtain $(\mathrm{d}I/\mathrm{d}V)^{\mathrm{eq}} = \nu e^2/h = 2e^2/h$ while without any scattering the conductance is $(\mathrm{d}I/\mathrm{d}V)^{\mathrm{ad}} = ge^2/h = e^2/h$. Thus we observe in Fig. 4b equilibrated transport ($\mathrm{d}I/\mathrm{d}V = 2e^2/h$) for positive voltages beyond a threshold while for voltages below that threshold and for negative polarity the conductance is $\mathrm{d}I/\mathrm{d}V = e^2/h$ and therefore scattering between the edge channels must be suppressed in this voltage range.

We can explain this observation using the schematics of the edge structure for different polarities displayed in Fig. 5. For negative polarity ($V < 0$, left) spin up electrons from the outer edge channel would have to tunnel into the inner one. But only spin down states are available and therefore the tunnelling has to be accompanied by a simultaneous spin flip. This process has a very low probability within the short interaction length ($1\,\mu\mathrm{m}$) – the scattering is strongly suppressed.

For positive bias ($V > 0$, right) spin down electrons have to tunnel from the inner into the outer edge channel. For small voltages $eV < \triangle E$ with

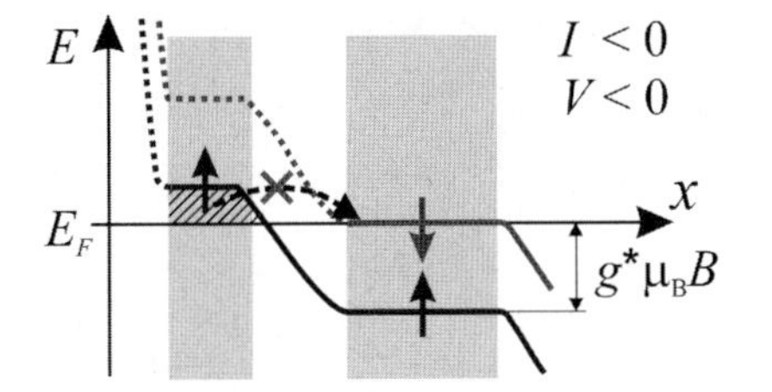

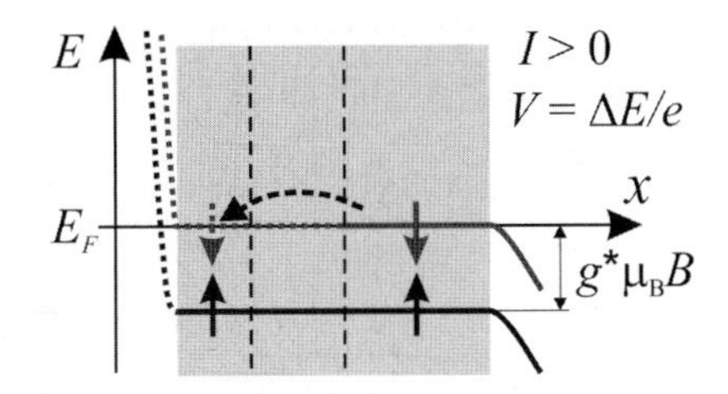

Fig. 5. *Left*: Scheme of the edge structure for negative polarity – tunnelling is blocked. *Right*: For positive polarity tunnelling is possible for $V \geq \triangle E$ with $\triangle E$ the Zeeman gap

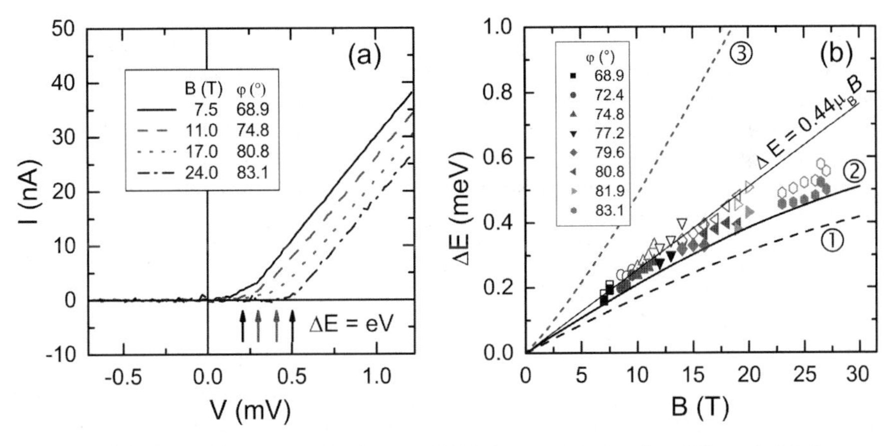

Fig. 6. (**a**) Tunnelling contribution to I–V characteristics for different total magnetic fields allow to extract the Zeeman gap $\triangle E$ as function of B (backward sweeps). (**b**) Compilation of all measured Zeeman gaps $\triangle E$. *Closed symbols* are determined from up ($\mathrm{d}I/\mathrm{d}T > 0$) sweeps, *open ones* from the down sweeps. Line (2) represents the calculation with 8-band $\mathbf{k} \cdot \mathbf{p}$ theory including inversion asymmetry and Bychkov–Rashba term but *without* exchange interaction; see text for other lines

$\triangle E$ the Zeeman gap the only available states are spin up and the scattering is still suppressed, $\mathrm{d}I/\mathrm{d}V = e^2/h$. But for $eV \geq \triangle E$ electrons can tunnel resonantly into the spin down level of the outer edge channel. This process conserves energy and spin and is therefore very likely even within the short interaction length - strong scattering sets in and the slope changes to $\mathrm{d}I/\mathrm{d}V = 2e^2/h$. The asymmetric diode like current-voltage characteristic due to the spin properties of the edge channels earns this device its name – inter-edge spin diode. This is even more obvious if the current $I = (e^2/h)V$ carried by the outer edge state is subtracted as done in Fig. 6a.

The onset of equilibrium between the edge states, $eV = \triangle E$, provides us with a nice tool to measure the Zeeman gap $\triangle E$ within the outer edge channel as function of magnetic field. To do so we have to change the total

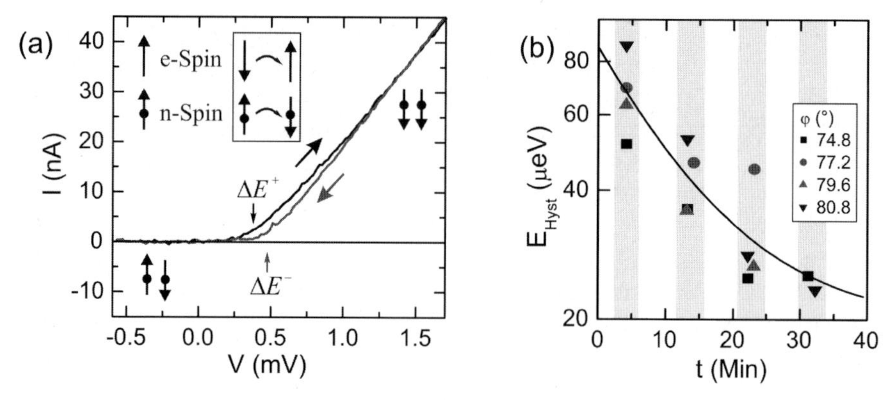

Fig. 7. (**a**) Hysteresis between up and down sweep caused by the nuclear spin polarisation. Spin down electrons relax their excess energy after tunnelling to the spin up ground state of the outer edge channel via a flip-flop process with the nuclei. The resulting Overhauser field shifts the measured gap $\triangle E$. (**b**) Compilation of the hysteresis $\Delta E_{\rm hyst} = \triangle E^- - \triangle E^+$ during the measurement sequence. Electrons can tunnelling and therefore the nuclear polarisation increases in the *grey areas*. The *line* acts as guide to the eye

magnetic field governing the Zeeman energy while keeping the perpendicular component within the bounds of the $\nu = 2$ quantum Hall plateau which is achieved by rotating the sample and adapting B. A selection of such measurements for various different rotations and different total fields is shown in Fig. 6a. For clarity we have subtracted the constant current $I = (e^2/h)V$ carried by the outer edge state even in the absence of scattering. For each rotation angle we also measure at different total fields within the range given by the width of the quantum Hall plateau.

The result the Zeeman gap measurements is compiled in Fig. 6b. For magnetic fields $B < 15\,\mathrm{T}$ the data agrees reasonably well with $\triangle E = |g^*|\mu_{\rm B}B$ with the bulk effective g-factor $g^* = -0.44$ (straight line in Fig. 6b). But the deviation at large B asks for a better description. The three curves in Fig. 6b marked (1),(2) and (3) were calculated by *R. Winkler* for our heterostructure using 8 band $\mathbf{k} \cdot \mathbf{p}$ theory [27]. Curve (1) includes effects of all bands and the compression of the wave function in the quantum well due to the large parallel field. In curve (2) the inversion asymmetry of the crystal (Dresselhaus term) and the built electric fields (Bychkov–Rashba term) were added. The third curve (3) adds exchange interaction. Obviously there is a very good agreement if we include all terms but exchange interaction.

If we now focus on the comparison of the data in Fig. 6b with curve (2) we notice that the data gained from down-sweeps (open symbols) are consistently lying further above the curve than the data from up-sweeps. Figure 7a displays one set of up and down-sweep from which this difference of the derived energy gaps $\triangle E^+$ and $\triangle E^-$ arises. We excluded any measurement artifacts and assured that the difference is caused by the time spent in the regime of scattering between the edge channels. We can relate this to a dynamic polarisation of the nuclei within the edge state scattering region [10, 28–30]: as sketched in Fig. 5 spin down electrons tunnel from the inner into the outer edge channel conserving their spin. But in a second step they have to relax their energy down to states at the Fermi energy which are spin up. The spin conservation in this process is achieved by a flip-flop process involving a nuclei as sketched in Fig. 7a: the electron spin flips from down to up and a nuclei spin flips from up to down. Starting from an unpolarised ensemble of nuclei this process builds up a local polarisation within the scattering volume. This generates an additional magnetic field (Overhauser effect) that increases the Zeeman gap of the electrons and thereby shifts the data points upwards of curve (2) in Fig. 6b.

The role of the nuclei is further confirmed by the time dependence of the effect: As described earlier we swept the current up and down several times for each angle. Before doing so the nuclei have sufficient time to establish an equilibrium. But between the sweeps the polarisation decays only partially. Thus each consequent sweep starts with a higher polarisation of the nuclei and the hysteresis between up and down sweep becomes smaller as can be seen in Fig. 7b. Using a model derived elsewhere [31] we can estimate the relaxation time of the nuclei spin to about 10 min.

4 Summary

In the first part we demonstrated a selective detection of edge states for time resolved current measurements. This allowed us to unambiguously identify two decoupled modes propagating along the edge of a $\nu = 3$ quantum Hall system: The strongly coupled outer two edge channel carry an edge magnetoplasmon mode while the decoupled inner edge channel supports an inter-edge magnetoplasmon. The analysis of their individual propagation velocities reveals details of the edge structure.

In the second part we used DC-tunnelling between the independently contacted edge states of a $\nu = 2$ QH system to determine the Zeeman gap in the edge channel. The observe magnetic field dependence agrees well with single particle calculations. In addition we observed a dynamic nuclear polarisation due to a spin flip between electron an nuclei.

Acknowledgements

We would like to thank R. Winkler for the calculation of the Zeeman gap, D. K. Maude for the help with the measurements at the Grenoble High Magnetic Field Laboratory and D. Reuter and A. D. Wieck for the heterostructure used in these experiments. We acknowledge financial support by the DFG and BMBF.

References

[1] K. v. Klitzing, G. Dorda, and M. Pepper, Phys. Rev. Lett. **45**, 494 (1980).
[2] B. I. Halperin, Phys. Rev. B **25**, 2185 (1982).
[3] R. J. Haug, Semicond. Sci. Technol. **8**, 131 (1993).
[4] D. B. Chklovskii, B. I. Shklovskii, and L. I. Glazman, Phys. Rev. B **46**, 4026 (1992).
[5] R. J. Haug, A. H. MacDonald, P. Streda, and K. von Klitzing, Phys. Rev. Lett. **61**, 2797 (1988).
[6] O. Kirichek, *Edge Excitations of Low-Dimensional Charged Systems*, in *Horizons in World Physics* Vol. 236 (Nova Science Publishers, Inc, Huntington, New York, 2000).
[7] G. Sukhodub, F. Hohls, and R. J. Haug, Phys. Rev. Lett. **93**, 196801 (2004).
[8] P. K. H. Sommerfeld, P. P. Steijaert, P. J. M. Peters, and R. W. van der Heijden, Phys. Rev. Lett. **74**, 2559 (1995).
[9] O. I. Kirichek and I. B. Berkutov, Low Temp. Phys. **21**, 305 (1995).
[10] D. C. Dixon, K. R. Wald, P. L. McEuen, and M. R. Melloch, Phys. Rev. B **56**, 4743 (1997).
[11] A. Würtz, R. Wildfeuer, A. Lorke, E. V. Deviatov, and V. T. Dolgopolov, Phys. Rev. B **65**, 075303 (2002).
[12] G. Sukhodub, F. Hohls, R. J. Haug, D. K. Maude, D. Reuter, and A. D. Wieck, Int. J. Mod. Phys. B **18**, 3649 (2004).
[13] G. Sukhodub, F. Hohls, R. J. Haug, D. K. Maude, D. Reuter, and A. D. Wieck, Physica E in press (2006).
[14] J. S. J. Allen, H. L. Störmer, and J. C. M. Hwang, Phys. Rev. B **28**, 4875 (1983).
[15] V. A. Volkov and S. A. Mikhailov, Sov. Phys. JETP **67**, 1639 (1988).
[16] O. G. Balev and P. Vasilopoulos, Phys. Rev. Lett. **81**, 1481 (1998).
[17] V. I. Talyanskii, A. V. Polisskii, D. D. Arnone, M. Pepper, C. G. Smith, D. A. Ritchie, J. E. Frost, and G. A. C. Jones, Phys. Rev. B **46**, 12427 (1992).
[18] N. B. Zhitenev, R. J. Haug, K. v. Klitzing, and K. Eberl, Phys. Rev. Lett. **71**, 2292 (1993).
[19] G. Ernst, R. J. Haug, J. Kuhl, K. von Klitzing, and K. Eberl, Phys. Rev. Lett. **77**, 4245 (1996).
[20] G. Sukhodub, F. Hohls, and R. J. Haug, Physica E **22**, 189 (2004).
[21] S. A. Mikhailov and V. A. Volkov, J. Phys. Condens. Matter **4**, 6523 (1992).
[22] S. A. Mikhailov, JETP Lett. **61**, 418 (1995).
[23] V. I. Talyanskii, D. R. Mace, M. Y. Simmons, M. Pepper, A. C. Churchill, J. E. Frost, D. A. Ritchie, and G. A. C. Jones, J. Phys. Condens. Matter **7**, L435 (1995).
[24] N. B. Zhitenev, R. J. Haug, K. v. Klitzing, and K. Eberl, Phys. Rev. B **52**, 11277 (1995).
[25] S. Takaoka, K. Oto, H. Kurimoto, K. Murase, K. Gamo, and S. Nishi, Phys. Rev. Lett. **72**, 3080 (1994).
[26] M. Büttiker, Phys. Rev. B **38**, 9375 (1988).
[27] R. Winkler, *Spin-Orbit Coupling Effects in Two-Dimensional Electron and Hole Systems*, in *Springer Tracts in Modern Physics* Vol. 191 (Springer, Berlin, 2003).
[28] B. E. Kane, L. N. Pfeiffer, and K. W. West, Phys. Rev. B **46**, 7264 (1992).

[29] T. Machida, T. Yamazaki, and S. Komiyama, Appl. Phys. Lett. **80**, 4178 (2002).
[30] A. Würtz, T. Müller, A. Lorke, D. Reuter, and A. D. Wieck, prl **95**, 056802 (2005).
[31] G. Sukhodub, Dissertation, Universität Hannover, 2005.

Decoherence of Fermions Subject to a Quantum Bath

Florian Marquardt

Sektion Physik, Center for NanoScience, and Arnold-Sommerfeld-Center for Theoretical Physics, Ludwig-Maximilians-Universität München, Theresienstr. 37, 80333 München, Germany
Florian.Marquardt@physik.lmu.de

Abstract. The destruction of quantum-mechanical phase coherence by a fluctuating quantum bath has been investigated mostly for a single particle. However, for electronic transport through disordered samples and mesoscopic interference setups, we have to treat a many-fermion system subject to a quantum bath. Here, we review a novel technique for treating this situation in the case of ballistic interferometers , and discuss its application to the electronic Mach–Zehnder setup. We use the results to bring out the main features of decoherence in a many-fermion system and briefly discuss the same ideas in the context of weak localization.

1 Introduction

There are two main messages of this brief review. First, regarding the physics of decoherence, we will argue that decoherence processes depend strongly on the type of system (single particle vs. many particles) and the type of noise (classical vs. quantum). Electronic interference experiments at low temperatures require a treatment of a many-fermion system coupled to a quantum bath. In that case, true many-body features come into play. This includes, in particular, the influence of Pauli blocking (that tends to restrain decoherence), and the fact that both hole- and particle-scattering processes contribute equally to the full decoherence rate. Second, regarding theoretical methods, we review a novel technique for treating ballistic interferometers subject to a quantum bath, which is based on the ideas behind the quantum Langevin equation (as it is known for the Caldeira–Leggett model). This is more efficient than generic methods (like Keldysh diagrams), and we will discuss the physical meaning of its ingredients. We apply it to the interference contrast and the current noise in an electronic Mach–Zehnder interferometer. We will also mention how the same ideas (if not the same technical methods) help to understand decoherence in weak localization within a path-integral framework.

The reasons for studying decoherence range from fundamental aspects of quantum mechanics to possible applications. On the fundamental level, the

R. Haug (Ed.): Advances in Solid State Physics,
Adv. in Solid State Phys. **46**, 161–172 (2008)

transition from the quantum world (with interference effects) to the classical world is due to the unavoidable fluctuations of the environment that tend to destroy macroscopic superpositions very rapidly (see, e.g., [1, 2]). These issues have been studied mostly with simplified single-particle models of decoherence.

In solid-state physics, decoherence first was investigated in the field of spin resonance, where straightforward Markoff master equation treatments are often sufficient [3, 4]. The quantum dissipative two-level system (spin-boson model) was studied in more detail at the beginning of the eighties [5]. During the preceding decade, interest in these questions has seen a revival due to the prospect of quantum information applications [6], where the decoherence time has to be at least ten thousand times longer than the time of elementary operations in order for error correction to work.

Regarding electronic transport phenomena, which will be our focus in the following, decoherence effects became important for the first time during the study of interference effects in disordered conductors, such as universal conductance fluctuations and weak localization (for a review see, e.g., [7, 10]). Later on, man-made interference structures were produced in metals and semiconductors, including Aharonov–Bohm rings, double quantum dot interferometers, and (most recently) Mach–Zehnder interferometers. These setups are also important in the quantum information context, both for generating, transporting, or detecting entanglement, and as highly sensitive measurement devices. The main nontrivial dependence of the interference contrast on temperature or transport voltage is produced by decoherence.

2 Single Particle Decoherence

Let us look at a a single particle traversing a two-way interferometer (Fig. 1). Its wavepacket has been split into two packets $\psi_{\mathrm{L/R}}$, going along the two arms (left/right). After these packets recombine, they form an interference pattern. This consists of a classical part (sum of probabilities) and an interference term, which is sensitive to a relative phase:

$$|\psi(x)|^2 = |\psi_{\mathrm{L}}(x)|^2 + |\psi_{\mathrm{R}}(x)|^2 + \psi_{\mathrm{L}}^*(x)\psi_{\mathrm{R}}(x)e^{i\varphi} + c.c.\,. \tag{1}$$

What happens once the particle is subjected to (classical) noise, i.e., a fluctuating potential $V(x,t)$? Even before acceleration/deceleration effects are noticeable, a random relative phase φ between the two paths is introduced. The actual pattern is obtained by averaging $|\psi|^2$ over many experimental runs. Since $\left|\langle e^{i\varphi}\rangle\right| \leq 1$, the interference term is suppressed.

At low temperatures ($k_{\mathrm{B}}T < \hbar\omega$) we have to consider a quantum bath, for which there exists an alternative description of decoherence: The bath acts as a kind of which-way detector, with its initial state evolving towards either one of two states, $|\chi_{\mathrm{R}}\rangle$ or $|\chi_{\mathrm{L}}\rangle$, depending on the path of the particle [8–10].

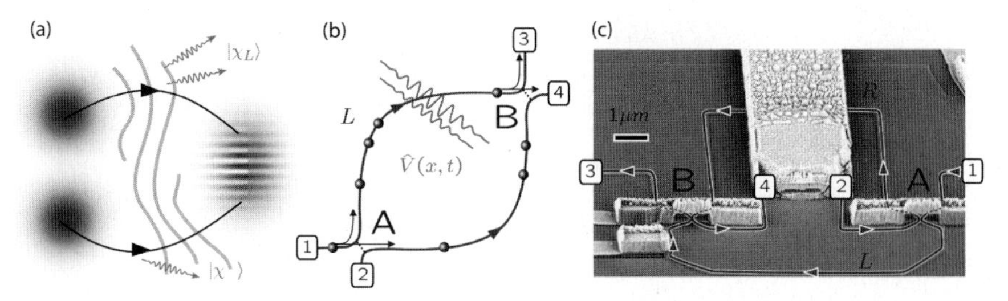

Fig. 1. (**a**) Decoherence in a two-way interferometer. The fluctuations (*wiggly lines*) introduce a random relative phase between the two paths. The quantum bath is left behind in two different states. This blurs the interference pattern (*right part* of pattern). (**b**) Schematic Mach–Zehnder setup. (**c**) SEM picture of electronic QHE Mach–Zehnder at the Weizmann institute (courtesy of I. Neder and M. Heiblum)

Now it is the overlap of these bath states that determines the suppression of the interference term. That overlap is nothing but the Feynman–Vernon influence functional [11].

Even if the particle is coupled to a quantum bath, decoherence may still be described using a classical noise spectrum, if the particle's energy is high and its motion is semiclassical (Fig. 2). To understand this, consider a simple weak coupling situation, where the total decoherence rate is given by the sum of downward and upward scattering rates, calculated using Fermi's Golden Rule. These rates can be related to the spectrum $\left\langle \hat{V}\hat{V} \right\rangle_\omega \equiv \int \mathrm{d}t e^{i\omega t} \left\langle \hat{V}(t)\hat{V}(0) \right\rangle$ of the quantum noise potential $\hat{V}$. We have $\Gamma_\downarrow \propto \left\langle \hat{V}\hat{V} \right\rangle_\omega \propto n(\omega)+1$ and $\Gamma_\uparrow \propto \left\langle \hat{V}\hat{V} \right\rangle_{-\omega} \propto n(\omega)$, where ω is the frequency transfer and $n(\omega)$ the thermal occupation. Obviously, the sum of these rates does not change if we replace $\hat{V}$ by classical noise with a symmetrized correlator $\langle VV \rangle_\omega = \left(\left\langle \hat{V}\hat{V} \right\rangle_\omega + \left\langle \hat{V}\hat{V} \right\rangle_{-\omega} \right) \Big/ 2$ (red curve in Fig. 2b). This can also be seen in a more general treatment, using a semiclassical evaluation of the Feynman–Vernon influence functional. We note that such a replacement is impermissible near the ground state of the system, where downward transitions are blocked.

3 Many Particles

Up to now, we have considered a single particle subject to classical or quantum noise. This has been the mainstay of research in quantum dissipative systems for a long time, with paradigmatic models such as the spin-boson model or the Caldeira–Leggett model of a single particle coupled to a bath of harmonic oscillators. However, in electronic transport experiments (and other setups, e.g., cold atom BEC interferometers) we are invariably dealing with a many-particle system. What are the new features arising in that case?

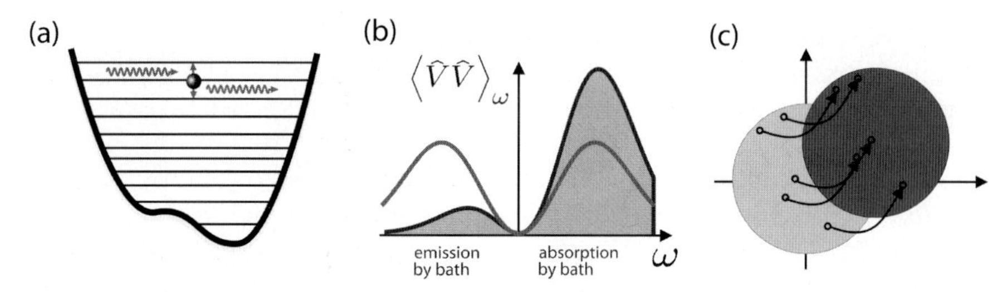

Fig. 2. (**a**) The scattering rates $\Gamma_{\downarrow/\uparrow}$ of a highly excited particle are related to the quantum noise spectrum (**b**). (**c**) A many-fermion system subject to classical noise

Everything remains straightforward if the noise is classical. Then, the many-particle problem reduces to the single-particle case: The wave function of each particle evolves according to the single-particle Schrödinger equation with a given noise field $V(x,t)$. In the case of many fermions, all the single-particle wave functions remain orthogonal, forming a Slater determinant (in the absence of intrinsic interactions). Pauli blocking is then completely unimportant.

We now discuss the one remaining combination: a many-fermion system coupled to a quantum bath. Unlike all the previous cases, this cannot be reduced to "single particle + classical noise": True many-body effects come into play (and appropriate methods are needed). Up to now, comparatively few quantum-dissipative many-particle systems have been studied. Examples include open Luttinger liquids [12], many-electron Aharonov–Bohm rings subject to quantum charge [13] or flux [14] noise, many-fermion generalizations of the Caldeira–Leggett model [14, 16, 17], and double quantum dot interferometers coupled to a quantum bath [15]. Here, we are going to review a recently developed general method of solution for ballistic interferometers [18, 26], and then briefly discuss the same physics in the context of disordered systems (weak localization).

4 The Mach–Zehnder Interferometer

The Mach–Zehnder (MZ) interferometer arguably represents the simplest kind of two-way interference setup (Fig. 1). Tuning the relative phase (via the magnetic flux ϕ) yields sinusoidal interference fringes in the currents at the two output ports.

Recently, this model has been realized in electronic transport experiments. The group of *Moty Heiblum* at the Weizmann institute managed to employ edge channels of the integer quantum Hall effect in a two-dimensional electron gas to build an ideal MZ setup with single-channel transport and without backscattering [19, 20]. The group measured the decrease of visibility (interference contrast in $I(\phi)$) as a function of rising temperature and transport voltage. No complete explanation for the results has been provided up to now, especially for the oscillations in the visibility [20]. Here, we will explore the possibility that at least part of the decrease in visibility is due to decoherence processes.

The effects of classical noise $V(x,t)$ onto a MZ setup have been studied intensively: The suppression of interference contrast to lowest order in the noise correlator was first calculated in the work of *Seelig* and *Büttiker* [21]. Building on this result, we treated the model to all orders in the interaction, calculating both the interference contrast and the effects on the shot noise in the output port of the interferometer [22, 23]. The shot noise has been measured and suggested as a tool to diagnose different sources of the loss in visibility [19]. These studies have recently been extended to the full counting statistics [24] and a renewed analysis of the dephasing terminal model [25]. However, as pointed out above, the situation is more involved for quantum noise, which is needed to account for the loss of visibility with rising bias voltage.

5 Equations of Motion Approach to Decoherence in Ballistic Interferometers

Recently, we have introduced a novel equations of motion technique for a many-particle system subject to a quantum bath [18], inside a ballistic interferometer (a detailed discussion may be found in [26]). It is similar in spirit to the quantum Langevin equation that can be employed to solve the Caldeira–Leggett model [4, 27]. Briefly, the idea of the latter is the following (when formulated on the level of Heisenberg equations). The total quantum force $\hat{F}$ acting on the given particle, due to the bath particles, can be decomposed into two parts:

$$\hat{F}(t) = \hat{F}_{(0)}(t) + \int_{-\infty}^{t} D^R(t-t')\hat{x}(t')\mathrm{d}t' . \tag{2}$$

The first describes the intrinsic fluctuations. It derives from the solution to the free equations of motion of the bath oscillators, with thermal and quantum (zero-point) fluctuations due to the stochastic initial conditions. The second part of the force is due to the response of the bath to the particle's motion. We will call it the "back-action" term, and it gives rise to features such as mass renormalization and friction. Equation (2) is valid on the operator level (not only for averages). In this way, one has "integrated out" the bath by solving for its motion. Plugging the force $\hat{F}$ into the right-hand-side (rhs) of the Heisenberg equation of motion for $\hat{x}$ yields the quantum Langevin equation, which in practice can only be solved for a free particle or a harmonic oscillator (linear equations).

In the case of a many-particle system, it is the density $\hat{n}(x) = \hat{\psi}^\dagger(x)\hat{\psi}(x)$ that couples to a scalar noise potential $\hat{V}(x)$. The place of $\hat{x}$ and $\hat{F}$ in the

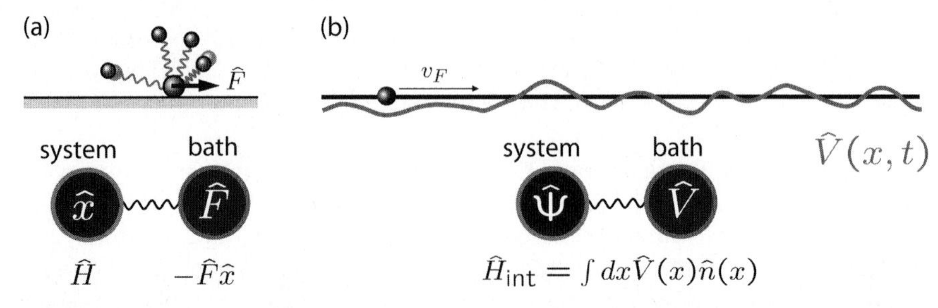

Fig. 3. (**a**) The Caldeira–Leggett model (single particle and oscillator bath) and (**b**) a ballistic many-particle system subject to a quantum noise potential $\hat{V}(x,t)$

quantum Langevin equation for a single particle is thus taken by the particle field $\hat{\psi}$ and $\hat{V}$, respectively. Let us now specialize to the case of fermions traveling ballistically inside the arm of an interferometer. We will assume chiral motion and use a linearized dispersion relation, as this is sufficient to describe decoherence. Then the fermion field obeys the following equation (with a slight approximation [18, 26]; we set $\hbar = 1$):

$$i(\partial_t + v_{\mathrm{F}}\partial_x)\hat{\psi}(x,t) = \hat{V}(x,t)\hat{\psi}(x,t)\,. \tag{3}$$

The formal solution of this equation is straightforward and analogous to the version for classical noise $V(x,t)$. The particle picks up a fluctuating "quantum phase" inside a time-ordered exponential:

$$\begin{aligned}\hat{\psi}(x,t) = \hat{T}\exp\left[-i\int_{t_0}^{t}\mathrm{d}t_1\,\hat{V}(x - v_{\mathrm{F}}(t-t_1),t_1)\right] \\ \times\,\hat{\psi}(x - v_{\mathrm{F}}(t-t_0),t_0)\,.\end{aligned} \tag{4}$$

In contrast to the case of classical noise, the field $\hat{V}$ contains the response to the fermion density, in addition to the intrinsic fluctuations $\hat{V}_{(0)}$:

$$\hat{V}(x,t) = \hat{V}_{(0)}(x,t) + \int_{-\infty}^{t}\mathrm{d}t'\,D^R(x,t,x',t')\hat{n}(x',t')\,. \tag{5}$$

Here D^R is the unperturbed retarded bath Green's function, $D^R(1,2) \equiv -i\theta(t_1 - t_2)\left\langle[\hat{V}(1),\hat{V}(2)]\right\rangle$, where $\hat{V}$-correlators refer to the free field. With these two equations, it becomes possible to calculate correlators of the fermion field (such as current and shot noise).

6 Decoherence Rate in a Many-Fermion System

Employing the formal solution from above (and using a lowest-order Markoff approximation [18, 26]), we find that the contribution of each electron to the interference term in the current is multiplied by a factor

$$1 - \Gamma_\varphi(\epsilon)\tau + i\delta\bar{\varphi}(\epsilon)\,, \tag{6}$$

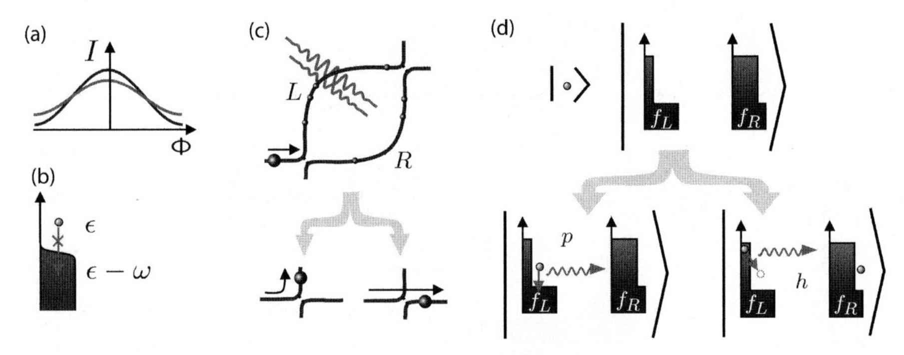

Fig. 4. (**a**) Decoherence suppresses the visibility of the interference pattern (*red curve (online color)*). (**b**) Pauli blocking restrains decoherence. (**c**) A particle arriving at the first beam splitter turns into a coherent superposition $t\,|R\rangle + r\,|L\rangle$. (**d**) For a many-fermion system, both particle- and hole-scattering contribute to decoherence: $\Gamma_\varphi = (\Gamma_p + \Gamma_h)/2$

with a phase shift $\delta\bar{\varphi} \propto \tau$. We focus on the suppression brought about by a decoherence rate $\Gamma_\varphi(\epsilon)$ that depends on the energy $\epsilon(k)$ of the incoming electron:

$$\Gamma_\varphi(\epsilon) = \int_0^\infty \frac{\mathrm{d}\omega}{v_\mathrm{F}} \mathrm{DOS}_q(\omega)[\underbrace{2n(\omega)+1}_{\substack{\text{thermal \& zeropoint} \\ \text{fluctuations}}} - \underbrace{(\bar{f}(\epsilon-\omega) - \bar{f}(\epsilon+\omega))}_{\substack{\text{from ``back -- action''} \\ \Rightarrow \text{Pauli blocking}}}] \tag{7}$$

The rate is an integral over all possible energy transfers ω. They are weighted by the bath spectral "density of states" $\mathrm{DOS}_q(\omega) = -\mathrm{Im}D_q^R(\omega)/\pi$, where $q = \omega/v_\mathrm{F}$ for ballistic motion. The first term in brackets stems from the $\hat{V}_{(0)}$ in the quantum phase. By itself, this would give rise to an energy-independent rate and a visibility independent of bias voltage (in contrast to experimental results [19, 20]). Thus, the second term is crucially important: It contains the average nonequilibrium distribution $\bar{f} = (f_\mathrm{L} + f_\mathrm{R})/2$ inside the arms (for equal coupling to both arms) and implements the physics of Pauli blocking. At $T = 0$, it suppresses all transitions that would take the electron into an occupied state (when $\bar{f}(\epsilon - \omega) = 1$ and this cancels against the 1 from the zero-point fluctuations).

The other main difference (vs. the case of a single particle) is less obvious but equally important. The decoherence rate Γ_φ is not simply given by the particle-scattering rate, but contains a contribution from hole scattering processes, where a particle at another energy $\epsilon + \omega$ is scattered into the given state at ϵ (with a factor $f(\epsilon + \omega)$ associated). This is a generic feature for decoherence of fermionic systems coupled to a quantum bath, and we now discuss the physical reason. In a single-particle language, the first beam

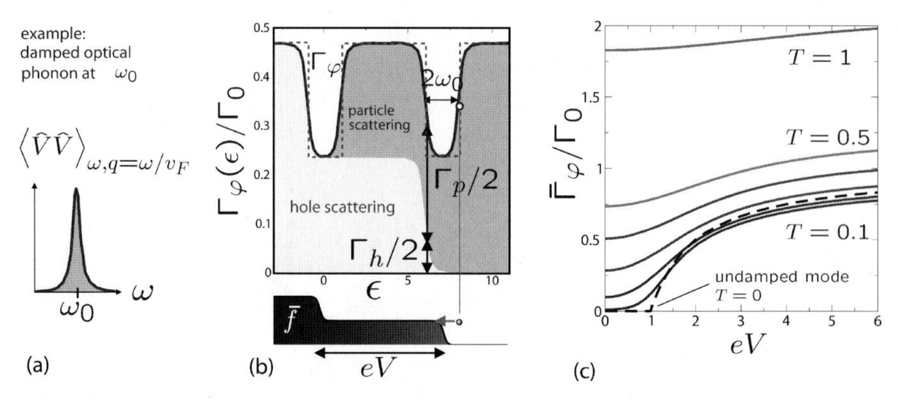

Fig. 5. The decoherence rate Γ_φ for the illustrative example of an optical phonon mode (**a**), as a function of energy of the incoming electron (**b**), and (**c**) the energy-averaged rate $\bar{\Gamma}_\varphi$ as a function of voltage V and temperature T (in units of ω_0). *Dashed curves* in (**b**) and (**c**) refer to an ideal undamped mode at $T = 0$

splitter creates a superposition of the form $t\,|R\rangle + r\,|L\rangle$, with t/r transmission and reflection amplitudes and R/L a packet inside the right/left arm. In the presence of a sea of other fermions, we should write instead a superposition of many-body states, for example:

$$t\,|1,1,0,\underline{0},0,0;\ 1,1,1,\underline{1},1,0\rangle + r\,|1,1,0,\underline{1},0,0;\ 1,1,1,\underline{0},1,0\rangle\ . \tag{8}$$

Here the occupations $|\text{left; right}\rangle$ of single-particle states in both arms are indicated, with a bar denoting the energy level ϵ of interest and the remaining particles (in the nonequilibrium distributions) playing the role of spectators. The interference is sensitive to the coherence $t\,|\ldots,\underline{0},\ldots;\ \ldots,\underline{1},\ldots\rangle + r\,|\ldots,\underline{1},\ldots;\ \ldots,\underline{0},\ldots\rangle$ that requires not only the presence of a particle in one arm but also the absence of a particle in the other respective arm.

This is why the many-body superposition can equally be destroyed by particle- and hole-scattering (leading to states with $|\ldots,\underline{0},\ldots;\ \ldots,\underline{0},\ldots\rangle$ or $|\ldots,\underline{1},\ldots;\ \ldots,\underline{1},\ldots\rangle$, respectively).

We have illustrated this in Fig. 5, where we have chosen a simple model bath spectrum (a broadened optical phonon mode at ω_0). Let us focus on small temperatures $T \ll \omega_0$. If the electron is far above the Fermi sea, it can easily undergo spontaneous emission and lose its coherence, thus the decoherence rate is maximal. For smaller energies, near the upper step in $\bar{f}$, it might end up in an occupied state, thus Γ_p is reduced, leading to a dip in Γ_φ. Inside the transport voltage window, Γ_p remains at 1/2 its previous value, but now Γ_h raises Γ_φ back to its maximal value. Finally, another dip is observed near the lower edge of the voltage window. The visibility is directly given by $1 - \bar{\Gamma}_\varphi$, with $\bar{\Gamma}_\varphi$ the energy-average of Γ_φ over the voltage window. At $T, V \to 0$, $\bar{\Gamma}_\varphi$ vanishes. Decoherence sets in only when the electron can emit phonons and thereby reveal its path through the MZ setup. At higher temperatures, the Fermi distribution becomes smeared, thereby easing the restrictions of Pauli blocking, and the thermal fluctuations of the bath grow, increasing Γ_φ. For $T \gg \omega_0$, the energy/voltage-dependence of Γ_φ becomes unimportant, and an approximate treatment becomes possible, replacing the quantum bath by classical noise.

7 Decoherence in Weak Localization

We now briefly discuss how the same concepts apply to weak localization [7, 29–31], where the constructive interference of time-reversed pairs of diffusive trajectories increases the electrical resistance of a disordered sample. One is interested in the linear response conductance, where the external perturbation (the electric field) induces a particle-hole excitation by lifting one of the particles above the Fermi sea. This creates a many-body state similar to the one above, $\sqrt{1-\delta^2}\,|\ldots,\underline{1},\ldots,\underline{0},\ldots\rangle + \delta\,|\ldots,\underline{0},\ldots,\underline{1},\ldots\rangle$, where δ is the small amplitude of the excited state. Following arguments analogous to those above [33], we see again why both particle- and hole-scattering processes contribute to the decoherence rate.

Many discussions of decoherence in weak localization have focussed on the thermal (classical) part of the Nyquist noise [29, 30]. This leads to a single-particle problem that can be treated using path integrals [30]. Diagrammatic calculations of the decoherence rate in the presence of a quantum bath [31] yielded results that can be interpreted in the manner discussed above. It is obviously desirable, though difficult, to cast these as well into the powerful path-integral framework. *Golubev* and *Zaikin* were the first to present a formally exact influence functional approach for many-fermion systems [35]. Their semiclassical evaluation yielded a decoherence rate that does not vanish at $T = 0$ and is independent of electron energy, in contrast to diagrammatic calculations and the ideas about Pauli blocking discussed above.

Recently, we have revisited this problem [33, 34] and have shown that the results of much more complicated diagrammatic calculations [32, 34] can be exactly reproduced by a rather simple prescription. The case "many particles + quantum bath" may be reduced to "single particle + classical noise", provided one uses an effective, modified noise spectrum of the following form [33]:

$$\left\langle \hat{V}\hat{V}\right\rangle_\omega \mapsto \langle VV\rangle_\omega^{\mathrm{eff}} \equiv \frac{1}{2}\left\langle\left\{\hat{V},\hat{V}\right\}\right\rangle_\omega + \frac{1}{2}\left\langle\left[\hat{V},\hat{V}\right]\right\rangle_\omega (f(\epsilon+\omega) - f(\epsilon-\omega)) \tag{9}$$

The first part is the symmetrized quantum correlator, containing the zero-point fluctuations. The second part incorporates Pauli blocking. These terms correspond to those in the equation of motion approach (7), with which this method is consistent. The resulting decoherence rate vanishes at $T = 0$ [35,

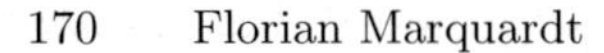

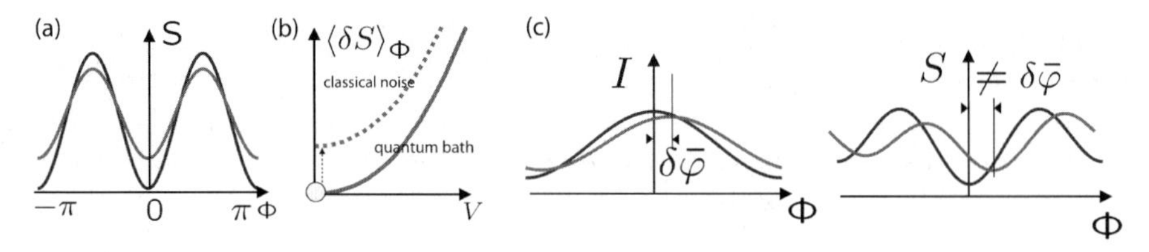

Fig. 6. Main effects of a quantum bath on shot noise in a MZ setup: (**a**) Suppression of visibility in $S(\phi)$. (**b**) No Nyquist noise correction, but classical conductance fluctuations $S \propto V^2$ at high voltages. (**c**) Different phase shifts in $I(\phi)$ and $S(\phi)$ for asymmetric setups

36]. Earlier similar ideas [30, 37, 38] represent approximations to the present approach.

8 Effects of a Quantum Bath on Shot Noise

We briefly return to the MZ setup. Using our approach, it is possible to discuss the influence of the quantum bath on the shot noise [28] power S in the output port [18, 26]. The visibility of the interference pattern $S(\phi)$ is reduced, although this cannot be described by the same decoherence rate as for the current. The most important feature refers to the phase shifts observed in asymmetric setups, for the current and the shot noise. These can become different: $I(\phi) = \bar{I} + \delta I \cos(\phi - \delta\bar{\varphi})$ and $S(\phi) = \bar{S} + S_1 \cos(\phi - \delta\phi_1) + S_2 \cos(2(\phi - \delta\phi_2))$, with $\delta\bar{\varphi} \neq \delta\phi_1 \neq \delta\phi_2$ in general. This prediction is in contrast to all simpler models (involving classical noise etc.), which usually do not give rise to phase shifts at all. Something like this seems to have been observed in recent experiments at the Weizmann institute. Another equally important avenue of current research is the application to nonlinear (non-Gaussian) environments.

9 Conclusions

Decoherence in transport interference situations can often be reduced to the case of a single particle subject to classical noise. However, for a many-fermion system subject to a quantum bath, true many body features remain and have to be taken into account via suitable techniques. The two main physical features are Pauli blocking and the importance of both hole- and particle-scattering processes. The technical innovation reviewed here is an equation of motion approach that is well suited to describe decoherence of many particles moving in ballistic interferometers. We have discussed its application to the MZ interferometer setup, the loss of visibility in the current and (briefly) the effects on shot noise. In addition, we have pointed out that the same kind of physics applies to decoherence in weak localization.

Acknowledgements

I thank M. Heiblum, I. Neder, J. v. Delft, C. Bruder, D. S. Golubev, M. Büttiker, Y. Imry, R. Smith, V. Ambegaokar, S. M. Girvin, and all the other people with whom I discussed and worked on these topics over the years.

References

[1] E. Joos et al.: *Decoherence and the Appearance of a Classical World in Quantum Theory* (Springer, Heidelberg, 2003)
[2] W. H. Zurek: Decoherence and the transition from quantum to classical, Phys. Today **44**, 36 (1991)
[3] K. Blum, *Density Matrix Theory and Applications* (Plenum, New York, 1996)
[4] U. Weiss: *Quantum Dissipative Systems* (World Scientific, Singapore, 2000)
[5] A. J. Leggett, S. Chakravarty, A. T. Dorsey, M. P. A. Fisher, A. Garg, W. Zwerger: Dynamics of the dissipative two-state system, Rev. Mod. Phys. **59**, 1-85 (1987)
[6] C. H. Bennett, D. P. DiVincenzo: Quantum information and computation, Nature **404**, 247-255 (2000)
[7] C. W. J. Beenakker, H. van Houten: Quantum transport in semiconductor nanostructures, Solid State Phys. **44**, 1-228 (1991)
[8] A. Stern, Y. Aharonov, Y. Imry: Phase uncertainty and loss of interference: A general picture, Phys. Rev. A **41**, 3436-3448 (1990)
[9] D. Loss, K. Mullen: Dephasing by a dynamic asymmetric environment, Phys. Rev. B **43**, 13252 (1991)
[10] Y. Imry: *Introduction to Mesoscopic Physics*, 2nd ed. (Oxford University Press, 2002)
[11] R. P. Feynman, F. L. Vernon: The theory of a general quantum mechanical system interacting with a linear dissipative system, Ann. Phys. (NY) **24**, 118-173 (1963)
[12] A. H. C. Neto, C. D. Chamon, C. Nayak: Open Luttinger liquids, Phys. Rev. Lett. **79**, 4629 (1997)
[13] P. Cedraschi, V. V. Ponomarenko, M. Büttiker: Zero-point fluctuations and the quenching of the persistent current in normal metal rings, Phys. Rev. Lett. 84, 346-349 (2000)
[14] F. Marquardt, C. Bruder: Aharonov–Bohm ring with fluctuating flux, Phys. Rev. B **65**, 125315 (2002)
[15] F. Marquardt, C. Bruder: Dephasing in sequential tunneling through a double-dot interferometer, Phys. Rev. B **68**, 195305 (2003)
[16] F. Marquardt, D. S. Golubev: Relaxation and dephasing in a many-fermion generalization of the caldeira–leggett model, Phys. Rev. Lett. **93**, 130404 (2004)
[17] F. Marquardt, D. S. Golubev: Many-fermion generalization of the Caldeira-Leggett model, Phys. Rev. A **72**, 022113 (2005)
[18] F. Marquardt: Fermionic Mach–Zehnder interferometer subject to a quantum bath, Europhys. Lett. **72**, 788 (2005)

[19] Y. Ji, Y. Chung, D. Sprinzak, M. Heiblum, D. Mahalu, H. Shtrikman: An electronic Mach–Zehnder interferometer, Nature **422**, 415 (2003)
[20] I. Neder, M. Heiblum, Y. Levinson, D. Mahalu, V. Umansky: Unexpected behavior in a two-path electron interferometer, Phys. Rev. Lett. **96**, 016804 (2006)
[21] G. Seelig, M. Büttiker: Charge-fluctuation-induced dephasing in a gated mesoscopic interferometer, Phys. Rev. B **64**, 245313 (2001)
[22] F. Marquardt, C. Bruder: Influence of dephasing on shot noise in an electronic Mach–Zehnder interferometer, Phys. Rev. Lett. **92**, 056805 (2004)
[23] F. Marquardt, C. Bruder: Effects of dephasing on shot noise in an electronic Mach–Zehnder interferometer, Phys. Rev. B **70**, 125305 (2004)
[24] H. Förster, S. Pilgram, M. Büttiker: Decoherence and full counting statistics in a Mach–Zehnder interferometer, Phys. Rev. B **72**, 075301 (2005)
[25] S. Pilgram, P. Samuelsson, H. Förster, M. Büttiker: Full counting statistics for voltage and dephasing probes, cond-mat/0512276
[26] F. Marquardt: Equations of motion approach to decoherence and current noise in ballistic interferometers coupled to a quantum bath, cond-mat/0604458 (2006)
[27] A. O. Caldeira, A. J. Leggett: Path integral approach to quantum Brownian motion, Physica **121A**, 587 (1983)
[28] Ya. M. Blanter, M. Büttiker: Shot noise in mesoscopic conductors, Phys. Rep. **336**, 1 (2000)
[29] B. L. Altshuler, A. G. Aronov, D. E. Khmelnitsky: Effects of electron–electron collisions with small energy transfers on quantum localization, J. Phys. C Solid State **15**, 7367 (1982)
[30] S. Chakravarty, A. Schmid: Weak localization: The quasiclassical theory of electrons in a random potential, Phys. Rep. **140**, 195 (1986)
[31] H. Fukuyama, E. Abrahams: Inelastic scattering time in two-dimensional disordered metals, Phys. Rev. B **27**, 5976 (1983)
[32] I. Aleiner, B. L. Altshuler, M. E. Gershenzon: Interaction effects and phase relaxation in disordered systems, Waves in Random Media **9**, 201 (1999) [cond-mat/9808053]
[33] F. Marquardt, J. v. Delft, R. Smith, V. Ambegaokar: Decoherence in weak localization I: Pauli principle in influence functional, cond-mat/0510556
[34] J. v. Delft, F. Marquardt, R. Smith, V. Ambegaokar: Decoherence in weak localization II: Bethe–Salpeter calculation of Cooperon, cond-mat/0510557; J. v. Delft: Influence functional for decoherence of interacting electrons in disordered conductors, cond-mat/0510563
[35] D. S. Golubev, A. D. Zaikin: Quantum decoherence and weak localization at low temperatures, Phys. Rev. B **59**, 9195 (1999)
[36] In contrast, the decoherence rate given in Ref. [35] is equal to that obtained for a single particle, with $f = 0$ in our formula, thus saturating at $T = 0$. We should note that the authors of Ref. [35] disagree with our conclusions, see their comment: D. S. Golubev and A. D. Zaikin, cond-mat/0512411.
[37] W. Eiler: Electron–electron interaction and weak localization, J. Low Temp. Phys. **56**, 481 (1984)
[38] D. Cohen, Y. Imry: Dephasing at low temperatures, Phys. Rev. B **59**, 11143 (1999)

Correlation Effects on Electronic Transport through Dots and Wires*

V. Meden

Institut für Theoretische Physik, Universität Göttingen,
Friedrich-Hund-Platz 1,
37077 Göttingen, Germany
meden@theorie.physik.uni-goettingen.de

Abstract. We investigate how two-particle interactions affect the electronic transport through meso- and nanoscopic systems of two different types: quantum dots with local Coulomb correlations and quasi one-dimensional quantum wires of interacting electrons. A recently developed functional renormalization group scheme is used that allows to investigate systems of complex geometry. Considering simple setups we show that the method includes the essential aspects of Luttinger liquid physics (one-dimensional wires) as well as of the physics of local correlations, with the Kondo effect being an important example. For more complex systems of coupled dots and Y-junctions of interacting wires we find surprising new correlation effects.

1 Introduction

In recent years the effect of electron correlations on the physics of meso- and nanoscopic systems has attracted growing interest. This led to an increasing overlap of the two communities working on mesoscopic physics and strongly correlated electron systems. In this article we focus on electronic transport properties, in particular the linear response conductance G. As discussed below two very fundamental correlation effects should be observable in transport through mesoscopic systems of relatively simple structure: 1. the Kondo effect [1] in single quantum dots [2,3] and 2. Luttinger liquid (LL) physics in quasi one-dimensional (1d) quantum wires with a single impurity [4].

The progress in nanostructuring techniques makes it now possible to design more complex geometries such as double- and triple-dot systems [5, 6] and junctions of several quasi 1d wires [7]. In these systems one can expect to find even more interesting correlation physics. In the near future complex setups might be used in conventional devices as well as for quantum information processing which provides a second reason to investigate the role of correlations. The theoretical tools commonly used to study many-body physics in dots and wires are rather specific to simple setups and cannot directly be applied to more complex geometries. Thus, there is need for novel techniques which can properly describe correlations, but are flexible and simple enough such that they can be used to investigate complex systems. In

* In memory of Xavier Barnabé-Thériault, who passed away in a tragic accident on August 15, 2004.

R. Haug (Ed.): Advances in Solid State Physics,
Adv. in Solid State Phys. **46**, 173–185 (2008)

this article, we show that an approximation scheme which is based on the functional renormalization group [8] (fRG) provides such a method. We will first apply it to the two simple setups mentioned above and show that it contains the essential physics. We then proceed and study a system of parallel double-dots and a Y-junction of three 1d quantum wires. In both examples the electron correlations lead to surprising new effects.

To exemplify the importance of correlations we first study transport through a quantum dot with spin degenerate levels. For simplicity we only consider a single level (e.g., described by the single impurity Anderson model [1]) and equal couplings to the left and right lead. The level position can be moved by a gate voltage V_{g}. At small T and for noninteracting dot electrons $G(V_{\mathrm{g}})$ shows a Lorentzian resonance of unitary height $2e^2/h$. The full width 2Γ of the resonance sets an energy scale Γ which is associated with the strength of the tunneling barriers. Including a Coulomb interaction U between the spin up and down dot electrons the line shape is substantially altered as can be seen from the exact $T = 0$ Bethe ansatz solution [9]. For increasing U/Γ it is gradually transformed into a box-shaped resonance of unitary height with a plateau of width U and a sharp decrease of G to the left and right of it [2, 3, 10, 11]. For gate voltages within the plateau the dot is half-filled implying a local spin-1/2 degree of freedom on the dot. Thus, the Kondo effect [1] leads to resonant transport throughout this so-called Kondo regime. The appearance of the plateau can be understood by studying the characteristics of the one-particle spectral function of the dot. For the present setup $G(V_{\mathrm{g}})$ is proportional to the spectral weight at the chemical potential μ [12]. For sufficiently large U/Γ the spectral function shows a sharp Kondo resonance (and additional Hubbard bands at higher energies). Its width sets an energy scale – the Kondo temperature T_{K} [1, 10]. At half-filling ($V_{\mathrm{g}} = 0$) the peak is located at the chemical potential, but even for gate voltages away from the particle-hole symmetric point it is pinned at μ and its height barely changes. This holds for $-U/2 < V_{\mathrm{g}} < U/2$, which explains the appearance of the plateau of width U in the conductance. Experimentally the appearance of Kondo physics in transport through quantum dots was demonstrated clearly [13, 14].

Although local correlations in a quantum dot (i.e., a zero-dimensional system) already have a strong effect, the system remains a Fermi liquid. However, the low-energy physics of correlated 1d metals (quantum wires) is not described by the Fermi liquid theory. Such systems fall into the LL universality class [4] which is characterized by power-law scaling of a variety of correlation functions and a vanishing quasi-particle weight. For spin-rotational invariant interactions and spinless models, on which we focus here, the exponents of the different correlation functions can be parametrized by a single number, the interaction dependent LL parameter $K < 1$ (for repulsive interactions; $K = 1$ in the noninteracting case). Instead of being quasi-particles the low lying excitations of LLs are collective density excitations. This implies that impurities, or more generally inhomogeneities, have a dramatic effect on the physical properties of LLs [15–18]. In the presence of only a single impurity on asymptotically small energy scales observables behave as if the 1d system was cut in two halfs at the position of the impurity, with open boundary conditions at the end points (open chain fixed point) [19–21]. Within a renormalization group approach the impurity increases from weak to strong. For a weak impurity and decreasing energy scale s – say the temperature T – the deviation of the linear conductance G from the impurity-free value first scales as $(s/s_0)^{2(K-1)}$, with K being the scaling dimension of the perfect chain fixed point and s_0 a characteristic energy scale (e.g., the band width). This holds as long as $|V_{\mathrm{back}}/s_0|^2 (s/s_0)^{2(K-1)} \ll 1$, with V_{back} being a measure for the strength of the $2k_{\mathrm{F}}$ backscattering of the impurity and k_{F} the Fermi momentum. For smaller energy scales or larger bare impurity backscattering this behavior crosses over to another power-law scaling $G(s) \sim (s/s_0)^{2(1/K-1)}$, with the scaling dimension of the open chain fixed point $1/K$. This scenario was verified within an effective field theoretical model for infinite LLs [19–21] as well as finite LLs connected to Fermi liquid leads [22]. In the latter case the scaling holds as long as the contacts are modeled to be "perfect", that is free of any bare and effective single-particle backscattering, and the impurity is placed in the bulk of the interacting quantum wire. Indications of power-law scaling of $G(T)$ were obtained in experiments on quasi 1d wires, but the results are ambiguous [23–26]. One reason for this is that experimentally the power-law behavior is restricted to less than one order of magnitude and often only achieved after a somewhat uncontrolled background subtraction on the data.

2 The RG Method and its Application to Simple Systems

The fRG was recently introduced as a powerful new tool for studying interacting Fermi systems [8]. It provides a systematic way of resumming competing instabilities and goes beyond simple perturbation theory even in problems which are not plagued by infrared divergences [27]. In our applications the dot(s) as well as the interacting quantum wire(s) will be coupled to noninteracting leads as it is the case in systems which can be realized in experiments. Before setting up the fRG scheme we integrate out the leads [28]. In the fRG procedure the noninteracting propagator $\mathcal{G}_0$ (now including self-energy contributions from the leads) is replaced by a propagator depending on an infrared cutoff Λ. Specifically, we use

$$\mathcal{G}_0^{\Lambda}(i\omega) = \Theta(|\omega| - \Lambda)\mathcal{G}_0(i\omega) \tag{1}$$

with Λ running from ∞ down to 0. Using $\mathcal{G}_0^{\Lambda}$ in the generating functional of the irreducible vertex functions and taking the derivative with respect to Λ one can derive an exact, infinite hierarchy of coupled differential equations for the vertex functions, such as the self-energy and the irreducible 2-particle interaction. In particular, the flow of the self-energy Σ^Λ (1-particle vertex) is determined by Σ^Λ itself and the 2-particle vertex Γ^Λ, while the flow of Γ^Λ is determined by Σ^Λ, Γ^Λ, and the flowing 3-particle vertex Γ_3^Λ. The latter could be computed from a flow equation involving the 4-particle vertex, and so on. At the end of the fRG flow $\Sigma^{\Lambda=0}$ is the self-energy Σ of the original, cutoff-free problem we are interested in [27]. A detailed derivation of the fRG flow equations for a general quantum many-body problem which only requires a basic knowledge of the functional integral approach to many-particle physics and the application of the method for a simple toy problem are presented in [29].

In practical applications the hierarchy of flow equations has to be truncated and $\Sigma^{\Lambda=0}$ only provides an approximation for the exact Σ. As a first approximation we here neglect the 3-particle vertex. The contribution of Γ_3^Λ to Γ^Λ is small as long as Γ^Λ is small, because Γ_3^Λ is initially (at $\Lambda = \infty$) zero and is generated only from terms of third order in Γ^Λ. Furthermore, Γ^Λ stays small for all Λ if the bare interaction is not too large. Below we will clarify the meaning of "not-too-large" in the cases of interest. This approximation leads to a closed set of equations for Γ^Λ and Σ^Λ [30]. We here do not give these equations but instead implement a second approximation: the frequency-dependent flow of the renormalized 2-particle vertex Γ^Λ is replaced by its value at vanishing (external) frequencies, such that Γ^Λ remains frequency independent. Since the bare interaction is frequency independent, neglecting the frequency dependence leads to errors only at second order for the self-energy, and at third order for the 2-particle vertex at zero frequency. For the approximate flow equations we then obtain

$$\frac{\partial}{\partial \Lambda} \Sigma^\Lambda_{1',1} = -\frac{1}{2\pi} \sum_{\omega=\pm\Lambda} \sum_{2,2'} e^{i\omega 0^+} \mathcal{G}^\Lambda_{2,2'}(i\omega)\, \Gamma^\Lambda_{1',2';1,2} \tag{2}$$

and

$$\frac{\partial}{\partial \Lambda} \Gamma^\Lambda_{1',2';1,2} = \frac{1}{2\pi} \sum_{\omega=\pm\Lambda} \sum_{3,3',4,4'} \left\{ \frac{1}{2} \mathcal{G}^\Lambda_{3,3'}(i\omega)\, \mathcal{G}^\Lambda_{4,4'}(-i\omega) \Gamma^\Lambda_{1',2';3,4}\, \Gamma^\Lambda_{3',4';1,2} \right.$$
$$\left. + \,\mathcal{G}^\Lambda_{3,3'}(i\omega)\, \mathcal{G}^\Lambda_{4,4'}(i\omega) \left[-\Gamma^\Lambda_{1',4';1,3}\, \Gamma^\Lambda_{3',2';4,2} + \Gamma^\Lambda_{2',4';1,3}\, \Gamma^\Lambda_{3',1';4,2} \right] \right\}, \tag{3}$$

where the lower indexes 1, 2, etc. stand for the single-particle quantum numbers and

$$\mathcal{G}^\Lambda(i\omega) = \left[\mathcal{G}_0^{-1}(i\omega) - \Sigma^\Lambda \right]^{-1} . \tag{4}$$

At the initial cutoff $\Lambda = \infty$ the flowing 2-particle vertex $\Gamma^\Lambda_{1',2';1,2}$ is given by the antisymmetrized interaction and the self-energy by the single-particle terms of the Hamiltonian not included in $\mathcal{G}_0$ (e.g., impurities).

2.1 The Single-Level Quantum Dot

Now the set of flow equations can be used to study the two-lead single impurity Anderson model [1, 31]. After integrating out the leads the only relevant single-particle quantum number is the spin σ of the dot electrons. If we take the level energies to be $\varepsilon_\uparrow = V_g - \mathcal{H}/2$ and $\varepsilon_\downarrow = V_g + \mathcal{H}/2$, where $\mathcal{H}$ denotes a magnetic field which lifts the spin-degeneracy, the projected noninteracting propagator in the so-called wide band limit is given by [1]

$$\mathcal{G}_{0,\sigma}(i\omega) = \frac{1}{i\omega - (V_g + \sigma\mathcal{H}/2) + i\Gamma\,\mathrm{sgn}(\omega)}\,, \tag{5}$$

where on the right-hand side $\sigma = \uparrow = +1$ and $\sigma = \downarrow = -1$.

During the fRG flow Σ^Λ remains frequency independent and $V_\sigma^\Lambda = V_g + \sigma\mathcal{H}/2 + \Sigma_\sigma^\Lambda$ can be interpreted as an effective, flowing level position, whose flow equation reads

$$\frac{\partial}{\partial\Lambda} V_\sigma^\Lambda = -\frac{U^\Lambda}{2\pi} \sum_{\omega=\pm\Lambda} \mathcal{G}_{\bar\sigma}^\Lambda(i\omega) = \frac{U^\Lambda\, V_{\bar\sigma}^\Lambda/\pi}{(\Lambda+\Gamma)^2 + (V_{\bar\sigma}^\Lambda)^2}\,, \tag{6}$$

with the initial condition $V_\sigma^{\Lambda=\infty} = V_g + \sigma\mathcal{H}/2$ and $\bar\sigma$ denoting the complement of σ. The cutoff-dependent propagator $\mathcal{G}_\sigma^\Lambda(i\omega)$ follows from $\mathcal{G}_{0,\sigma}(i\omega)$ by replacing $V_g + \sigma\mathcal{H}/2 \to V_\sigma^\Lambda$. Using symmetries, the flow of the 2-particle vertex can be reduced to a single equation for the effective interaction between spin up and spin down electrons

$$\begin{aligned}\frac{\partial}{\partial\Lambda} U^\Lambda &= \frac{(U^\Lambda)^2}{2\pi} \sum_{\omega=\pm\Lambda} \left[\tilde{\mathcal{G}}_\uparrow^\Lambda(i\omega)\,\tilde{\mathcal{G}}_\downarrow^\Lambda(-i\omega) + \tilde{\mathcal{G}}_\uparrow^\Lambda(i\omega)\,\tilde{\mathcal{G}}_\downarrow^\Lambda(i\omega)\right] \\ &= \frac{2\left(U^\Lambda\right)^2 V_\uparrow^\Lambda V_\downarrow^\Lambda/\pi}{\left[(\Lambda+\Gamma)^2 + (V_\uparrow^\Lambda)^2\right]\left[(\Lambda+\Gamma)^2 + (V_\downarrow^\Lambda)^2\right]}\,,\end{aligned} \tag{7}$$

with $U^\Lambda = \Gamma^\Lambda_{\sigma,\bar\sigma;\sigma,\bar\sigma}$ and the initial condition $U^{\Lambda=\infty} = U$.

Within our approximation the dot spectral function at the end of the fRG flow is given by

$$\rho_\sigma(\omega) = \frac{1}{\pi}\,\frac{\Gamma}{\left(\omega - V_\sigma\right)^2 + \Gamma^2}\,, \tag{8}$$

with $V_\sigma = V_\sigma^{\Lambda=0}$, that is a Lorentzian of full width 2Γ and height $1/(\pi\Gamma)$ centered around V_σ. We first consider $\mathcal{H} = 0$. Although this spectral function neither shows the narrow Kondo resonance nor the Hubbard bands, for $U/\Gamma \gg 1$ the pinning of the spectral weight at the chemical potential is included in our approximation. We thus reproduce the line shape of the conductance quantitatively up to very large U/Γ. This is shown in Fig. 1a where we plot $V^{\Lambda=0}$ as a function of V_g (dashed-dotted line) as well as $G(V_g)$ (dashed line) for $U/\Gamma = 4\pi$. For comparison the exact result for $G(V_g)$ obtained by Bethe ansatz [9, 11] is shown as the solid line. Neglecting the flow of the two-particle vertex the exponential pinning of spectral weight can even be shown analytically [31].

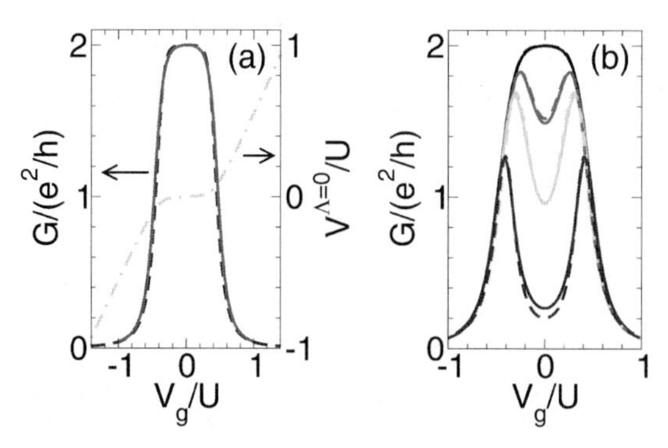

Fig. 1. (**a**) Gate Voltage V_g dependence of the effective level position $V^{\Lambda=0}$ (*dashed-dotted line*) and the linear conductance G (*solid line*: Bethe ansatz of [11]; *dashed line*: fRG) of a single level quantum dot for $U/\Gamma = 4\pi$, $\mathcal{H} = 0$. (**b**) $G(V_g)$ for $U/\Gamma = 3\pi$ and $\mathcal{H} = 0, 0.5T_K, T_K$, and $5T_K$, from *top* to *bottom*. Here $T_K = 0.116\Gamma$. *Solid line*: NRG data from [32]. *Dashed line*: fRG approximation

Even without the sharp Kondo resonance in the spectral function, which is usually used to define T_K, our approximation contains the Kondo temperature. In Fig. 1b we show a comparison of fRG data (dashed lines) and high precision numerical renormalization group (NRG) data [32] for $G(V_g)$ and different $\mathcal{H}$. The T_K given in the caption was obtained from the width of the Kondo resonance at $V_g = 0$ using NRG. Within NRG at $\mathcal{H} = T_K$ (third solid curve from top) $G(V_g = 0) = e^2/h$. This exemplifies that T_K can equally be defined as the magnetic field which is necessary to suppress G down to e^2/h. This criterion can be used for the fRG data and the excellent agreement of the fRG curves and the NRG results in Fig. 1b shows that our approximation (in contrast to other simple approximation schemes) indeed contains T_K.

We note that using a fRG based truncation scheme in which the full frequency dependence of the 2-particle vertex is kept (leading to a frequency dependent self-energy) it was shown that one can also reproduce the Kondo resonance and Hubbard bands of the spectral function, however with a much higher numerical effort [27].

2.2 A Quantum Wire with a Single Impurity

Next we show that the fRG based approximation scheme is also able to produce power-law scaling of the conductance, which is characteristic for inhomogeneous LLs. This requires fairly long chains of interacting electrons – say lattice systems with $N = 10^4$ sites corresponding to a length in the μmrange. Therefore the fRG flow equations have to be simplified further. For a spinless lattice model of the interacting chain with nearest-neighbor hopping t and a nearest-neighbor interaction U (see Fig. 2) we achieve this by parameterizing the Λ dependent 2-particle vertex by a flowing nearest-neighbor interaction U^{Λ}. Then Σ^{Λ} is a tridiagonal matrix in the real space

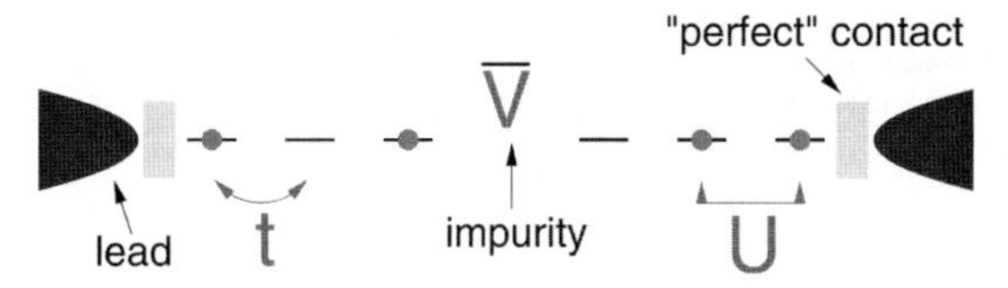

Fig. 2. Sketch of the model for an interacting quantum wire with nearest-neighbor hopping t, nearest-neighbor interaction U, a local impurity of strength V, and N lattice sites. The wire is connected to noninteracting leads by "perfect" contacts

basis of Wannier states and large systems can be treated [30]. A similar approximation has also been implemented for the (spinful) extended Hubbard model [33]. For the spinless model the LL parameter K is known exactly from Bethe ansatz [34]. Besides U it also depends on the filling n. To compute $G(T)$ one has to generalize the fRG to $T > 0$. This is described in [28] and [33]. We obtain the conductance from the one-particle Green function (and thus the self-energy) using a generalized Landauer–Büttiker relation [28]. Within our approximation current vertex corrections vanish (because $\Sigma^{\Lambda=0}$ is real) and using this relation does not require any additional approximations [28]. As we are only interested in the effect of a single impurity we model the contacts between the leads and the interacting wire to be "perfect", that means for $V = 0$ (see Fig. 2) the $T = 0$ conductance through the wire is e^2/h. This requires that the interaction is turned on and off smoothly close to the contacts [28].

The power-law scaling of $G(T)$ close to the perfect and open chain fixed points can very elegantly be shown in a single plot using a one-parameter scaling ansatz. Plotting $G(T)$ as a function of T/T_0, with an appropriately chosen nonuniversal scale $T_0(U, n, V)$, all data obtained for different V, U, and n – with the restriction that $K(U, n)$ is fixed – collapse on a single K dependent curve [19, 21]. This is shown in Fig. 3. The open symbols were obtained for $n = 1/2$, $U = 0.5$ while the filled ones were computed for $n = 1/4$, $U = 0.851$, both parameter sets leading to $K = 0.85$. Different symbols stand for different V. For convenience $(T/T_0)^{K-1}$ is chosen as the x-axis such that $G/(e^2/h)$ for $x \ll 1$ scales as $1 - x^2$ and for $x \gg 1$ as $x^{-2/K}$. Within our approximation the scaling exponents are correct to leading order in U [28, 30, 33]. During the flow the self-energy remains frequency independent and $\Sigma^{\Lambda=0}$ can be interpreted as an effective impurity potential. Scattering off the spatially long-ranged oscillations of the self-energy generated during the fRG flow leads to the observed power-law behavior [28, 30].

We note that in the limits of weak and strong impurities the scaling can be shown analytically with our fRG scheme [35]. Up to now the fRG is the only method which allows to study the entire crossover from weak to strong impurities for microscopic lattice models. Studying such models is important as similar to experimental systems, they contain energy scales that set upper and lower bounds for scaling. They are absent if field theoretical models are used.

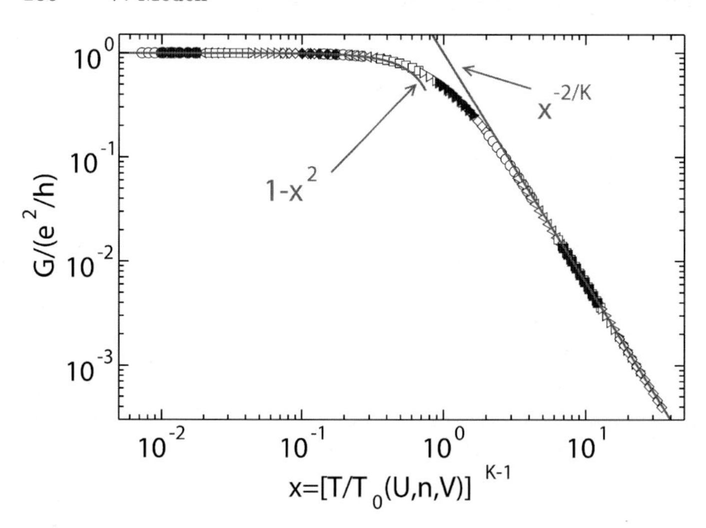

Fig. 3. One-parameter scaling of $G(T)$ for a quantum wire with $N = 10^4$ lattice sites and a single impurity. For details see the text

3 Correlation Effects in More Complex Geometries

After showing that our method contains the essential physics in the case of locally correlated systems as well as for inhomogeneous LLs we proceed and study more complex setups.

3.1 Novel Resonances in a Parallel Double-Dot System

As a first example we consider a parallel single-level double-dot system as sketched in Fig. 4a. The dots are coupled to common leads by tunneling barriers and the electrons interact by an inter-dot interaction U. The ring geometry is pierced by a magnetic flux ϕ. We here neglect the spin and thus suppress the spin Kondo effect. Experimentally, the contribution of spin physics may be excluded by applying a strong magnetic field. We focus on temperature $T = 0$. For this setup the flow equations (2) and (3) can directly be applied. G is computed from a generalized Landauer–Büttiker relation [31].

Figure 4b shows the evolution of $G(V_g)$ for degenerate levels $\varepsilon_j = V_g$, generic parameters[36] Γ_j^l, ϕ, and increasing U. At $U = 0$, $G(V_g)$ is a Lorentzian at large $|V_g|$, while it shows a dip at $V_g = 0$. With increasing U the height of the two peaks resulting from the dip at $V_g = 0$ increases and the maximum flattens. At a critical $U = U_c(\{\Gamma_j^l\}, \phi)$ each peak splits into two. For the present example the fRG approximation is $U_c/\Gamma \approx 4.69$, with Γ being the sum over all Γ_j^l. Further increasing U the two outer most peaks move towards larger $|V_g|$ and become the Coulomb blockade peaks located at $V_g \approx \pm U/2$. The other two peaks at $\pm V_{\mathrm{CIR}}$, where $V_{\mathrm{CIR}} > 0$ decreases

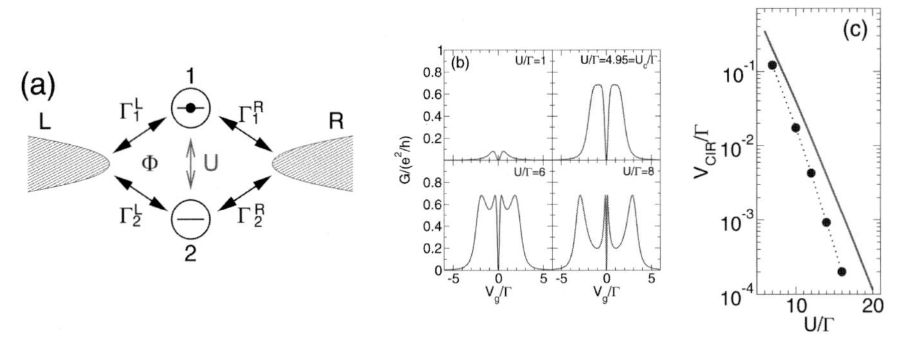

Fig. 4. (**a**) Sketch of the double-dot setup. (**b**) Generic results for $G(V_\mathrm{g})/(e^2/h)$ at different U obtained from the fRG with $\Gamma_1^L = 0.27\Gamma$, $\Gamma_1^R = 0.33\Gamma$, $\Gamma_2^L = 0.16\Gamma$, $\Gamma_2^R = 0.24\Gamma$, and $\phi = \pi$. The two novel correlation induced resonances are visible in the *lower panels* (large U), near $V_\mathrm{g} = 0$. (**c**) U dependence of the position V_CIR of the novel resonances (*line*: fRG; *circles*: NRG)

with increasing U, are novel resonances following from the interplay of quantum interference and correlations. For $U > U_c$ the height of all four peaks is equal and does not change with U. Note that in contrast to [36] here the flow of the 2-particle vertex is included which leads to improved results. The above scenario was confirmed using NRG [36]. Furthermore, the appearance of the novel resonances is robust: It appears for almost arbitrary combinations of the four tunnel couplings and the magnetic flux, also remains visible for a small detuning of the dot level energies [36], and small temperatures. In Fig. 4c we show the dependence of V_CIR on U obtained from fRG (line) and NRG (circles) on a linear-log scale. Both curves follow a straight line which shows that

$$V_\mathrm{CIR}/\Gamma \propto \exp\left[-C\left(\{\Gamma_j^l\},\phi\right)U/\Gamma\right], \tag{9}$$

with $C > 0$. The fRG apparently underestimates C. Within the fRG for a certain class of $\{\Gamma_j^l\},\phi$ the appearance of an energy scale depending exponentially on U/Γ can be shown analytically [36]. We note that the above correlation effect is unrelated to spin and orbital Kondo physics.

Double-dot geometries that could form the basis to verify the above predictions have been experimentally realized in [5].

3.2 Novel Fixed Points in a Y-Junction of Three Quantum Wires

As a second example for the application of the fRG to more complex geometries we study the conductance through a Y-junction of interacting quantum wires. The three wires are assumed to be equal and described by the model of spinless fermions with nearest-neighbor hopping t and nearest-neighbor interaction U as already used above. Each wire is connected to noninteracting leads by "perfect" contacts. The junction is assumed to be symmetric and

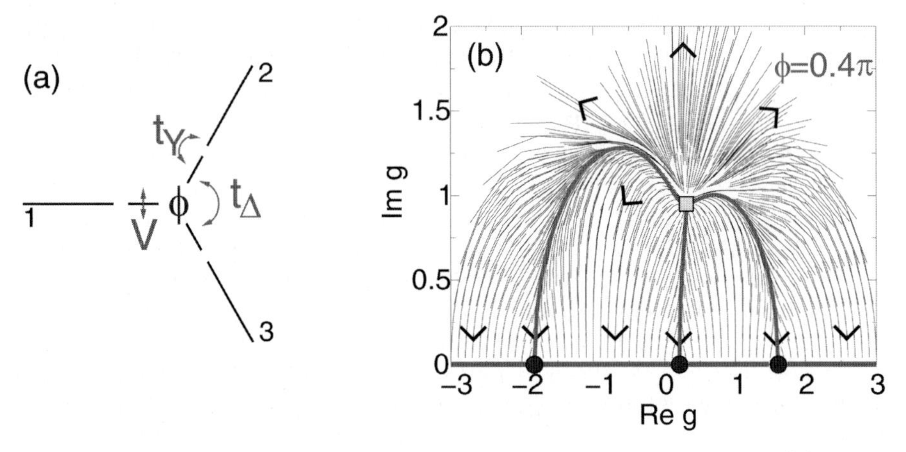

Fig. 5. (a) Setup of the Y-junction of three interacting quantum wires. (b) $U > 0$ fRG flow within the complex g plane. For each point in the g plane the conductance follows using Eq. (10). The symbols indicate fixed points. The condition $\mathrm{Im}\, g = 0$ defines a line of fixed points. For further details see the text

characterized by three parameters t_Y, t_Δ, and V as sketched in Fig. 5a. The ring structure is furthermore pierced by a magnetic flux ϕ and for $\phi \neq n\pi$, with an integer n, time-reversal symmetry is broken. For generic junction parameters in the non-interacting case this leads to an asymmetry of the conductance from wire ν to wire ν' (with $\nu, \nu' = 1, 2, 3$) and vice versa $G_{\nu,\nu'} \neq G_{\nu',\nu}$ and the breaking of time-reversal symmetry is indicated by the conductance. This can be seen using single-particle scattering theory. $G_{\nu',\nu}$ can be expressed in terms of the single complex and junction parameter dependent number $g = (-V - t_Y^2\, \tilde{\mathcal{G}}_{1,1})/|t_\Delta|$ as

$$G_{\nu,\nu'} = \frac{4\,(\mathrm{Im}\, g)^2 \left|e^{-i\phi} - g\right|^2}{\left|g^3 - 3g + 2\cos\phi\right|^2}\,, \qquad (10)$$

with ν, ν' in cyclic order [37]. $G_{\nu',\nu}$ follows by replacing $\phi \to -\phi$. Here $\tilde{\mathcal{G}}$ denotes the (wire index independent) single particle Green function of one of the semi-infinite wires obtained after setting $t_Y = 0$ and taking the energy $\varepsilon + i0$ with $\varepsilon \to 0$. The index $1, 1$ stands for its diagonal matrix element taken at the first site. At $U = 0$ it is given by $\tilde{\mathcal{G}}_{1,1} = i/t$ [28].

At $T = 0$ Eq. (10) also holds for $U > 0$. In this case $\tilde{\mathcal{G}}_{1,1}(\Sigma)$ becomes self-energy and thus U dependent. As for the single interacting wire with one impurity Σ can approximately be computed using the fRG. As scaling variable we this time use an energy scale $\delta_N = \pi v_{\mathrm{F}}/N$, with v_{F} being the $U = 0$ Fermi velocity, set by the length of the three interacting wires. As $\Sigma^{\Lambda=0}$ becomes δ_N dependent, also g and thus $G_{\nu,\nu'}$ will depend on the length of the interacting wires. Increasing N (that is lowering the infrared cutoff δ_N) for fixed $U, \phi \neq n\pi, t_Y, t_\Delta, V$ then leads to a flow in the complex g plane and

$G_{\nu,\nu'}$ changes according to (10). The flow for $U = 1$ and $\phi = 0.4\pi$ is shown in Fig. 5b for a variety of different t_Y, t_Δ, V. Each line stands for a fixed set of the junction parameters and varying δ_N. The direction of the flow is indicated by the arrows.

The square indicates a fixed point at which the asymmetry is maximal, that is $G_{1,2} = e^2/h$ and $G_{2,1} = 0$. For $U > 0$ it is unstable [37, 38]. For generic junction parameters the flow is directed towards the line of fixed points defined by the $\mathrm{Im}\, g = 0$ line. As is apparent from (10) on this line and for generic $\mathrm{Re}\, g$, both $G_{1,2}$ and $G_{2,1}$ vanish. This line of fixed points is the analogue of the open chain fixed point obtained for one impurity in a single wire. Analyzing the scaling close to the line of fixed points in more detail we find that for $\delta_N \to 0$, $|G_{1,2} - G_{2,1}|/(G_{1,2} + G_{2,1}) \to 0$, i.e. $G_{1,2}$ and $G_{2,1}$ become equal faster than they go to zero. Even more surprising behavior is found if the bare parameters are tuned such that the flow starts on the thick line. In that case it leads to one of the three fixed points indicated by the circles. On these fixed points we find $G_{1,2} = G_{2,1} = (4/9)(e^2/h)$, which is the conductance maximally allowed by the unitarity of the scattering matrix for a symmetric Y-junction of noninteracting wires. Combined, these two observations show that due to the interaction on very low energy scales (i.e. for long interacting wires and at low temperatures) the conductance no longer indicates a broken time-reversal symmetry. In that sense the electron correlations restore the time-reversal symmetry [37].

Associated to the novel $(4/9)(e^2/h)$ fixed points is a novel scaling dimension that should show up in e.g. the finite temperature corrections of G with respect to the fixed point conductance [37].

4 Summary

In the present article we have shown that electron correlations in nano- and mesoscopic systems can lead to a variety of surprising effects. While for simple geometries established methods are available to investigate important aspects of the many-body problem, they cannot directly be applied to more complex setups, such as systems of locally correlated quantum dots and junctions of interacting quantum wires. To investigate such systems, which in the near future will shift towards the focus of experimentally activities, we introduced a reliable, simple to implement, and numerically very fast approximation scheme which is based on the functional renormalization group. In certain limiting cases it can also be used to obtain analytical results.

Acknowledgements

The author would like to thank T. Costi and J. von Delft for providing their NRG and Bethe ansatz data, S. Andergassen, T. Enss, C. Karrasch, F. Marquardt, W. Metzner, U. Schollwöck, K. Schönhammer, and A. Sedeki for collaboration on the issues presented, and P. Wächter for useful comments on the manuscript. This work was supported by the Deutsche Forschungsgemeinschaft (SFB 602).

References

[1] A. C. Hewson, *The Kondo Problem to Heavy Fermions*, (Cambridge University Press, Cambridge, UK, 1993)
[2] L. Glazman and M. Raikh, JETP Lett. **47**, 452 (1988)
[3] T. Ng and P. Lee, Phys. Rev. Lett. **61**, 1768 (1988)
[4] K. Schönhammer in D. Baeriswyl (Ed.): *Interacting Electrons in Low Dimensions,* (Kluwer Academic Publishers, 2005)
[5] A. W. Holleitner *et al.,* Phys. Rev. Lett. **87**, 256802 (2001); Science **297**, 70 (2002)
[6] N. J. Craig *et al.,* Science **304**, 565 (2004)
[7] M. Terrones *et al.,* Phys. Rev. Lett. **89**, 075505 (2002)
[8] M. Salmhofer, *Renormalization,* (Springer, Berlin, 1998)
[9] A. M. Tsvelik and P. B. Wiegmann, Adv. Phys. **32**, 453 (1983)
[10] T. A. Costi, A. C. Hewson, and V. Zlatic, J. Phys.: Condens. Matter **6**, 2519 (1994)
[11] U. Gerland, J. von Delft, T. A. Costi, and Y. Oreg, Phys. Rev. Lett. **84**, 3710 (2000)
[12] Y. Meir and N. S. Wingreen, Phys. Rev. Lett. **68**, 2512 (1992)
[13] D. Goldhaber-Gordon *et al.,* Nature **391**, 156 (1998)
[14] W. van der Wiel *et al.,* Science **289**, 2105 (2000)
[15] A. Luther and I. Peschel, Phys. Rev. B **9**, 2911 (1974)
[16] D. C. Mattis, J. Math. Phys. **15**, 609 (1974)
[17] W. Apel and T. M. Rice, Phys. Rev. B **26**, 7063 (1982)
[18] T. Giamarchi and H. J. Schulz, Phys. Rev. B **37**, 325 (1988)
[19] C. L. Kane and M. P. A. Fisher, Phys. Rev. Lett. **68**, 1220 (1992); Phys. Rev. B **46**, 15233 (1992)
[20] A. Furusaki and N. Nagaosa, Phys. Rev. B **47**, 4631 (1993)
[21] P. Fendley, A. W. W. Ludwig, and H. Saleur, Phys. Rev. Lett. **74**, 3005 (1995)
[22] A. Furusaki and N. Nagaosa, Phys. Rev. B **54**, R5239 (1996)
[23] M. Bockrath *et al.,* Nature **397**, 598 (1999)
[24] Z. Yao, H. Postma, L. Balents, and C. Dekker, Nature **402**, 273 (1999)
[25] O. Auslaender *et al.,* Phys. Rev. Lett. **84**, 1764 (2000)
[26] R. de Picciotto *et al.,* Nature **411**, 51 (2001)
[27] R. Hedden, V. Meden, Th. Pruschke, and K. Schönhammer, J. Phys.: Condens. Matter **16**, 5279 (2004)
[28] T. Enss *et al.,* Phys. Rev. B **71**, 155401 (2005)
[29] V. Meden, lecture notes on the "Functional renormalization group", http://www.theorie.physik.uni-goettingen.de/~meden/funRG/
[30] S. Andergassen *et al.,* Phys. Rev. B **70**, 075102 (2004)
[31] C. Karrasch, T. Enss, and V. Meden, cond-mat/0603510
[32] T. A. Costi, Phys. Rev. B **64**, 241310(R) (2001)
[33] S. Andergassen *et al.,* Phys. Rev. B **73**, 045125 (2006)

[34] F. D. M. Haldane, Phys. Rev. Lett. **45**, 1358 (1980)
[35] V. Meden, W. Metzner, U. Schollwöck, and K. Schönhammer, J. Low Temp. Physics **126**, 1147 (2002)
[36] V. Meden and F. Marquardt, Phys. Rev. Lett. **96**, 146801 (2006)
[37] X. Barnabé-Thériault, A. Sedeki, V. Meden, and K. Schönhammer, Phys. Rev. Lett. **94**, 136405 (2005)
[38] C. Chamon, M. Oshikawa, and I. Affleck, Phys. Rev. Lett. **91**, 206403 (2003)

Part IV

Optoelectronics

Ultrafast Dynamics of Optically-Induced Charge and Spin Currents in Semiconductors

Torsten Meier[1], Huynh Thanh Duc[1,2], Quang Tuyen Vu[1], Bernhard Pasenow[1], Jens Hübner[1,⋆], Sangam Chatteryee[1], Wolfgang W. Rühle[1], Hartmut Haug[3], and Stephan W. Koch[1]

[1] Department of Physics and Material Sciences Center, Philipps University, Renthof 5, 35032 Marburg, Germany
torsten.meier@physik.uni-marburg.de

[2] Institute of Physics, Mac Dinh Chi 1, Ho Chi Minh City, Vietnam

[3] Institut für Theoretische Physik, Johann Wolfgang Goethe-Universität Frankfurt, Max-von-Laue-Strasse 1, 60438 Frankfurt, Germany

Abstract. The ultrafast dynamics of optical excitations in semiconductors and semiconductor nanostructures can be computed on a microscopic theoretical basis using the semiconductor Bloch equations . This set of equations allows one to nonperturbatively evaluate light-field-induced intraband and interband excitations and is also well suited for the analysis of many-body effects such as excitonic resonances and carrier-carrier and carrier-phonon scattering processes. In this article we start with a description of an experimental observation of coherent spin photocurrents, then we briefly introduce the theoretical approach and review some results of our theory concerning the microscopic description of the ultrafast coherent optical generation and the temporal decay of charge and spin currents in semiconductor nanostructures. The computed transients show an enhanced damping of the spin current relative to the charge current as a consequence of Coulomb scattering between carriers with opposite spin. The influence of quantum kinetic memory effects on the dynamics of the photocurrents is analyzed by evaluating the scattering terms beyond the Markov approximation.

1 Introduction

The optical and electronic properties of semiconductors and semiconductor nanostructures are of substantial current interest. In these systems, one can study on the one hand questions which are of relevance in the area of fundamental physics, i.e., many-body and non-equilibrium effects, ultrafast dy-

⋆ Present address: Institut für Festkörperphysik, Universität Hannover, Appelstraße 2, D-30167 Hannover, Germany.

R. Haug (Ed.): Advances in Solid State Physics,
Adv. in Solid State Phys. **46**, 189–200 (2008)

namics, coherent phenomena, influence of disorder, etc. On the other hand, such structures are useful for a great variety of applications including optoelectronic devices. To a certain degree, it is possible in semiconductor nanostructures to design the optoelectronic response by chosing suitable material combinations and growth conditions when producing such samples.

A proper microscopic theory of the optical properties of semiconductors has to describe the light field, the material excitations, and their interaction. By exciting a semiconductor with light, one raises the electrons energetically to the previously unoccupied conduction bands and generates holes in the initially occupied valence band. Whereas in atomic systems the energy spectrum of the electronic states is discrete which means that the optical response can in many situations be described by few-level systems, see, e.g., [1], in extended semiconductors and semiconductor nanostructures the continuous dispersion of the electronic states, i.e., the band structure, has to be incorporated into the theoretical analysis [2–4].

Another very important difference between atomic systems and semiconductors is that in semiconductors the Coulomb interaction among the photoexcited charged quasi-particles gives rise to strong many-body effects which significantly influence the optoelectronic properties. Therefore, any proper theory for the optical response of semiconductors has to treat this many-body problem adequately. Within the framework of the semiconductor Bloch equations (SBE) it is possible to compute a number of important many-body effects on a microscopic theoretical basis [2–4]. These equations describe the dynamical evolution of the electronic coherences between the relevant bands and of the electronic distributions during and after the photoexcitation.

If one is interested in the combined description of optical and transport properties, one has to take into account that besides exciting electrons between bands the electric field also accelerates electrons within a band [5–7]. As shown, e.g., in [7–9] this intraband acceleration can be incorporated through additional terms into the SBE. The resulting extended SBE have been used more than a decade ago for the description of coherent transport effects like Bloch oscillations [10, 11], i.e., an oscillating current in the presence of a static bias field, and dynamical localization [12, 13]. Using ultrafast optical excitations of biased semiconductor superlattices nanostructures it was possible to experimentally demonstrate in the early 1990's the existence of Bloch oscillations directly in the time domain [14–16]. The experimentally observed transients can be described well if one numerically solves the extended SBE including many-body effects and scattering events [7–9].

The observation of the coherent transport effects mentioned above, requires the photoexcitation of semiconductors or semiconductor nanostructures which are biased by an electric field. The possibility to generate photocurrents in semiconductors without any bias field, i.e., by purely optical excitation, was predicted in [17]. The proposed excitation scheme involves two light fields with frequencies ω and 2ω satisfying $2\hbar\omega > E_{\text{gap}} > \hbar\omega$, where E_{gap} is the band gap energy. Considering one- and two-photon transitions,

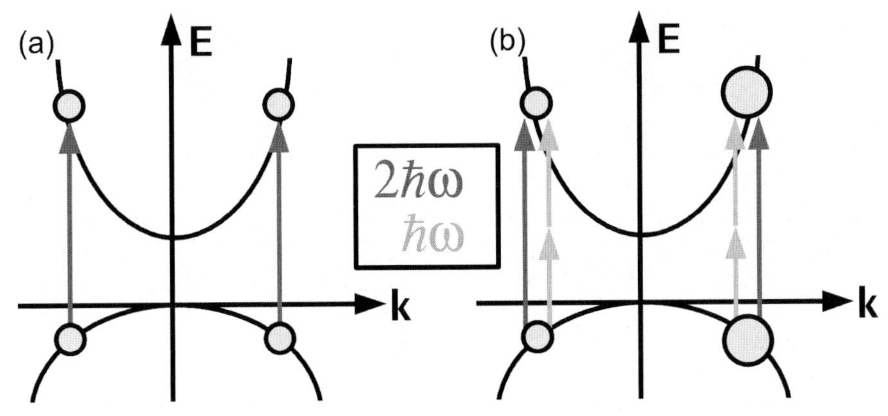

Fig. 1. Schematical illustration of optical interband excitations in a two-band model of a semiconductor. (**a**) A single optical field with frequency 2ω resonantly generates electrons and holes above the band gap E_{gap}. In this case, the photoexcited electron and hole distribution are symmetric in k-space and no current is generated. (**b**) The incident field consists of two frequencies ω and 2ω which satisfy $\hbar\omega < E_{\text{gap}} < 2\hbar\omega$. In this so-called two-color excitation scheme due to one- and two-photon transitions the same initial and final states are connected by two pathways. Therefore, it is possible to create excitations which are not symmetric in k-space, i.e., correspond to a finite current

in this two-color excitation scheme the initial and final states are coupled by two pathways which allows one to generate asymmetric electronic excitations in momentum, i.e., **k**, space, see Fig. 1. The direction of the asymmetry and thus of the accompanying current can be coherently controlled by varying the relative phase between the two incident ω and 2ω fields. Already one year after the prediction, the proposed purely optical coherent generation of currents was demonstrated experimentally [18]. Later, it has been predicted that in disordered semiconductors sequences of temporally delayed excitation pulses can be used to induce current echoes [19, 20]. Quite recently it has been shown that it is even possible to control the photocurrent via the carrier-envelope phase [21] which makes this effect interesting for potential applications in optical metrology.

The purely optically generated ultrafast photocurrents received additional attention after the prediction of [22] which showed that basically the same interference scheme with just different polarization directions of the incident light fields, see Fig. 2, can also be employed for the generation of pure spin currents which are not accompanied by any charge current. Such spin currents which have been observed [23, 24] could be of great interest for possible applications in the area of spintronics [25].

The charge generation process has been described theoretically on the basis of band structure calculations, in terms of nonlinear optical susceptibilities [17, 26]. In this approach, the optical transitions are treated on the level

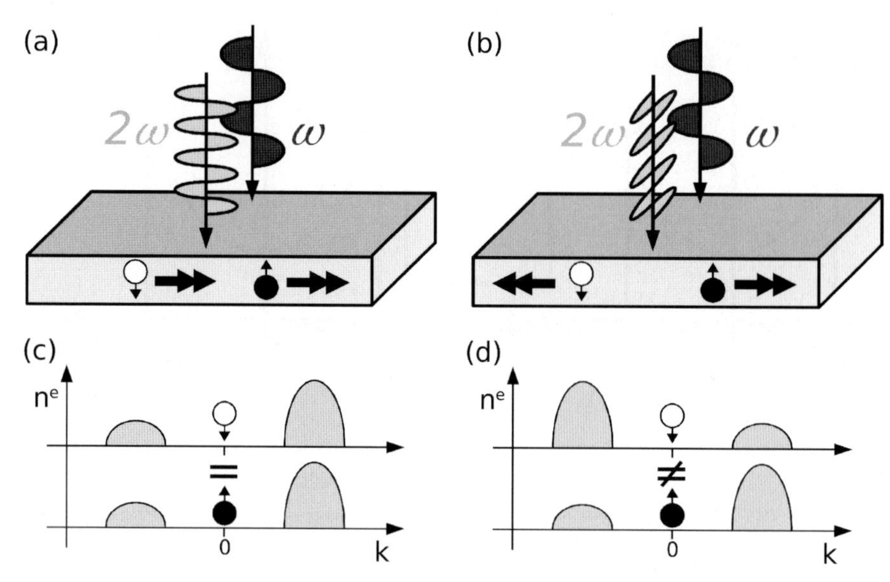

Fig. 2. Schematical illustration of the coherent optical generation of currents. (**a**) By using linear parallel polarization directions of the ω and the 2ω beams a charge current is injected. In this case, the electrons with both spin directions flow into the same direction, see (**c**) which shows the generated momentum distributions. (**b**) With linear perpendicular polarization directions of the ω and the 2ω beams a spin current is injected. In this case, the electrons with both spin directions flow into opposite directions, see (**d**)

of Fermi's Golden Rule. The dynamics of the generation process has been analyzed using Bloch equations for the relevant intra- and interband transitions and carrier distributions. This approach has been used to compute photocurrents in disordered semiconductors within a two-band model [19, 20] and in ordered semiconductor quantum wells within a multiband formalism [27]. Although the decay of the electronic current by the scattering with LO phonons has been studied [28], in most of the existing publications this decay is still modeled by phenomenological decay times. The Coulomb interaction among the photoexcited carriers has so far not been the focus of particular attention. However, in semiconductor quantum wires excitonic effects have been addressed on the Hartree–Fock level in [29] and photocurrents which oscillate with transition frequencies between different excitonic states have been computed. Furthermore, recently some interesting phase shifts originating from excitonic effects have been predicted [30].

In [31] we have presented and analyzed a microscopic many-body theory which is based on the extended SBE that is capable of describing the dynamical generation, the coherent evolution, and the decay of charge and spin currents. These equations nonperturbatively include light-field-induced intraband and interband excitations beyond the rotating wave approxima-

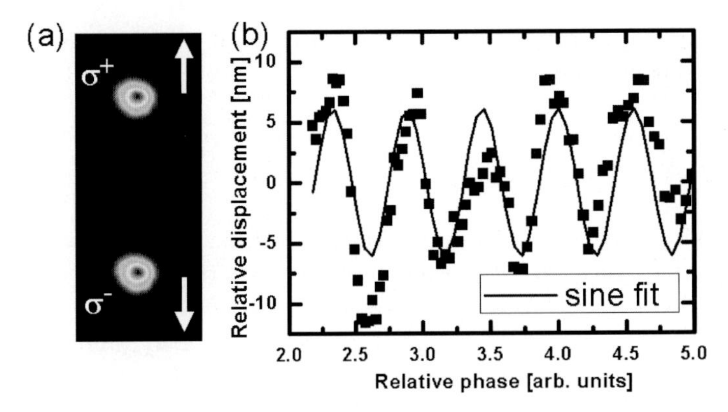

Fig. 3. (**a**) CCD image of the oppositely circularly polarized photoluminescence spots of a 290 nm ZnSe sample at a temperature of 100 K. The sample is excited using coherent cross-linearly polarized 100 fs laser pulses at 3.10 eV and 1.55 eV. (**b**) The displacement between the circularly polarized photoluminescence components varies sinusoidally as function of the relative phase between the two laser fields. After [24]

tion, contain excitonic effects and Coulombic nonlinearities due to energy and field renormalizations, and many-body correlation contributions due to the carrier LO-phonon coupling and the Coulomb interaction at the second Born–Markov level.

In this review, we start by describing in Sect. 2 an experiment which shows that one can generate coherent spin currents in semiconductors using optical laser pulses and that these currents can be detected by measuring the circularly polarized photoluminescence. Then we briefly describe our theoretical approach in Sect. 3 and present numerical results on the ultrafast dynamics of coherent charge and spin currents in Sect. 4. The most important results are summarized in Sect. 5.

2 Experimental Observation of Coherent Spin Currents

In [24] coherent spin currents have been generated via two-color optical excitation of a bulk ZnSe sample. In order to measure that the electrons with opposite spin move into opposite directions, in this experiment the two oppositely circularly polarized photoluminescence components were separated by a Wollaston prism and measured spatially-resolved. The luminescence spots were imaged onto a CCD, see Fig. 3a. When the relative displacement between the oppositely circularly polarized photoluminescence spots is recorded as function of the phase difference between the ω and the 2ω components of the laser field one finds a sinusoidal variation, see Fig. 3b. These findings [24] and also the results of other experiments [23] are in agreement with the predicted behavior of coherently photogenerated spin currents [22] and show that it is indeed possible to generate observable spin currents in semiconductors using purely optical excitation.

3 Theoretical Approach

Within a two-band approach, the dynamical optoelectronic response is obtained by solving the Heisenberg equations of motion for the electron distribution $\hat{n}^e_{\sigma\mathbf{k}} = \hat{a}^\dagger_{c\sigma\mathbf{k}}\hat{a}_{c\sigma\mathbf{k}}$, the hole distribution $\hat{n}^h_{\sigma\mathbf{k}} = 1 - \hat{a}^\dagger_{v\sigma\mathbf{k}}\hat{a}_{v\sigma\mathbf{k}}$, and the interband polarization $\hat{p}_{\sigma\mathbf{k}} = \hat{a}^\dagger_{v\sigma\mathbf{k}}\hat{a}_{c\sigma\mathbf{k}}$. Here, the operator $\hat{a}^\dagger_{\lambda\sigma\mathbf{k}}$ creates and the operator $\hat{a}_{\lambda\sigma\mathbf{k}}$ destroys an electron with wave vector $\mathbf{k}$ and spin σ in band λ. Commuting the dynamical quantities with the Hamiltonian and taking expectation values, i.e., $\eta_{\sigma\mathbf{k}} = \langle\hat{\eta}_{\sigma\mathbf{k}}\rangle$ for $\eta = p, n^e, n^h$, the resulting extended SBE read [7–9, 31]

$$\left(\frac{\mathrm{d}}{\mathrm{d}t} + \frac{e}{\hbar}\mathbf{E}(t)\cdot\nabla_\mathbf{k}\right) p_{\sigma\mathbf{k}} = -\frac{i}{\hbar}\left[\epsilon_{c\mathbf{k}} - \epsilon_{v\mathbf{k}} - \sum_{\mathbf{q}\neq 0} V_\mathbf{q}(n^e_{\sigma\mathbf{k}+\mathbf{q}} + n^h_{\sigma\mathbf{k}+\mathbf{q}})\right] p_{\sigma\mathbf{k}}$$
$$- \frac{i}{\hbar}(n^e_\mathbf{k} + n^h_\mathbf{k} - 1)\left(\mathbf{d}^{cv}_{\sigma\mathbf{k}}\cdot\mathbf{E}(t) + \sum_{\mathbf{q}\neq 0} V_\mathbf{q} p_{\sigma\mathbf{k}+\mathbf{q}}\right) + \left.\frac{\mathrm{d}}{\mathrm{d}t}p_{\sigma\mathbf{k}}\right|_{\mathrm{coll}}, \tag{1}$$

$$\left(\frac{\mathrm{d}}{\mathrm{d}t} + \frac{e}{\hbar}\mathbf{E}(t)\cdot\nabla_\mathbf{k}\right) n^\alpha_{\sigma\mathbf{k}} = -\frac{2}{\hbar}\,\mathrm{Im}\left[\left(\mathbf{d}^{cv}_{\sigma\mathbf{k}}\cdot\mathbf{E}(t) + \sum_{\mathbf{q}\neq 0} V_\mathbf{q} p_{\sigma\mathbf{k}+\mathbf{q}}\right) p^*_{\sigma\mathbf{k}}\right]$$
$$+ \left.\frac{\mathrm{d}}{\mathrm{d}t}n^\alpha_{\sigma\mathbf{k}}\right|_{\mathrm{coll}}. \tag{2}$$

In (1)–(2) $V_\mathbf{q}$ denotes the matrix element of the Coulomb interaction in k-space, i.e., the Fourier transform of the Coulomb potential in real space, $\mathbf{d}^{cv}_{\sigma\mathbf{k}}$ is the interband transition dipole, and the index $\alpha = e, h$ refers to electrons and holes, respectively, The terms given explicitely in (1)–(2) describe the coherent field-induced dynamics including many-body effects in time-dependent Hartree–Fock approximation. The collision terms, which describe the carrier LO-phonon and the carrier-carrier scattering [2, 31] are denoted by $|_{\mathrm{coll}}$.

Except for the spin indices, equations similar to (1)–(2) have been used to describe the optoelectronic response of semiconductor superlattices in the presence of static and Terahertz fields [7–9]. In these studies, however, the optical fields have been considered to solely generate interband transitions via the terms proportional to $\mathbf{E}(t)\cdot\mathbf{d}^{cv}_{\sigma\mathbf{k}}$ whereas the static and Terahertz fields were taken into account only in the intraband acceleration terms which are proportional to $\mathbf{E}(t)\cdot\nabla_\mathbf{k}$. In the spirit of a rotating wave approximation, such a distinction is useful if the involved fields are characterized by very different frequencies. However, since we here describe the generation of currents by the interference of optical fields with frequencies ω and 2ω, this simplification is not useful and therefore the total field is considered for both the inter- and the intraband excitations.

Numerical solutions of (1)–(2) provide the time-dependent polarization and electron and hole distributions. The carrier distributions determine the charge and spin current densities which are given by $\mathbf{J} = e\sum_{\sigma\mathbf{k}} \mathbf{v}^c n^e_{\sigma\mathbf{k}} - e\sum_{\sigma\mathbf{k}} \mathbf{v}^v n^h_{\sigma\mathbf{k}}$ and $\mathbf{S} = \frac{\hbar}{2}\sum_{\sigma\mathbf{k}} \sigma\mathbf{v}^c n^e_{\sigma\mathbf{k}} - \frac{\hbar}{2}\sum_{\sigma\mathbf{k}} \sigma\mathbf{v}^v n^h_{\sigma\mathbf{k}}$, respectively. Here, $\mathbf{v}^\lambda = \nabla_{\mathbf{k}}\epsilon_{\lambda\mathbf{k}}/\hbar$ is the group velocity which is linear in $\boldsymbol{k}$ for a parabolic band.

4 Numerical Results

In this section we present some computed numerical results on the dynamics of coherent charge and spin photocurrents.

4.1 Coherent Generation and Decay of Charge and Spin Currents

Using the approach outlined in Sect. 3, we have computed the coherent generation and the decay of charge and spin photocurrents including carrier LO-phonon and carrier-carrier scattering on the second Born–Markov level [31]. Figure 4 shows the time-dependence of the electron and heavy-hole distributions of a GaAs quantum well in k-space. The short 20 fs two-color laser pulse initially generates carriers with a combined excess energy of about 150 meV above the band gap. Due to their smaller mass, most of this kinetic energy is given to the electrons and only little to the heavy holes. Directly after the excitation, the electron and heavy-hole distributions are very similar since the optical interband transitions are diagonal in k-space. As a consequence of the quantum interference between the ω and 2ω field components, the distributions are larger for positive k_x than for negative k_x. Therefore, this situation corresponds to a current that flows in x-direction. In the course of time, the electron and heavy-hole distributions relax to energetically lower states and thermalize towards quasi-equilibrium distributions at the band edges in the limit of long times. Due to their larger mass, the distribution of the heavy holes is wider than that of the electrons.

Figure 5(a) clearly demonstrates that for the considered excitation conditions the dynamical evolution of the charge and of the spin current is influenced by both carrier LO-phonon and carrier-carrier scattering. If calculations are performed which artificially include only the carrier LO-phonon scattering, the charge and spin currents decay similarly. This decay is not exactly exponential, however, its onset can be approximated by an exponential decay with a time constant of 240 fs. Adding carrier-carrier scattering to the analysis, speeds up the decay of both currents. Surprisingly, we find in our complete model that the spin current decays more rapidly than the charge current even though none of the considered scattering events changes

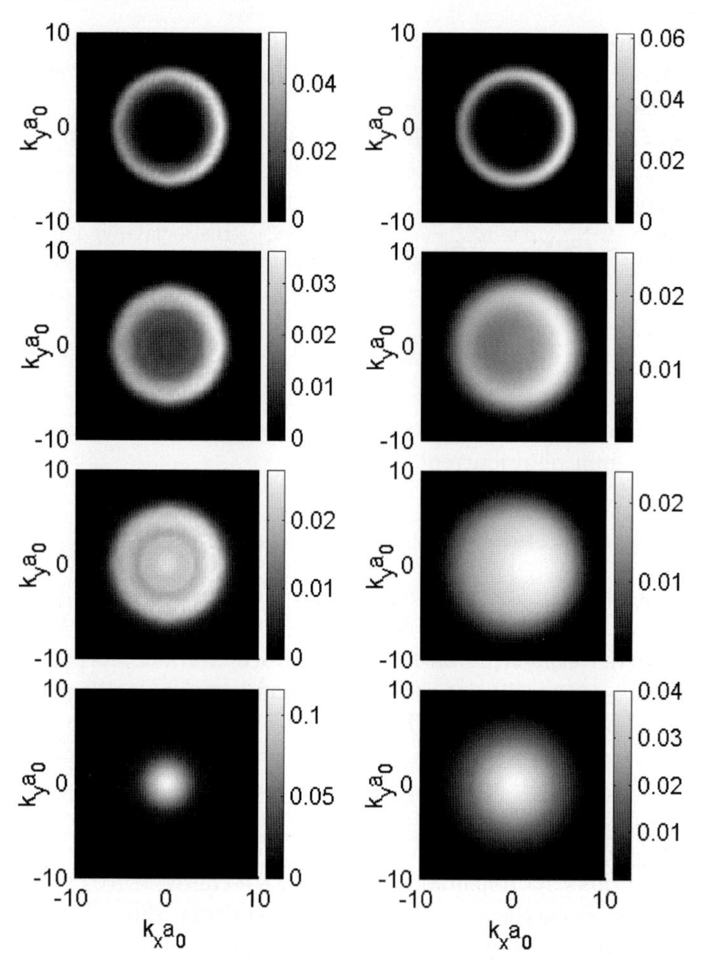

Fig. 4. *Left (right) column*: Grey-scale contour plots of the electron (heavy-hole) distribution n^e (n^h) in the two-dimensional **k**-space of a GaAs quantum well. Shown are snapshots of the relaxation dynamics from *top* to *bottom* at $t = 50, 100, 150$, and 400 fs, respectively. The incident Gaussian two-color laser pulse has its maximum at $t = 0$, a duration of 20 fs, and the amplitudes of the two fields are $A_\omega = 2A_{2\omega} = 108A_0$, with $A_0 = E_0/ea_0 \approx 4\,\mathrm{kV/cm}$, where E_0 is the exciton Rydberg and a_0 the exciton Bohr radius. The frequency $2\hbar\omega$ is about 150 meV above the band gap, the density of photoinjected carriers is $N = 10^{11}\ \mathrm{cm}^{-2}$, and the temperature is $T = 50\,\mathrm{K}$. Taken from [31]

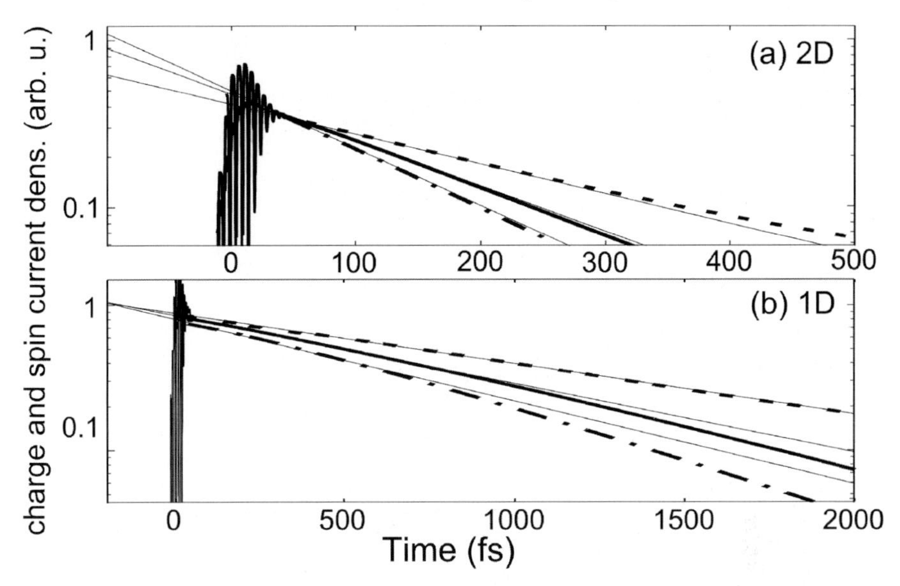

Fig. 5. (**a**) Time dependence of the charge (*solid*) and spin (*dash-dot*) currents computed for a quantum well using the same parameters as in Fig. 4. Also shown is the identical decay of both currents if only carrier LO-phonon scattering is considered (*dashed*). The *thin solid lines* represent exponential decays with time constants of 240, 155, and 125 fs, respectively. (**b**) Same as (a) for a quantum wire. In this case, the density of the photoinjected carriers is $N = 5\times\ 10^5\,\text{cm}^{-1}$ and the other parameters are the same as in (a). The *thin solid lines* represent exponential decays with time constants of 1250, 900, and 740 fs, respectively. Taken from [31]

the electronic spin. This effect can be understood by recalling that the excitation of a charge current corresponds to identical electron distributions for the different spins, see Fig. 2(c). Therefore, in this case the average momentum of the electron system is finite. Since carrier-carrer scattering only exchanges momenta among the carriers, a finite average momentum cannot be reduced by this process. The situation is, however, different when a spin current is excited. In this case, the electron distributions for the two spins add up to a vanishing total electronic momentum, see Fig. 2(d). Consequently, carrier-carrer scattering events between electrons with opposite spin can relax the photoexcited momenta of the spin-up and spin-down electrons.

As is shown in Fig. 5(b), we obtain qualitatively similar results also for quantum wires. However, due to the one-dimensional phase space the scattering is reduced and therefore the computed decay times are longer than in two dimensions. Additional investigations on the dependence of the currents on the intensities of the incident laser pulses can be found in [31].

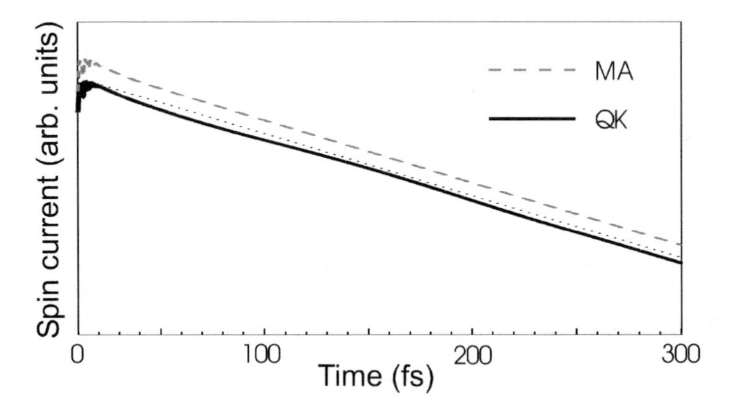

Fig. 6. Time-dependent spin current for a GaAs quantum wire computed for a carrier density of $N = 10^5\,\mathrm{cm}^{-1}$. The incident Gaussian laser pulses have a duration of 20 fs and the frequency $2\hbar\omega$ is 90 meV larger than the band gap. Compared are the results of calculations which use the Markov approximation (MA) with fully quantum-kinetic (QK) computations. The *dashed line* represents an exponential decay with a time constants of 360 fs

4.2 Influence of Quantum-Kinetic Memory Effects on the Current Transients

When the dynamics of the photoexcited charge and spin currents is computed on the second-Born Markov level, the transients decay nearly exponentially, see Fig. 5 and the thin line in Fig. 6. The result of a fully quantum-kinetic non-Markovian treatment of the carrier LO-phonon and Coulomb scattering [5] is shown by the thick line in Fig. 6. Comparing the results of the two approaches clearly demonstrates that the time scale of the overall decay of the transients is unaffected by quantum-kinetic effects, i.e., is described well already by the Markovian approximation. In the quantum-kinetic treatment the magnitude of the current is just slightly smaller and as a consequence of the memory effects, the transients show some weak oscillatory structure. Additional calculations (not shown in figure) demonstrate that the quantum-kinetic oscillations appear also when a charge current is excited. These oscillations originate from the contributions of the heavy holes to the currents. This explains why they are rather weak since the main contribution to the current comes from the lighter electrons. Furthermore, the numerical results show that oscillations are present only if carrier LO-phonon coupling is included. For the parameters considered in Fig. 6, the excess energy of the heavy-holes is smaller than the LO-phonon energy. Therefore, in a Markovian treatment no LO-phonon emission is possible. However, in the quantum-kinetic calculations the virtual emission of LO-phonons is included and thus influences the dynamics of the coupled carrier-phonon system, in particular, on ultrashort time scales. It can be shown [32] that the oscillations of the currents can be understood as beats between a state involving a virtual LO-phonon emission process and the heavy-hole band edge.

5 Summary

In this review, we describe an experiment which shows that one can generate coherent spin currents in semiconductors using optical laser pulses and that these currents can be detected by measuring the circularly polarized photoluminescence. We have then introduced our microscopic theoretical approach and presented numerical results on the ultrafast dynamics of coherent charge and spin currents. Our numerical results predict that due to Coulomb scattering the spin current decays more rapidly than the charge current even though none of the considered scattering processes changes the electronic spin. Furthermore, interesting quantum-kinetic oscillations may appear in the current transients due to non-Markovian memory effects.

Presently, we are extending our two-band model and incorporate coupled heavy- and light-hole transitions into our approach. This leads in quantum wells to additional currents which flow in unusual directions [26, 33]. Also further experimental investigations on photocurrents in semiconductor nanostructures are in progress.

Acknowledgements

This work is supported by the Deutsche Forschungsgemeinschaft (DFG), by the Ministry of Education and Research (BMBF), and by the Interdisciplinary Research Center Optodynamics, Philipps-University Marburg. T.M. thanks the DFG for support via a Heisenberg fellowship (ME 1916/1). We thank the John von Neumann Institut für Computing (NIC), Forschungszentrum Jülich, Germany, for continued grants of computer time on their supercomputer systems.

References

[1] L. Allen, J. H. Eberly: *Optical Resonance and Two-Level Atoms* (Wiley, New York, 1975).
[2] H. Haug, S. W. Koch: *Quantum Theory of the Optical and Electronic Properties of Semiconductors*, 4th ed. (World Scientific, Singapore, 2004).
[3] W. Schäfer, M. Wegener: *Semiconductor Optics and Transport Phenomena* (Springer, Berlin, 2002).
[4] T. Meier, P. Thomas, S. W. Koch: *Coherent Semiconductor Optics: From Basic Concepts to Nanostructure Applications* (Springer, Berlin), scheduled for June 2006.
[5] H. Haug, A.-P. Jauho: *Quantum Kinetics in Transport and Optics of Semiconductors* (Springer, Berlin, 1996).

[6] E. Schöll, (Ed.): *Theory of Transport Properties of Semiconductor Nanostructures* (Chapman and Hall, London, 1997).
[7] T. Meier, P. Thomas, S. W. Koch: Coherent dynamics of photoexcited semiconductor superlattices with applied homogeneous electric fields, in K. T. Tsen (Ed.) *Ultrafast Phenomena in Semiconductors*, (Springer, New York, 2001) pp. 1–92.
[8] T. Meier, G. von Plessen, P. Thomas, S. W. Koch: Phys. Rev. Lett. **73**, 902 (1994).
[9] T. Meier, F. Rossi, P. Thomas, S. W. Koch: Phys. Rev. Lett. **75**, 2558 (1995).
[10] F. Bloch: Z. Phys. **52**, 555 (1928).
[11] C. Zener: Proc. Roy. Soc. A **145**, 523 (1934).
[12] D. H. Dunlap, V. M. Kenkre: Phys. Rev. B **34**, 3525 (1986).
[13] M. Holthaus: Phys. Rev. Lett. **69**, 351 (1992).
[14] J. Feldmann, K. Leo, J. Shah, D. A. B. Miller, J. E. Cunningham, T. Meier, G. von Plessen, A. Schulze, P. Thomas, S. Schmitt-Rink: Phys. Rev. B **46**, 7252 (1992).
[15] K. Leo, P. Haring Bolivar, F. Brüggemann, R. Schwedler, K. Köhler: Sol. State. Comm. **84**, 943 (1992).
[16] C. Waschke, H. G. Roskos, R. Schwedler, K. Leo, H. Kurz, K. Köhler: Phys. Rev. Lett. **70**, 3319 (1993).
[17] R. Atanasov, A. Haché, J. L. P. Hughes, J. E. Sipe, H. M. van Driel: Phys. Rev. Lett. **76**, 1703 (1996).
[18] A. Haché, Y. Kostoulas, R. Atanasov, J. L. P. Hughes, J. E. Sipe, H. M. van Driel: Phys. Rev. Lett. **78**, 306 (1997).
[19] J. Stippler, C. Schlichenmaier, A. Knorr, T. Meier, M. Lindberg, P. Thomas, S. W. Koch: Pphys. Stat. Sol. (B) **221**, 379 (2000).
[20] C. Schlichenmaier, I. Varga, T. Meier, P. Thomas, S. W. Koch: Phys. Rev. B **65**, 085306 (2002).
[21] T. M. Fortier, P. A. Ross, D. J. Jones, S. T. Cundiff, R. D. R. Bhat, J. E. Sipe: Phys. Rev. Lett. **92**, 147403 (2004).
[22] R. D. R. Bhat, J. E. Sipe: Phys. Rev. Lett. **85**, 5432 (2000).
[23] M. J. Stevens, A. L. Smirl, R. D. R. Bhat, A. Najmaie, J. E. Sipe, H. M. van Driel: Phys. Rev. Lett. **90**, 136603 (2003).
[24] J. Hübner, W. W. Rühle, M. Klude, D. Hommel, R. D. R. Bhat, J. E. Sipe, H. M. van Driel: Phys. Rev. Lett. **90**, 216601 (2003).
[25] I. Žutić, J. Fabian, S. Das Sarma: Rev. Mod. Phys. **76**, 323 (2004).
[26] A. Najmaie, R. D. R. Bhat, J. E. Sipe: Phys. Rev. B **68**, 165348 (2003).
[27] D. H. Marti, M.-A. Dupertuis, B. Deveaud: Phys. Rev. B **69**, 035335 (2004).
[28] P. Král, J. E. Sipe: Phys. Rev. B **61**, 5381 (2000).
[29] D. H. Marti, M.-A. Dupertuis, B. Deveaud: Phys. Rev. B **72**, 075357 (2005)
[30] R. D. R. Bhat, J. E. Sipe: Phys. Rev. B **72**, 075205 (2005).
[31] H. T. Duc, T. Meier, S. W. Koch: Phys. Rev. Lett. **95**, 086606 (2005).
[32] Q. T. Vu, et al.: to be published.
[33] B. Pasenow, et al.: to be published.

From a Fundamental Understanding of Phase Change Materials to Optimization Rules for Nonvolatile Optical and Electronic Storage

C. Steimer, Henning Dieker, Wojciech Welnic, Ralf Detemple,
Daniel Wamwangi, and Matthias Wuttig

Institute of Physics (IA), RWTH-Aachen University,
Sommerfeldstr. 14–18,
52056 Aachen, Germany
steimer@physik.rwth-aachen.de

Abstract. Phase change materials are commercially used in rewritable optical storage and investigated as non-volatile electronic storage. A short laser or current pulse of high intensity melts a sub-micron sized spot of crystalline material before quenching it to the amorphous state. A second pulse of lower intensity but longer duration may recrystallise and erase that bit. Since reflectivity and conductivity of the amorphous state are lower, a third even weaker laser or current pulse can be used to read out the state of the bit without changing it. As recrystallisation is the slowest process involved, materials with a small structural difference between the crystalline and amorphous phase promise higher data transfer rates. Such structural similarity however limits the optical and electronic contrast between the phases and the stability against spontaneous recrystallisation. This contradiction makes the development of phase change media a challenge that despite commercial applications still heavily relies on empirical approaches. This summary of recent experiments and ab-initio calculations reflects first steps toward an atomistic understanding of phase change materials.

1 Introduction

Phase change (PC) materials have been in commercial use in rewritable optical storage (DVD-RW, e.g.) for a decade and are currently investigated as non-volatile electronic storage to replace conventional FLASH-memory. Figure 1 explains the operation principle.

A ns-duration laser or current pulse of high intensity is used to melt a sub-micron sized spot of crystalline material and quench it to the amorphous state. A second pulse of lower intensity but longer duration leads to a less pronounced spatial temperature profile and lower cooling rates but provides sufficient thermal energy for crystallisation. Since reflectivity and conductivity of the amorphous state are lower than the crystalline values, a third laser or current pulse, even lower in intensity and thus not changing the material, can be used to read out the current state of the bit [1-3]. As recrystallisation is the slowest process involved, materials with a small structural difference between the crystalline and amorphous phase promise higher data transfer

R. Haug (Ed.): Advances in Solid State Physics,
Adv. in Solid State Phys. **46**, 201–211 (2008)

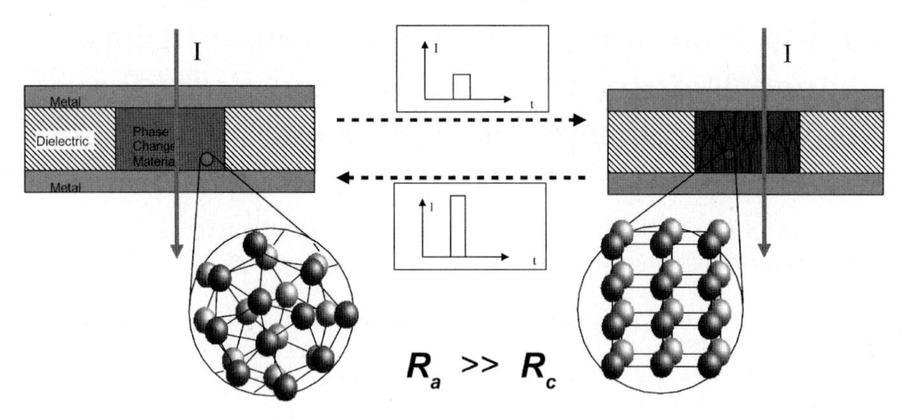

Fig. 1. Principle of phase change storage, here for nonvolatile electronic memory

rates. Such structural similarity however limits the optical and electronical contrast between both phases, i.e., readability, and the stability of the amorphous phase against spontaneous recrystallisation, i.e., data retention, as well. With these restrictions given it is clear that a fundamental understanding is needed to find suitable PC-materials, that usually consist of binary up to quaternary alloys of Te with group IV-V elements and noble metals. Surprisingly, despite their commercial application material development of PC-media still heavily relies on try and error. After this introduction, we will show how stoichiometry determines the structure of the crystalline phase in the second chapter. The amorphous short range order as the remaining loose end will be treated in the third chapter before we try to come up with a conclusion and outlook summarizing our current understanding and future challenges.

2 Stoichiometry and Crystalline Structures

Based on their different compositions and their different crystallisation behaviour common PC-materials can be divided into a Sb-rich and Te-rich class, which are based on Sb_2Te or Sb_2Te_3 [12,13]. Two of the phases formed upon annealing of AgIn doped Sb_2Te are $AgInTe_2$ and $AgSbTe_2$, which were chosen as model-systems [4]. After molecular beam deposition on glass-substrates, the samples were crystallised by annealing in a N_2-atmosphere. A static tester was used to laser amorphise a spot of a diameter of about 0.3 µm to start out from identical conditions.

The reflectivity of the amorphous spot was measured before exposure to a crystallisation pulse of defined intensity and duration and a second measurement of the reflectivity with the same laser. The procedure was repeated for an array of spots with varying intensities and durations of the crystallisation pulse. The results are visualized in figure 2: As expected for low intensities and long pulse durations $AgSbTe_2$ shows an increase in reflectivity due to

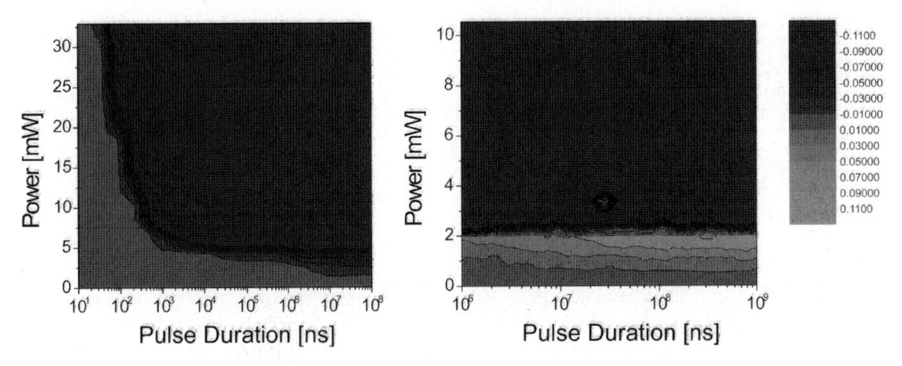

Fig. 2. Power-time-effect diagrams for $AgInTe_2$ (*left*) and $AgSbTe_2$ (*right*). Both, reflectivity measurements and crystallisation pulses were performed with the same laser-diode and optics at 830 nm. The colour coded in- or decreases in reflectivity are attributed to crystallisation or ablation, respectively [10]

crystallisation. For $AgInTe_2$ however, the low intensity wings for low power or short duration are followed by immediate ablation instead of crystallisation. Thus despite their chemical similarity these model alloys show that not all Te-alloys exhibit a pronounced optical contrast on crystallisation needed for successful applications as phase change materials. To understand that apparently constant reflectivity of $AgInTe_2$, the sheet-resistance of as deposited samples was monitored upon annealing in a N_2-atmosphere. Figure 3 shows the drop beyond 100°C together with the optical contrast, calculated from the optical constants determined before and after annealing. To relate the different contrasts to structure, XRD measurements were taken at the annealed samples. They confirmed crystallisation in a chalcopyrite-structure, which is similar to a zinc-blende structure, for $AgInTe_2$. $AgSbTe_2$ forms a rock salt-like structure, where the Te-atoms occupy one of the sub-lattices while the remaining places are taken by Ag and Sb at random.

The formation of a chalcopyrite structure for chalcogenides with an average valence electron number up to four and of rock salt structures beyond that value can be understood by density functional theory (DFT): For a structure given DFT varies the electron density until a minimum in energy is reached. Assuming a chalcopyrite and a rock salt structure for $AuInTe_2$ (N_{sp} = 4.0) and $AuSbTe_2$ (N_{sp} = 4.5) respectively and allowing the lattice constants to relax yields the electron density plots in Fig. 4 [5]. Up to four valence electrons can be hosted in bonding sp^3-orbitals leading to tetrahedral coordination. Beyond that value occupation of orthogonal p-orbitals is preferred to avoid the anti bonding sp^3-states and thus octahedral coordinations are assumed as found in simple cubic or rock salt arrangements for instance. The energy difference between a rock salt - and a chalcopyrite structure for various chalcogenide alloys as a function of the valence electron numbers obtained from DFT in Fig. 6 shows that this structural transition is a general

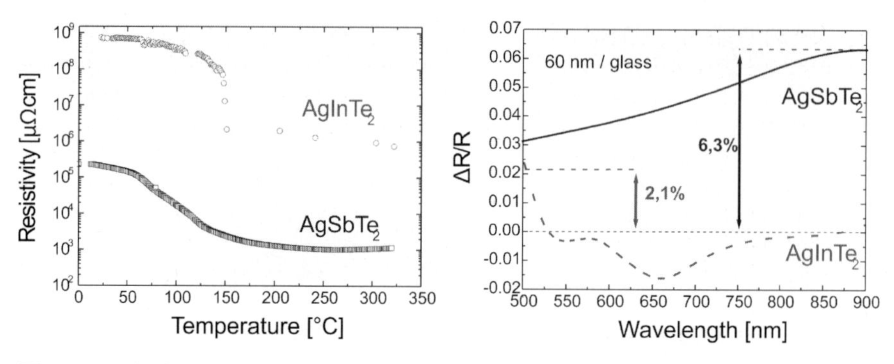

Fig. 3. *Left*: Sheet resistance of a unit60nm film on glass measured on annealing. *Right*: Optical contrast between amorphous and crystalline phase, calculated from ellipsometric measurements [4]

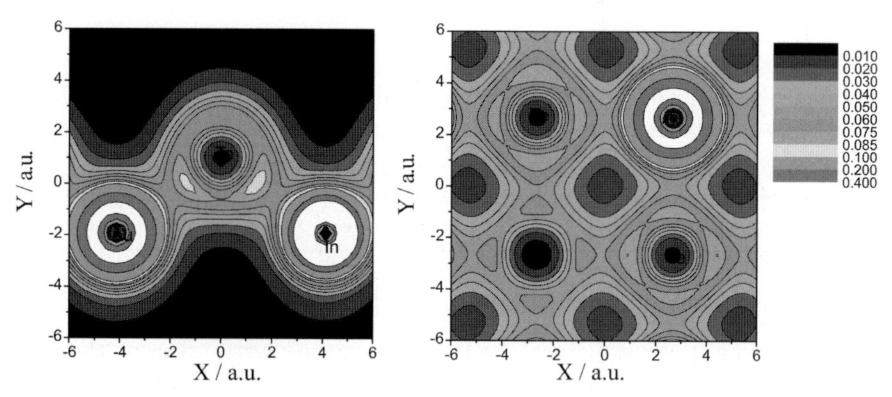

Fig. 4. Calculated electron-densities in arbitrary units for the (110)- and (100)-plane of $AuInTe_2$ (*left*) and $AuSbTe_2$ (*right*) with average valence electron numbers of 4.0 and 4.5 [5]

rule for chalcogenides rather than a peculiarity of these very alloys [5]. This trend is reconfirmed by a compilation of experimental structural data, which also shows an insufficient electrical and optical contrast for all tetrahedral arrangements as opposed to the octahedral coordinations given in Table 1. The lower optical contrast on amorphisation of chalcopyrite-alloys as compared to the octahedral arrangements had been attributed to the lower density contrast between the crystalline and amorphous phase. However this picture reduces contrast to an average over all wavelengths and structure to density. To gain a atomistic understanding of the switching mechanism that accounts for different wavelengths and can be turned into design rules for phase change media the a combination of the known crystalline structures and in particular an idea about the unknown amorphous structure have to be combined. We will present such a model for the amorphous short range order of PC-

Table 1. Crystalline structures and average valence-electron numbers for some chalcogenides, showing sufficient or insufficient contrast for optical storage, depending on their crystalline structures or average valence electron numbers N_{sp} respectively [5]. For alloys like $Ge_1Sb_2Te_4$ with a rock salt-like metastable structure, where one sub-lattice is occupied by Te and the other one shared by 2 Sb, 1 Ge and 1 vacancy, the vacancies were counted in the denominator as atoms to determine average electron numbers N_{sp}. Note that a) not all Te-rich chalcogenides are suitable and b) not all successful materials contain Ge

	Material	Structure	N_{sp}
Unsuccessful samples	$AgInTe_2$	Chalcopyrite	4.00
	$AuInTe_2$	Chalcopyrite	4.00
	In_3SbTe_2	Rock Salt	4.30
Successful samples	$Au_{25}Ge_4Sn_{11}Te_{60}$	Cubic	4.45
	$AgSbTe_2$	Rock Salt	4.50
	$AuSbTe_2$	Cubic	4.50
	$GeSb_2Te_4$	Rock Salt (metastable)	4.75
		Hexagonal (stable)	5.43
	$GeSb_2Te_4$	Rock Salt (metastable)	4.80
	$Ag_3In_4Sbn_{76}Te_{17}$	Cubic	4.93
	GeTe	Rock Salt (metastable)	5.00
	$Ge_4Sb_1Te_5$	Rock Salt (metastable)	5.10

materials with compositions along the GeTe-Sb_2Te_3 pseudo-binary line in the next chapter.

3 Amorphous Short-Range Order and Contrast

$Ge_2Sb_2Te_5$ is the most widely used phase change material. Its pronounced optical contrast between the amorphous and crystalline phase is compared with GaAs as a prototype for covalently bonded alloy in Fig. 6. The amorphous phase of GaAs follows the dielectric functions of the crystalline phase except for a damping of the sharper features that are smeared-out to some extent by deviations from perfect order. Nevertheless the relative variation is on the order of 25 %. The situation is totally different for the phase change alloy $Ge_2Sb_2Te_5$ where differences as large as 150 % appear. Such an extraordinary difference in optical properties of PC-alloys can only be explained in terms of large structural changes appearing on short length-scales as they determine the electronic and optical structure. Figure 7 shows the Fourier-transformed EXAFS-Signal for GaAs and $Ge_2Sb_2Te_5$ in the amorphous and crystalline phase. The signal measured at the k-resonance of Ga, As, Ge or Te gives the pair-correlation and thus the short-range order in the vicinity of the respective atoms. The difference in the short-range order between the amorphous and crystalline phase is much lower for the GaAs-reference than

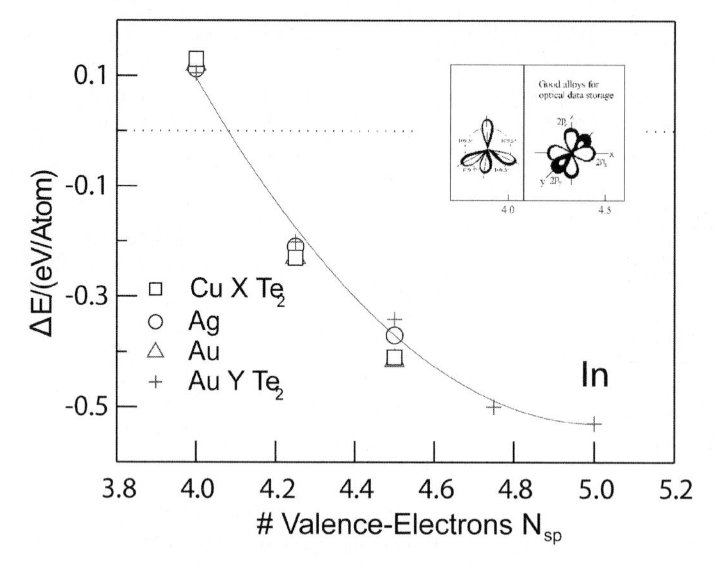

Fig. 5. Energy difference between rock salt - and chalcopyrite structure, calculated by DFT for different model-alloys with varying number of valence electrons but constant ratio of constituents. X = Cd, In, Sn, Sb; Y = Ga, Ge, As, Sb; see [5] for details

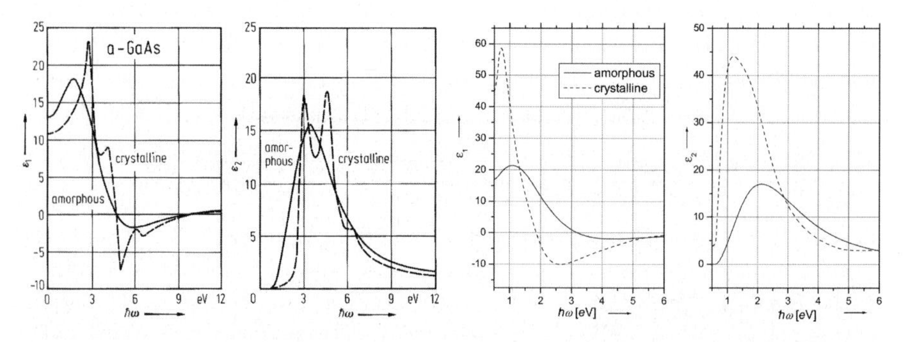

Fig. 6. Real and imaginary part of the dielectric functions over energy for GaAs from Landolt Börnstein and $Ge_2Sb_2Te_5$ (*right*) in the crystalline and amorphous phase, respectively. To compare the difference for both states note the different scales

for the recent measurements of *Kolobov* et al. on $Ge_2Sb_2Te_5$ [6]. Kolobov suggested, that Ge switches from an octahedral to a tetrahedral coordination on amorphisation. Based on that model, we chose $Ge_1Sb_2Te_4$ for calculation because of the smaller unit-cell as compared to $Ge_2Sb_2Te_5$. It can be modeled by an array of 64 sites as shown in Fig. 8. We started out from a rock salt structure as described above for the crystalline- and a spinel structure to represent the short range ordering of the amorphous phase. That spinel structure

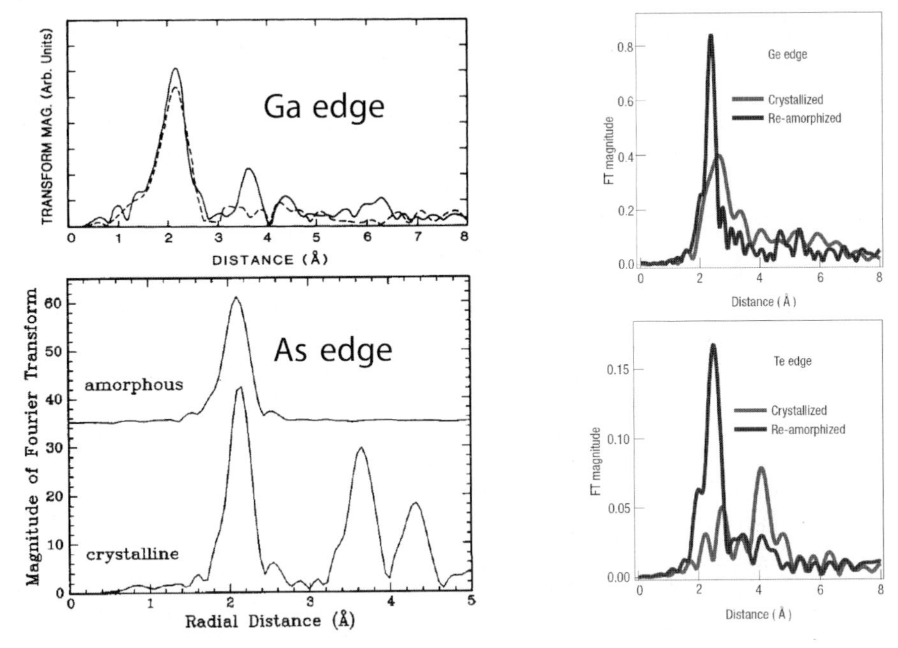

Fig. 7. Compilations of Fourier-transformed EXAFS-signals for the Ga- and As-edge of GaAs (*left*) [7,8] and the Ge- and Te- edge of $Ge_2Sb_2Te_5$ (*right*) [6] in the crystalline and amorphous phase from *Rigday*, *Bouldin*, *Kolobov* and their respective coworkers

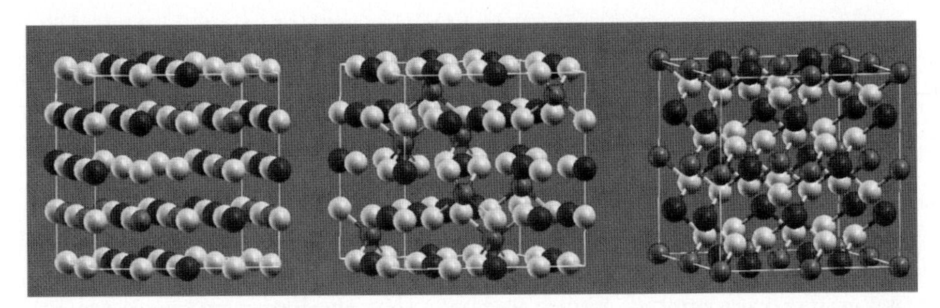

Fig. 8. Structures obtained after molecular dynamics relaxation of the ideal rock salt-, spinel- and chalcopyrite structures used as an input for the DFT-calculations of the ground-state energy [9]

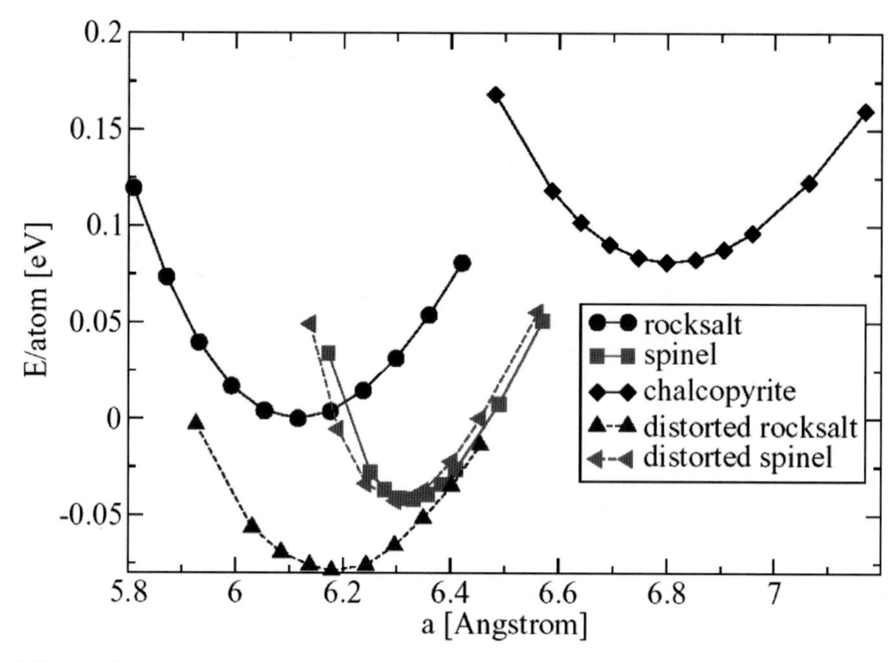

Fig. 9. Ground-state energy vs. lattice constant for the ideal and relaxed rock salt-, spinel- and chalcopyrite structure as given in Fig. 8. Due to its negligible energetic and structural relaxation the ideal chalcopyrite-values are shown only

was chosen because it can be thought of as a mixture of the rock salt- and the chalcopyrite structures typical for the chalcogenides investigated. It offers tetrahedral arrangements for the Ge atoms like a chalcopyrite- but maintains octahedral coordinations for Sb and Te like a rock salt cell.

We varied the lattice constants and allowed the atoms to relax from their ideal initial positions for all arrangements to minimise the DFT-energies [9] and plotted the results in Fig. 9. The chalcopyrite arrangement is far off in energy. As relaxation did not show a visible effect for the chalcopyrite that points were omited. Figure 9 shows, that the lowest energy is reached for a slightly relaxed rock salt structure at a lattice constant of about 6.2 while the relaxed spinel structure has a energy minimum at about 6.3 that is 40 meV/atom higher only in energy. The chalcopyrite arrangement is far off in energy. As relaxation did not show a visible effect for the chalcopyrite its relaxation is not plotted in Fig. 9. The shift to higher lattice constants and energies in the spinel-like arrangement is well in line with the comparatively large density loss of the cubic crystalline structures of phase change alloys of about 5 % as compared to 2 % observed for the chalcopyrite alloys. It fits to recent differential scanning calorimetry-results as well [11]. Finally that density change combined with the transition from octahedral to tetrahedral coordination of the Ge-atoms on amorphisation gives a plausible explanation

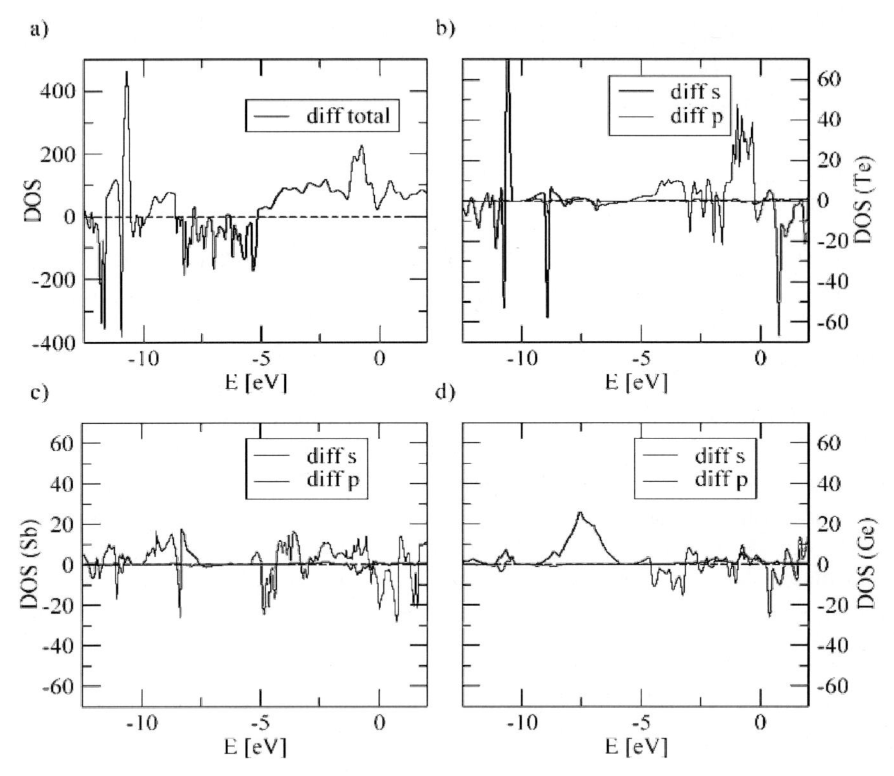

Fig. 10. Difference of the density of states between relaxed rock salt- and spinel structure for $Ge_1Sb_2Te_4$ for all elements (**a**) and the *s*- and *p*-bands of Te (**b**), Sb (**c**) and Ge (**d**)

for the strong optical contrast for the Ge-containing phase change alloys with cubic crystalline structures in Table 1. Interestingly, the structural change of the Ge-coordination is related to differences in the electronic structure of the Te- and Sb-bands rather than in Ge. Figure 10 shows the difference of the density of states between the relaxed rock salt- and spinel structures for all atoms and for the *s*- and *p*-bands for each element. The difference of the overall density of states shows an increase just below the Fermi-level (0 down to -5 eV), which fits to the semiconductor-metal-transition and the optical contrast in the visible range observed on crystallisation. This difference can be attributed mostly to the *p*-orbitals of Te, as Ge- shows a broad change in the *s*-band well below the energies relevant to optical properties only but almost no difference in the *p*-band at all. The changes in the Sb *s*- and *p*-band are in between the extremes Ge and Te. However the stronger differences in Te are increased by its higher concentration and thus deliver the main contributions. The increase in the Te-*p*-band is easily identified in the overall difference in Fig. 10a and is responsible for the electronic and optical changes upon crystallisation/amorphisation. The outcome of our DFT-calculations agrees well with *Kolobovs* results on tetrahedral Ge-coordination in the amorphous phase [6] and gives insight into the role of the electronic states involved.

4 Conclusion

Crystalline chalcogenide alloys for PC-storage either assume chalcopyrite structures with a low- or cubic structures with a high optical contrast to their amorphous phase. Both crystalline structures can be understood in terms of their coordination numbers and predicted from stoichiometry by their average number of valence electrons. For the Ge-containing alloys that crystallise in octahedral coordination we suggest a spinel-like short range order for the amorphous phase with the Ge-atoms in fourfold coordination, that does not involve long-range diffusion on switching. That change in the Ge coordination on crystallisation to the cubic phases explains the pronounced difference in the opto-electronic properties. This model is backed up by the lattice expansion of the spinel phase, that could account for the higher density loss of the crystalline cubic phases of about 5 % as compared to about 2 % for the chalcopyrite phases on amorphisation. An important point to note for this system is the influence of the Te p-bands that prefer a second structure and thus push the system just to the onset of an instability between two arrangements that may be responsible for the balance between fast switching and stability. Future work should focus on kinetics to establish a similar understanding of the influence of stoichiometry on that field. From the structural point of view, other classes of phase change materials without Ge, e.g., are the next promising field to be investigated in their amorphous phase, as they might show a different switching mechanism.

References

[1] S. Ovshinsky: Phys. Rev. Lett. **21**, 1453 (1968)
[2] V. Weidenhof, I. Friedrich, S. Ziegler, M. Wuttig: J. Appl. Phys. **89**, 3168 (2001)
[3] M. Wuttig, R. Detemple, I. Friedrich, W. Njoroge, I. Thomas, V. Weidenhof, H. Woeltgens, S. Ziegler: J. Magn. Magn. Mat. **249**, 492 (2002)
[4] R. Detemple, D. Wamwangi, M. Wuttig, G. Bihlmayer: Appl. Phys. Lett. **83**, 2572 (2003)
[5] M. Luo, M. Wuttig: Adv. Mat. **16**, 439 (2004)
[6] A. Kolobov, P. Fons, A. Frenkel, A. Ankudinov, J. Tominaga, T. Uruga: Nat. Mat. **03**, 703 (2004)
[7] M. Rigday, C. Glover, G. Foran, M. Yu: J. Appl. Phys. **83**, 4613 (1998)
[8] C. Bouldin, R. Forman, M. Bell et al.: Phys. Rev. B **35**, 1429 (1987)
[9] W. Welnic, R. Detemple, C. Steimer, M. Wuttig: Nat. Mat. **05**, 56 (2006)
[10] R. Detemple: Dissertation RWTH-Aachen Germany (2003)

[11] M. Klein, RWTH-Aachen Germany, private communication (2006)
[12] J. Kalb et al.: Appl. Phys. Lett. **84**, 5240 (2004)
[13] J. Kalb et al.: J. Appl. Phys. **98**, 054910 (2005)

GaInAs/AlAsSb Quantum Cascade Lasers: A New Approach towards 3-to-5 μm Semiconductor Lasers

Quankui Yang, Christian Manz, Wolfgang Bronner, Christian Mann, Klaus Köhler, and Joachim Wagner

Fraunhofer-Institut für Angewandte Festkörperphysik (IAF)
Tullastrasse 72, 79108 Freiburg, Germany
quankui.yang@iaf.fraunhofer.de

Abstract. At present GaInAs/AlInAs based quantum cascade (QC) lasers represent the state-of-the-art with respect to the short-wavelength ($< 5\,\mu$m) performance of the QC laser concept. This performance, however, is intrinsically limited by the available conduction band offset of 0.5–0.7 eV, thus motivating research on materials combinations with larger band offsets, such as GaN/AlN and InAs/AlSb. A particularly attractive materials combination is GaInAs/AlAsSb grown lattice-matched on InP. It offers a Γ-point conduction band offset of ~ 1.6 eV, while the mature growth and processing technologies available for InP-based lasers can be used and the favorable properties of InP as a waveguide cladding material can be exploited. In this paper recent advances in GaInAs/AlAsSb QC laser research will be reviewed, leading to a maximum pulsed operating temperature of > 400 K for devices emitting at 4.6 μm and an impressive maximum peak output power of 8 W at 77 K (corresponding to a total power efficiency of 23 %) for a QC laser emitting at 3.7 μm. Furthermore, current limitations of the GaInAs/AlAsSb QC laser concept and challenges for future research are discussed.

1 Introduction

Conventional semiconductor diode lasers are based on the radiative recombination of two types of charge carriers, namely electrons and holes, making optical transitions across the fundamental bandgap of the semiconductor. Quantum cascade (QC) lasers [1], in contrast, are unipolar semiconductor light sources based on only one type of carriers, e. g., electrons, making transitions between confined energy levels within the conduction band. As the emission wavelength of a QC laser is determined not by the semiconductor bandgap, but by quantum confinement in a potential well formed by two different semiconductors acting as the well material and the barrier material, respectively, the emission of the QC lasers can be adjusted over a wide wavelength range for a given material system.

There exists a gap regarding the overall performance of practical infrared (IR) semiconductor lasers in the 3–5 μm wavelength range. This gap arises on one hand from the fact that the performance of diode lasers, based on

R. Haug (Ed.): Advances in Solid State Physics,
Adv. in Solid State Phys. **46**, 213–227 (2008)

interband transitions, rolls off for wavelengths exceeding 3 μm due to fundamental parasitic loss mechanisms such as Auger recombination and intervalence band absorption [2–4]. State-of-the-art QC lasers, on the other hand are not affected by the above loss mechanisms and exhibit an impressive overall performance for wavelengths $> 5\,\mu m$. But they show a decrease in maximum output power and operating temperature towards shorter wavelengths ($\lambda < 5\,\mu m$). The 3–5 μm wavelength interval, however, is of particular interest for molecular spectroscopy since characteristic absorption bands of, e. g., CO, N_2O, HCl, and CH_2O lie within that wavelength range, as well as for security applications such as directed infrared countermeasures.

The shortest possible emission wavelength of a QC laser is ultimately limited by the band offset between the barrier and quantum well material. In order to obtain sufficient electron confinement for the upper laser level in the active region, a higher conduction band offset, i. e., a larger quantum well depth, is necessary for short-wavelength (large energy separation) QC lasers. The above mentioned inferior performance of QC lasers in the short-wavelength regime ($\lambda < 5\,\mu m$) [5–8] compared to that of QC lasers emitting at longer wavelengths ($6 - 10\,\mu m$) [9–11] is mainly due to the lack of a sufficient confinement of the electrons in the upper laser level. Therefore there is ongoing research on materials combinations with a larger conduction-band-offset for the realization of high-performance QC lasers operating at wavelengths shorter than 5 μm.

The first material system used to fabricate QC lasers is the $Ga_{0.47}In_{0.53}As/Al_{0.48}In_{0.52}As$ material combination grown lattice-matched on InP substrates [1], offering a conduction-band offset (quantum well depth) of 0.51 eV. In order to increase the conduction-band offset to achieve sufficient electron confinement for short-wavelength QC lasers, *Faist* et al. [5] adopted compressively strained $Ga_xIn_{1-x}As$ ($x < 0.47$) layers as the quantum wells, compensated by tensile strained $Al_yIn_{1-y}As$ ($y > 0.48$) layers for the barriers. By doing so, as well as adjusting the thickness of the individual layers, these authors obtained a conduction-band offset of up to 0.74 eV while strain balancing the whole QC layer sequence to achieve a net strain close to zero. As a result, they succeeded in fabricating QC lasers emitting at a wavelength of 3.5 μm and operating up to 280 K [5]. The concept of strain-compensation has been used subsequently also by other groups for the fabrication of short-wavelength QC lasers [6–8, 11–14]. Another approach to achieve strain compensation, very similar to that described above, is to insert highly strained InAs and AlAs layers into the otherwise lattice-matched GaInAs and AlInAs layers of an overall strain-balanced GaInAs/AlInAs layer sequence [15–17]. This way the effective quantum well depth can be increased up to ~ 0.7 eV while keeping the majority of the whole structure lattice-matched as well as the total net strain close to zero.

Searching for an alternative well/barrier material combination with a larger conduction-band offset than the above mentioned strain-compensated GaInAs/AlInAs heterostructures, several groups investigated the GaN/AlGaN material combination, which theoretically offers a conduction-band offset of up to several eV. *Gmachl* et al. have reported intersubband absorption in GaN/AlGaN quantum wells (QW), peaking at wavelengths which range from 4.2 μm down to 1.35 μm [18–20]. Independently, *Hofstetter* et al. have achieved intersubband absorption ranging from 1.4 μm up to 5.1 μm for the same material combination [21–23]. However, due to the difficulty in the growth of high-quality GaN/AlGaN heterostructures as well as lack of appropriate substrates for this material system, no QC lasers have been demonstrated up to now based on group III-nitrides.

Another material combination offering a high conduction-band offset is InAs/AlSb, grown either on GaSb substrates [24] or on InAs substrates [25–29]. InAs/AlSb offers a conduction-band offset of 2.1 eV, which is in principle sufficient for the fabrication of QC lasers covering the 3–5 μm atmospheric transparency window. In 2001, Becker et al. [24] reported electroluminescence in the wavelength range 3.7–5.3 μm for InAs/AlSb quantum-cascade structures grown on GaSb substrates. The first InAs/AlSb-based QC lasers, demonstrated by *Ohtani* and *Ohno* [27, 28], however, operated at a much longer wavelength of 10 μm. These lasers were grown on InAs substrates. Recently, the emission wavelength of InAs/AlSb based QC lasers has been pushed towards short wavelengths (4.5 μm). These lasers operate up to room temperature [26]. A major drawback of the InAs/AlSb material system is the narrow bandgap of InAs of only 0.354 eV at room temperature, which corresponds to a wavelength of $\lambda = 3.5$ μm. Thus parasitic band-to-band absorption hinders the realization of InAs/AlSb QC lasers emitting at a wavelength shorter than 3.5 μm. As a matter of fact, no InAs/AlSb QC lasers emitting at wavelengths shorter than 4 μm have been reported so far. Another challenge of the InAs/AlSb material system in particular when grown on InAs substrates is the lattice mismatch between the well material InAs and the barrier material AlSb. This problem, however, can be handled by growing specially designed strain-compensating interface layers between the well and the barrier. Regarding vertical waveguiding, the InAs/AlSb QC lasers demonstrated so far rely on the plasmon enhanced waveguide concept employing highly doped n-InAs cladding layers [26]. But, as discussed above, this approach limits the short-wavelength capability because of 1. strong band-to-band absorption in the InAs for wavelengths below 3.5 μm, and 2. the reduced effectiveness of this waveguiding concept at shorter wavelengths. An alternative approach is the use of dielectric AlGaAsSb cladding layers. But for this material achieving a sufficient n-type conductivity remains a major issue. The combination of these limitations will certainly have an adverse effect on the future development and eventual applications of InAs/AlSb based QC lasers.

Compared to all the above-mentioned alternative material systems, GaInAs/AlAsSb offers the following advantages for the fabrication of short wavelength QC lasers: 1. GaInAs/AlAsSb heterostructures can be grown lattice-matched on InP substrates, which helps to maintain a high crystalline quality of the whole epitaxial layer stack required for the fabrication of QC lasers, which typically contain several hundred of individual layers only a few nm in width. 2. The lattice-matched $Ga_{0.47}In_{0.53}As/AlAs_{0.56}Sb_{0.44}$ materials combination has a band offset at the direct conduction band edge (Γ-valley) of about 1.6 eV [30]. This large band offset is sufficient for the fabrication of short-wavelength QC lasers emitting in the 3–5 μm wavelength range. 3. An active region composed of GaInAs/AlAsSb can easily be combined with the established InP-waveguide technology, because the refractive index of InP is lower than the GaInAs/AlAsSb active region. Additionally, device fabrication benefits from the mature processing technology available for InP-based lasers. These three features make the GaInAs/AlAsSb materials combination almost an ideal candidate for the fabrication of QC lasers emitting in the 3–5 μm atmospheric transparency window.

The first intersubband emission from GaInAs/AlAsSb QC structures was demonstrated by *Revin* et al. [31]. They observed electroluminescence at 10 K from QC structures emitting at wavelengths ranging from 4 μm to 5.3 μm in late 2003, and pushed the minimum electroluminescence wavelength down to 3.1 μm several months later [32]. Subsequently, in 2004, the same group demonstrated the first GaInAs/AlAsSb-based QC lasers emitting at a wavelength of $\lambda \sim 4.5\,\mu m$, operating in pulsed mode up to 240 K [33]. For a device with the size of $20\,\mu m \times 1.5\,mm$, they obtained a maximum peak power per fact of around 200 mW at low temperature (100 K). The characteristic temperature T_0, which describes the increase of threshold current density as a function of temperature, was found to be around 150 K [33].

In the present paper recent progress in InP-based GaInAs/AlAsSb QC lasers [34–39], achieved since their above mentioned first demonstration, will be reviewed. The organization of the paper is as follows: in Sect. 2 certain aspects of the molecular beam epitaxy (MBE) growth of GaInAs/AlAsSb laser structures as well as device processing will be discussed. In Sect. 3 the design and the resulting performance of GaInAs/AlAsSb QC lasers will be described. Finally, in Sect. 4 we will summarize and discuss some important issues to be addressed in future work on GaInAs/AlAsSb QC lasers.

2 Epitaxial Growth and Processing

For the growth of GaInAs/AlAsSb QC lasers it is instrumental to achieve high quality interfaces between adjacent GaInAs and AlAsSb layers. This task is by far not trivial as there is a change in the group V constituents from just As in GaInAs to As and Sb in AlAsSb across these interfaces. There are reports [30, 40, 41] that an increased incorporation of Sb into the GaInAs quantum well layer alters the potential profile of GaInAs/AlAsSb QWs grown by molecular beam epitaxy. An often used procedure to improve the interface quality is to interrupt the growth at each hetero-interface, allowing the growth surface to smoothen as well as to change group III and in particular group V fluxes [30]. Several groups have reported on the growth of multiple-QW (MQW) structures including complete QC lasers with growth interruptions at the GaInAs/AlAsSb interface varying between 10 s and 120 s [31, 40]. On the other hand QC lasers are normally composed of several hundred individual layers, each a few nanometer in thickness, and the growth time for each of these layers lies in the range of a few seconds only (typically between 5 and 15 s). Considering the large number of interfaces and thus the unfavorably large ratio between the accumulated duration of all growth interruptions and the actual growth time, the frequent growth interruptions will at least triple the total growth time and consequently also at least triple the background impurity concentration due to pick-up of residual contamination from the growth chamber. We therefore concentrated on establishing a procedure for the MBE growth of GaInAs/AlAsSb heterostructures without growth interruption [42].

All QC laser structures presented in the following were grown in a Varian Gen-II solid-source MBE system equipped with valved As- and Sb-cracker cells. Elemental Ga, In, and Al were used as the group III sources, and Sb_2 and As_2 were used as the group V species. The appropriate adjustment of the alloy composition for both ternary compounds started with establishing the growth of lattice-matched $Ga_{0.47}In_{0.53}As$ at a growth rate of about 1 μm/h. The Sb flux was adjusted such that $Ga_{0.47}In_{0.53}As$ and $AlAs_{0.56}Sb_{0.44}$, both lattice-matched to InP, could be grown at a growth rate of about 1 μm/h with a constant As flux for both GaInAs and AlAsSb. This led to an As/Sb beam equivalent pressure (BEP) ratio of about 3.

After MBE growth, the wafers were transferred into a metal-organic vapor phase epitaxy (MOVPE) system to grow Si-doped InP layers ($n = 5 \times 10^{17}\,\mathrm{cm}^{-3}$, 20 nm; $2 \times 10^{17}\,\mathrm{cm}^{-3}$, 1300 nm; $7 \times 10^{18}\,\mathrm{cm}^{-3}$, 1300 nm) serving as the upper waveguide cladding and contact layers. The lower waveguide cladding is formed by the InP substrate.

Following the MOVPE growth mesa waveguide lasers with different ridge widths ranging from 7 to 34 μm were fabricated by etching double-channels of approximately 7 μm width through the active region using chemically-assisted ion-beam etching (CAIBE). After depositing 350 nm of SiN_x as an isolation layer, windows were opened for top contact formation, and Ge/Ni/Au followed by Ti/Au was evaporated as the top contact. The substrate was then thinned down to around 110 μm, and finally the backside metallization (Ge/Ni/Au) was deposited.

The lasers were cleaved into 1–3 mm long bars leaving the facets uncoated. The laser bars were mounted substrate-side down onto gold-plated copper heatsinks and wire bonded. The mounted lasers were placed either on a thermo-electrical cooler (temperature range 270 to 400 K) or inside a continuous-flow cryostat (temperature range 77 to 270 K) for electro-optical characterization. The devices were driven by current pulses of 100 ns width at a repetition rate between 1 kHz and 5 kHz. The emitted light was collimated by an $f/1.6$ off-axis parabolic mirror and focused either onto a Fourier

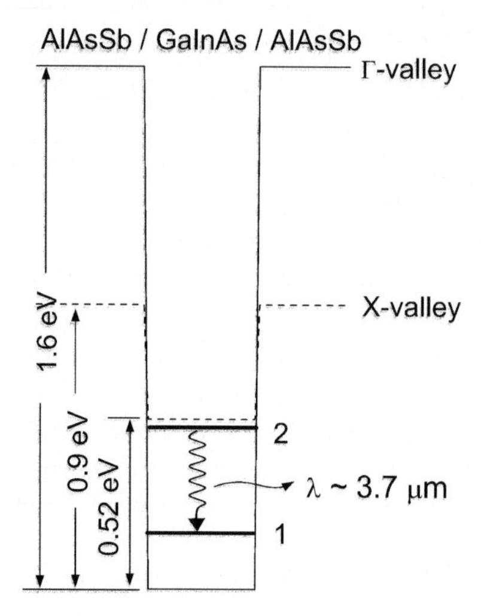

Fig. 1. Schematic conduction band diagram of an AlAsSb/GaInAs/AlAsSb quantum well lattice-matched to InP. The *solid* (*dashed*) line represents the Γ-valley (X-valley) energy position, respectively. Energy levels 2 and 1 are the lowest two confined Γ-states in the GaInAs quantum well. The 2-to-1 transition energy corresponds to an emission wavelength of $\lambda \sim 3.7\,\mu$m in the case level 2 is just below the X-valley position of the GaInAs quantum well

transform spectrometer equipped with a liquid-nitrogen cooled InSb detector for characterization of the emission spectra, or onto a calibrated room-temperature pyroelectric detector for direct power measurement.

3 GaInAs/AlAsSb Quantum Cascade Laser Design and Device Results

The demonstration of short-wavelength ($\lambda \sim 4.5 - 5.0\,\mu$m) GaInAs/AlAsSb and GaInAs/AlGaAsSb QC lasers by our group, operating in pulsed mode up to temperatures as high as 400 K [35,38], has encouraged us to explore the potential of this material system for the fabrication of QC lasers operating at even shorter wavelengths. Although the lattice-matched GaInAs/AlAsSb material system exhibits a large conduction band offset ΔE_c of about 1.6 eV at the direct conduction-band minimum (Γ-point), it is not yet clear to what extent this large Γ-valley conduction-band offset can be fully exploited for the confinement of electrons in the active region of practical QC lasers due to the following reasons (see Fig. 1): 1. The barrier material AlAsSb is an indirect bandgap semiconductor, with its indirect X-valley energy position lower than its direct Γ-valley energy position. This is expected to deteriorate the confinement of electrons once the initial state of the laser transition lies above the X-valley energy position of the AlAsSb barrier [43]. 2. The X-

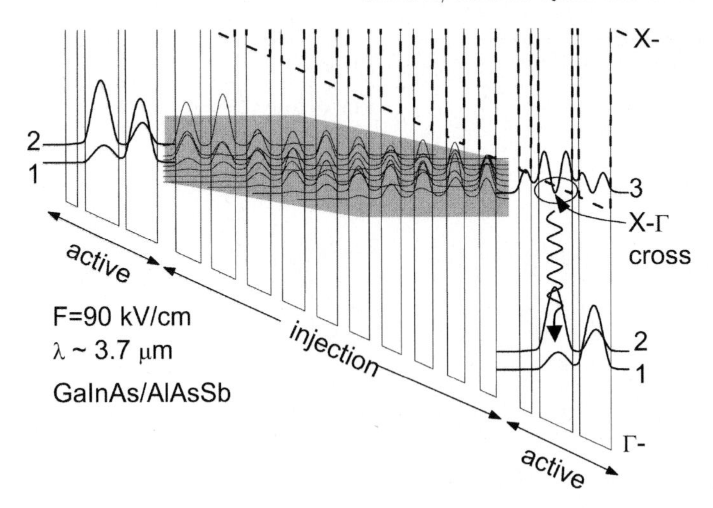

Fig. 2. Part of the conduction-band profile of two active regions connected by an injector and the squared moduli of the relevant wave functions (labeled as *3*, *2*, and *1*). The laser transition is indicated by the *wavy arrow* and the *shaded area* indicates the injection miniband. Levels 3 and 2 in this vertical-transition design essentially represent levels 2 and 1 in Fig. 1, respectively. Note that level 3 is exactly below the X-valley energy position in the transition-occurring quantum well

valley energy position of the GaInAs quantum well is even lower than the X-valley energy position of the AlAsSb barrier [44]; this may further limit the performance of GaInAs/AlAsSb QC lasers at shorter wavelengths as a loss of electrons from the initial laser level might occur through X–Γ scattering once this level is located above the X-valley energy position in the GaInAs quantum well [45]. Therefore, it appears to be crucial to demonstrate and analyze short-wavelength QC lasers, in which the initial laser level lies just below the X-valley energy position of the GaInAs quantum wells [17], prior to attempting GaInAs/AlAsSb QC laser operation at even shorter wavelengths. The corresponding emission wavelength at this critical point is, as illustrated in Fig. 1, calculated to be around $\lambda \sim 3.7\,\mu$m, when including conduction-band nonparabolicity [46].

The design of the present short-wavelength GaInAs/AlAsSb QC lasers is based on triple quantum-well vertical transition active regions [47] connected by appropriate injection regions. The latter also act as Bragg reflectors creating minigaps to prevent electrons escaping out of the laser initial state of the active regions [48]. Part of the conduction band profile containing two active regions connected by an injection region is shown in Fig. 2. The solid line and the dashed line in Fig. 2 represent the direct Γ-valley energy position and the indirect X-valley position, respectively. As discussed above, the initial state of the laser transition (level 3 in Fig. 2) is designed to be just below the X-valley energy position of the GaInAs quantum well where the lasing transition oc-

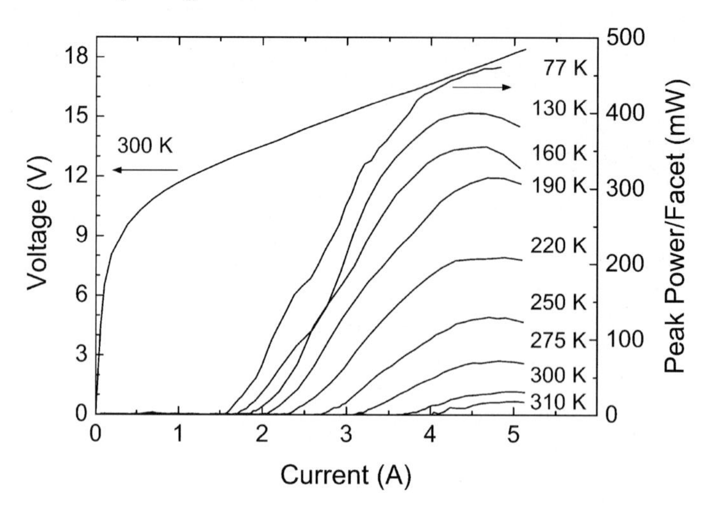

Fig. 3. Light output versus injection current characteristics at various heat-sink temperatures (*right-hand scale*), as well as voltage-current dependence at 300 K (*left-hand scale*) for a laser with the size of 14 μm × 3.0 mm operated in pulsed-mode (100 ns, 1 kHz)

curs, corresponding to the calculated emission wavelength $\lambda \sim 3.7\,\mu$m at the operating electric field of $F = 90\,$kV/cm. The layer sequence of one period of the GaInAs/AlAsSb structure, in nanometers, from left to right starting from the injection barrier (left-most layer) is: 2.5/1.4, 0.8/3.8, 0.7/3.6, 2.0/2.9, 1.1/2.8, 1.2/2.7, 1.2/2.6, 1.2/2.4, 1.2/**2.3**, 1.3/**2.2**, 1.4/**2.1**, 1.5/**2.0**, 1.6/1.9. The layers in bold face are Si doped to $n = 3 \times 10^{17}$ cm^{-3}. The dipole matrix element of the optical transition in this design amounts to 1.05 nm, and the calculated optical phonon scattering lifetimes are $\tau_{31} = 3.58$ ps and $\tau_{32} = 2.93$ ps. Energy level 1 is designed to be one longitudinal-optical phonon energy below energy level 2 to rapidly depopulate the final laser state. As a result τ_{21} is calculated to be only 0.38 ps $\ll \tau_{32}$, enabling population inversion between the initial and the final laser state. The entire gain region of the present QC lasers is composed of 30 periods of alternating active and injection regions.

The present short-wavelength QC lasers show pulsed-mode lasing up to 310 K at a wavelength of $\lambda \sim 3.7 - 3.9\,\mu$m. Figure 3 shows typical light output versus injection current dependencies for a device with a ridge size of 14 μm × 3.0 mm recorded at various heat-sink temperatures, as well as the voltage-current characteristic at 300 K. The maximum peak power per facet amounts to 460 mW (31 mW) at 77 K (300 K) .

The threshold current increases from 1.63 A at 77 K to 3.72 A at 300 K and 4.00 A at 310 K, corresponding to threshold current densities of 3.90 kA/cm^2, 8.83 kA/cm^2, and 9.56 kA/cm^2, at the respective temperature. The slope efficiency of the device, considering light emission from only one as-cleaved

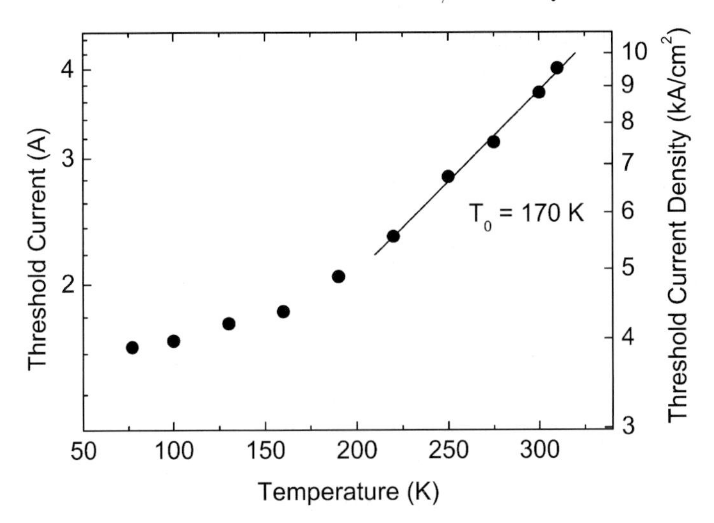

Fig. 4. Threshold current in pulsed-mode (100 ns, 1 kHz) as a function of heat-sink temperature at various temperatures between 77 K and 310 K (*dots*). The *solid line* represents an exponential fit according to the empirical relation $I_{\mathrm{th}} \sim \exp(T/T_0)$, yielding a characteristic temperature T_0 of 170 K in the temperature range 220 K to 310 K

facet, is 197 mW/A, 33 mW/A, and 24 mW/A at 77 K, 300 K, and 310 K, respectively. The threshold current as a function of heat-sink temperature is plotted in Fig. 4. The exponential fit to the measured data yields a characteristic temperature T_0 of 170 K in the temperature range between 220 K and 310 K. It has to be noted that although the lasers exhibit a relatively high characteristic temperature T_0, the threshold current density of the lasers remains still high compared to GaInAs/AlInAs based QC lasers emitting at comparable wavelengths [49].

Taking the computed values for τ_{32}, τ_{31}, τ_{21}, and z_{32} and estimating $2\gamma_{32} = 43\,\mathrm{meV}$ at 77 K from the experimentally determined electroluminescence linewidth [34], we calculate the gain coefficient g following [47], assuming 100 % injection efficiency into level 3,

$$g = \tau_3 \left(1 - \frac{\tau_{21}}{\tau_{32}}\right) \frac{4\pi e |z_{32}|^2}{\varepsilon_0 L_{\mathrm{p}} n_{\mathrm{eff}} \lambda} \cdot \frac{1}{2\gamma_{32}}\,, \tag{1}$$

where $\tau_3^{-1} = \tau_{32}^{-1} + \tau_{31}^{-1}$, e is the unit charge, ε_0 is the vacuum dielectric constant, $L_{\mathrm{p}} = 50.4\,\mathrm{nm}$ is the length of one period, $n_{\mathrm{eff}} = 3.3$ is the effective refractive index of the waveguide, and $\lambda = 3.8\,\mu\mathrm{m}$ is the emission wavelength. The above values result in $g = 8.1\,\mathrm{cm/kA}$. This relatively low gain coefficient is mainly due to the large electroluminescence linewidth. To derive values of the threshold current density from this gain coefficient, we assumed waveguide losses of around $8\,\mathrm{cm}^{-1}$ and calculated a mirror loss of $\alpha_{\mathrm{m}} = 4.2\,\mathrm{cm}^{-1}$ for a 3.0 mm long cavity with as-cleaved facets. The confinement factor Γ

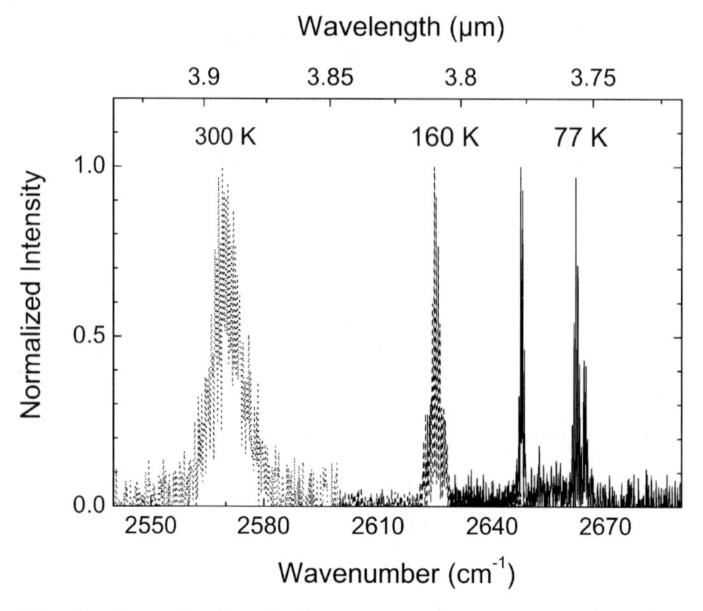

Fig. 5. Normalized, pulsed emission spectra at 77 K (*solid*), 160 K (*dashed*) and 300 K (*dotted*), respectively. The spectra were taken at an injection current slightly above threshold at the respective temperatures. The device was driven in pulsed mode (100 ns, 5 kHz)

of the lasing mode with the entire stack of 30 periods of active regions and injectors is calculated to be 0.728. These values allow us to estimate the threshold current density at low temperature, using $J_{\mathrm{th}} = (\alpha_{\mathrm{w}} + \alpha_{\mathrm{m}})/(g\Gamma)$, to be $J_{\mathrm{th}} = 2.1\,\mathrm{kA/cm^2}$, which is lower than the experimentally determined low-temperature value of $3.9\,\mathrm{kA/cm^2}$. A possible reason for this discrepancy could be carrier leakage out of the initial laser state (level 3) into GaInAs conduction band satellite-valleys through X–Γ scattering. In order to verify this assumption, we performed the same analysis for a QC laser structure emitting at a longer wavelength of $\lambda \sim 4.5\,\mu\mathrm{m}$ (see [35]), for which the upper laser state (level 3) is much lower in energy than the GaInAs X-valley. These longer-wavelength QC lasers have the same material quality as the shorter wavelength ones and are based on an identical design [35] as the $3.7\,\mu\mathrm{m}$ QC lasers. The calculated threshold current density for the 25-period $\lambda \sim 4.5\,\mu\mathrm{m}$ QC laser is $2.3\,\mathrm{kA/cm^2}$ ($g = 8.6\,\mathrm{cm/kA}$, $\Gamma = 0.612$) at 77 K, in excellent agreement with the experimental value of $2.2\,\mathrm{kA/cm^2}$. This indicates, as already discussed at the beginning of this section, that carrier leakage out of the upper laser level due to the X–Γ intervalley scattering may constitute an essential limitation for GaInAs/AlAsSb QC laser emitting at wavelengths below $3.7\,\mu\mathrm{m}$.

Figure 5 shows temperature dependent lasing spectra of the present short-wavelength QC laser. At a heatsink temperature of 77 K there appear two

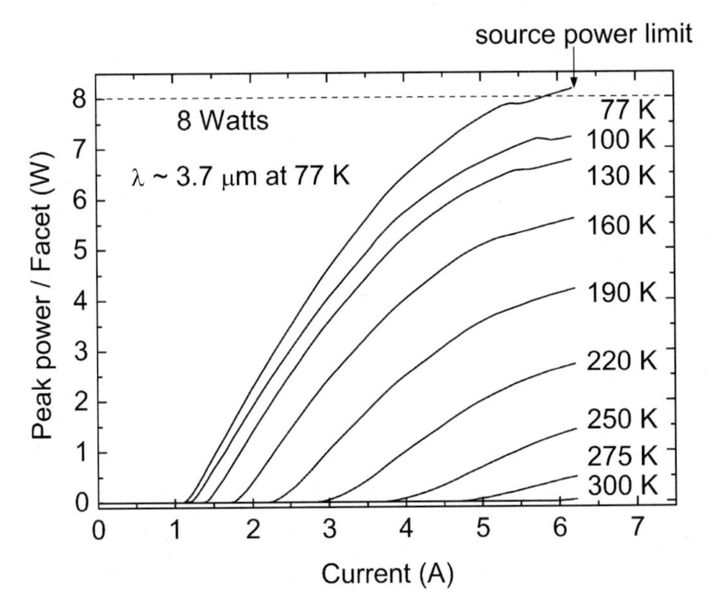

Fig. 6. Light output versus injection current characteristics at various heat-sink temperatures for a high-power $\lambda \sim 3.7\,\mu\text{m}$ GaInAs/AlGaAsSb QC laser with the size of $18\,\mu\text{m}\times\ 2.7\,\text{mm}$ operated in pulsed-mode (100 ns, 1 kHz). The maximum peak power exceeds 8 W per facet at 77 K

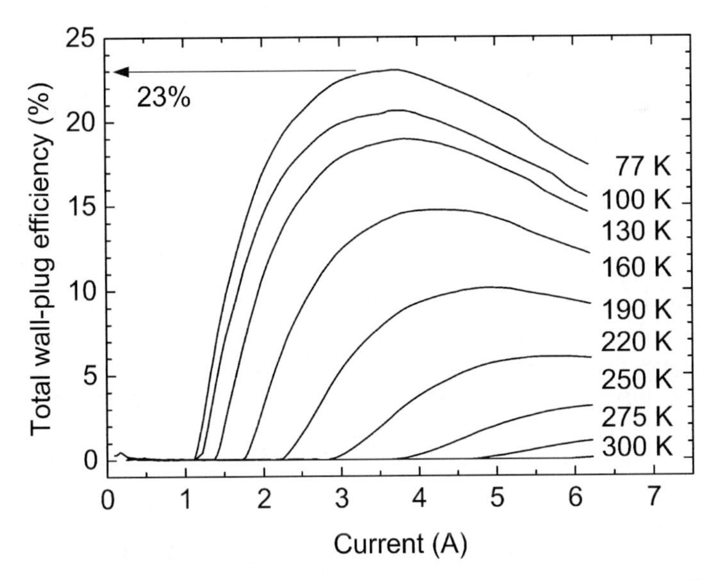

Fig. 7. Total wall-plug efficiency (taking into account emission from both uncoated facets) vs. current characteristics of a high-power $\lambda \sim 3.7\,\mu\text{m}$ GaInAs/AlGaAsSb QC laser. At 77 K, the maximum wall-plug efficiency is as high as 23 %

groups of lasing modes, with maximum positions of $2662\,\mathrm{cm}^{-1}$ ($\lambda \sim 3.76\,\mu\mathrm{m}$) and $2647\,\mathrm{cm}^{-1}$ ($\lambda \sim 3.78\,\mu\mathrm{m}$), respectively. The experimental peak wavelengths are close to the designed value of $\lambda \sim 3.7\,\mu\mathrm{m}$. At 160 K and 300 K the maximum position of the multi-mode spectrum shifts to $2625\,\mathrm{cm}^{-1}$ ($\lambda \sim 3.81\,\mu\mathrm{m}$) and to $2570\,\mathrm{cm}^{-1}$ ($\lambda \sim 3.89\,\mu\mathrm{m}$), respectively.

Using a slightly different design, short-wavelength ($\lambda \sim 3.7\,\mu\mathrm{m}$) GaInAs/AlGaAsSb QC lasers with a high peak output power ($> 8\,\mathrm{W/facet}$ at 77 K) have been realized. Figure 6 displays typical light output versus injection current dependencies for a device with a ridge size of $18\,\mu\mathrm{m}\times 2.7\,\mathrm{mm}$ recorded at various heat-sink temperatures. The maximum peak power per facet exceeds 8 W at 77 K. Even at the maximum drive current of 6.22 A (limited by the pulsed current source), the output power characteristics has not yet been reached the rollover point, as can be seen from Fig. 6.

In Fig. 7 the total wall-plug efficiency (taking into account light emitted from both uncoated facets) versus current characteristics is plotted at different heatsink temperatures. At 77 K, the maximum wall-plug efficiency reaches 23 %. The total differential quantum efficiency of amounts to 1605 %, which translates into a differential quantum efficiency of 54 % per stage, considering that the gain region is composed of 30 stages. With increasing temperature, however, the output power and the power efficiency drop rapidly and the threshold current increases more steeply for the high-power QC laser than for the device shown in Fig. 2 , resulting in a comparable high-temperature ($\sim 300\,\mathrm{K}$) performance of both QC laser structures.

4 Summary and Outlook

We have given a brief overview over various III-V semiconductor materials combinations capable of producing a sufficiently large conduction band offset to realize practical short-wavelength (3-to-5 µmspectral range) QC lasers. Then, focusing on the GaInAs/AlAsSb materials combination which can be grown lattice matched on InP substrates and thus allows one to take advantage of an already well developed InP-based waveguide and processing technology, recent advances in the area of GaInAs/AlAsSb-on-InP QC lasers have been reviewed. These advances include the demonstration of GaInAs/AlAsSb QC lasers emitting at a wavelength as short as 3.7 µmwith a maximum low-temperature (77 K) peak output power in pulsed operation of 8 W (single-ended output, uncoated facets) and a maximum total power efficiency of 23 %.

Up to now there are no reports on QC lasers emitting at wavelengths below 3.4 µm[5]. Therefore one of the major tasks for future work on GaInAs/AlAsSb QC lasers will be to explore the potential of this material combination for the realization of QC lasers emitting at $\lambda < 3.4\,\mu\mathrm{m}$. As pointed out in Sect. 3, GaInAs/AlAsSb QC lasers emitting at wavelengths $< 3.7\,\mu\mathrm{m}$might suffer from X–Γ scattering in the GaInAs quantum well layer. This issue will have to be clarified by designing and testing of $\lambda < 3.4\,\mu$mQC structures. If future work will confirm X–Γ scattering as a major source for carrier loss in the upper laser level, simultaneous engineering of both the Γ-valley and the X-valley conduction bandedge profile will become a major task for the realization of $\lambda < 3.4\,\mu$mGaInAs/AlAsSb QC lasers. This way it should become possible to extend the short-wavelength limit for GaInAs/AlAsSb to wavelengths close to $3\,\mu$m.

Another formidable task for the future will be to improve on the so far poor continuous-wave (cw) operating performance of GaInAs/Al(Ga)AsSb QC lasers [39]. The main factor which limits the cw performance of present GaInAs/AlAsSb based QC lasers is the much higher threshold current density compared to that of GaInAs/AlInAs based QC lasers (e. g., $0.4\,\mathrm{kA/cm^2}$ at 77 K, $1.2\,\mathrm{kA/cm^2}$ at 300 K, [49]). The high threshold current density mainly results from the large width of the gain spectrum, which can be inferred >from the comparatively broad intersubband electroluminescence spectrum observed for GaInAs/AlAsSb QC structures [34]. In order to reduce the high threshold current density, future work will have to focus on a further optimization of the MBE growth of GaInAs/Al(Ga)AsSb materials combination, aiming at the reduction of interface roughness and thus inhomogeneous broadening of the gain spectrum.

Acknowledgements

The authors would like to thank H. Menner for MOVPE overgrowth, P. Hiesinger, L. Kirste, M. Maier, T. Fuchs, F. Windscheid for material characterization, K. Schäuble and G. Kaufel for help with processing the lasers, B. Raynor, K. Schwarz, R. Moritz for technical support, F. Fuchs for help with laser characterization, and G. Weimann for encouragement and continuous support. Financial support by the European Community under the project "ANSWER" is gratefully acknowledged.

References

[1] J. Faist, F. Capasso, D. L. Sivco, C. Sirtori, A. L. Hutchinson, A. Y. Cho, Science **264**, 553 (1994).
[2] J. G. Kim, L. Shterengas, R. U. Martinelli, G. L. Belenky, D. Z. Garbuzov, W. K. Chan, Appl. Phys. Lett. **81**, 3146 (2002).
[3] E. P. O'Reilly and A. Adams, IEEE J. Quantum Electron. **30**, 366 (1994).
[4] C. Sirtori, H. Page, C. Becker, V. Ortiz, IEEE J. Quantum Electron. **38**, 547 (2002).
[5] J. Faist, F. Capasso, D. L. Sivco, A. L. Hutchinson, A. Y. Cho, Appl. Phys. Lett. **72**, 680 (1998).
[6] J. S. Yu, A. Evans, S. Slivken, S. R. Darvish, M. Razeghi, IEEE Photonics Technol. Lett. **17**, 1154 (2005).
[7] A. Evans, J. S. Yu, S. Slivken, M. Razeghi, Appl. Phys. Lett. **85**, 2166 (2004).

[8] J. S. Yu, S. Slivken, S. R. Darvish, A. Evans, B. Gokden, M. Razeghi, Appl. Phys. Lett. **87**, 041104 (2005).
[9] M. Beck, D. Hofstetter, T. Aellen, J. Faist, U. Oesterle, M. Ilegems, E. Gini, H. Melchior, Science **295**, 301 (2002).
[10] J. S. Yu, S. Slivken, A. Evans, S. R. Darvish, J. Nguyen, M. Razeghi, Appl. Phys. Lett. **88**, 091113 (2006).
[11] A. Friedrich, G. Scarpa, G. Boehm, M.-C. Amann, Electron. Lett. **40**, 1416 (2004).
[12] D. Hofstetter, M. Beck, T. Aellen, J. Faist, Appl. Phys. Lett. **78**, 396 (2001).
[13] R. Köhler, C. Gmachl, A. Tredicucci, F. Capasso, D. L. Sivco, S. N. G. Chu, A. Y. Cho, Appl. Phys. Lett. **76**, 1092 (2000).
[14] N. Ulbrich, G. Scarpa, A. Sigl, J. Roßkopf, G. Böhm, G. Abstreiter, M.-C. Amann, Electron. Lett. **37**, 1341 (2002).
[15] Q. Yang, Ch. Mann, F. Fuchs, R. Kiefer, K. Köhler, N. Rollbühler, H. Schneider, J. Wagner, Appl. Phys. Lett. **80**, 2048 (2002).
[16] Q. Yang, Ch. Mann, F. Fuchs, K. Köhler, W. Bronner, J. Crystal Growth **278**, 714 (2005).
[17] M. P. Semtsiv, M. Ziegler, S. Dressler, W. T. Masselink, N. Georgiev, T. Dekorsy, M. Helm, Appl. Phys. Lett. **85**, 1478 (2004).
[18] C. Gmachl, H. M. Ng, A. Y. Cho, Appl. Phys. Lett. **77**, 334 (2000).
[19] C. Gmachl, H. M. Ng, S. N. G. Chu, A. Y. Cho, Appl. Phys. Lett. **77**, 3722 (2000).
[20] C. Gmachl, H. M. Ng, A. Y. Cho, Appl. Phys. Lett. **79**, 1590 (2001).
[21] D. Hofstetter, L. Diehl, J. Faist, W. J. Schaff, J. Hwang, L. F. Eastman, C. Zellweger, Appl. Phys. Lett. **80**, 2991 (2002).
[22] D. Hofstetter S. S. Schad, H. Wu, W. J. Schaff, L. F. Eastman, Appl. Phys. Lett. **83**, 572 (2003).
[23] D. Hofstetter, E. Baumann, F. R. Giorgetta, M. Graf, M. Maier, F. Guillot, E. Bellet-Amalric, E. Monroy, Appl. Phys. Lett. **88**, 121112 (2006).
[24] C. Becker, I. Prevot, X. Marcadet, B. Vinter, C. Sirtori, Appl. Phys. Lett. **78**, 1029 (2001).
[25] R. Teissier, D. Barate, A. Vicet, D. A. Yarekha, C. Alibert, A. N. Baranov, X. Marcadet, M. Garcia, C. Sirtori, Electron. Lett. **39**, 1253 (2003).
[26] R. Teissier, D. Barate, A. Vicet, C. Alibert, A. N. Baranov, X. Marcadet, C. Renard, M. Garcia, C. Sirtori, D. Revin, J. Cockburn, Appl. Phys. Lett. **85**, 167 (2004).
[27] K. Ohtani and H. Ohno, Jpn. J. Appl. Phys. Part 2, **41**, L1279 (2002).
[28] K. Ohtani and H. Ohno, Appl. Phys. Lett. **82**, 1003 (2003).
[29] K. Ohtani, K. Fujita, H. Ohno, Appl. Phys. Lett. **87**, 211113 (2005).
[30] N. Georgiev and T. Mozume, J. Appl. Phys. **89**, 1064 (2001).
[31] D. G. Revin, L. R. Wilson, E. A. Zibik, R. P. Green, J. W. Cockburn, M. J. Steer, R. J. Airey, M. Hopkinson, Appl. Phys. Lett. **84**, 1447 (2004).
[32] D. G. Revin, M. J. Steer, L. R. Wilson, R. J. Airey, J. W. Cockburn, E. A. Zibik, R. P. Green, Electron. Lett. **40**, 874 (2004).
[33] D. G. Revin, L. R. Wilson, E. A. Zibik, R. P. Green, J. W. Cockburn, M. J. Steer, R. J. Airey, M. Hopkinson, Appl. Phys. Lett. **85**, 3992 (2004).
[34] Q. Yang, C. Manz, W. Bronner, K. Köhler, J. Wagner, Electron. Lett. **40**, 1339 (2004).

[35] Q. Yang, C. Manz, W. Bronner, Ch. Mann, L. Kirste, K. Köhler, J. Wagner, Appl. Phys. Lett. **86**, 131107 (2005).
[36] Q. Yang, C. Manz, W. Bronner, K. Köhler, J. Wagner, Appl. Phys. Lett. **88**, 121127 (2006).
[37] Q. Yang, W. Bronner, C. Manz, B. Raynor, H. Menner, Ch Mann, K. Köhler, J. Wagner (to be published in Appl. Phys. Lett., 2006).
[38] Q. Yang, C. Manz, W. Bronner, L. Kirste, K. Köhler, J. Wagner, Appl. Phys. Lett. **86**, 131109 (2005).
[39] Q. Yang, W. Bronner, C. Manz, R. Moritz, Ch. Mann, G. Kaufel, K. Köhler, J. Wagner, IEEE Photon. Technol. Lett. **17**, 2283 (2005).
[40] N. Georgiev and T. Mozume, Appl. Phys. Lett. **75**, 2371 (1999).
[41] N. Georgiev and T. Mozume, J. Cryst. Growth **209**, 247 (2000).
[42] C. Manz, Q. Yang, K. Köhler, M. Maier, L. Kirste, J. Wagner, W. Send, D. Gerthsen, J. Crystal Growth **280**, 75 (2005).
[43] The X-valley position in the AlAsSb barrier has been determined by photoluminescence (unpublished result).
[44] I. Vurgaftman, J. R. Meyer, L. R. Ram-Mohan, J. Appl. Phys. **89**, 5815 (2001).
[45] J. Feldmann, J. Nunnenkamp, G. Peter, E. Göbel, J. Kuhl, K. Ploog, P. Dawson, C. T. Foxon, Phys. Rev. B **42**, 5809 (1990).
[46] C. Sirtori, F. Capasso, J. Faist, S. Scandolo, Phys. Rev. B **50**, 8663 (1994).
[47] J. Faist, F. Capasso, C. Sirtori, D. L. Sivco, J. N. Baillargeon, A. L. Hutchinson, A. Y. Cho, Appl. Phys. Lett. **68**, 3680 (1996).
[48] C. Sirtori, F. Capasso, J. Faist, D. L. Sivco, S. N. G. Chu, A. Y. Cho, Appl. Phys. Lett. **61**, 898 (1992).
[49] J. S. Yu, S. R. Darvish, A. Evans, J. Nguyen, S. Slivken, M. Razeghi, Appl. Phys. Lett. **88**, 041111, (2006).

Part V

Thin Films and Materials

Hydrostatic Pressure Effects in the Magnetocaloric Compounds $R_5(Si_xGe_{1-x})_4$

Cesar Magen[1], Luis Morellon[2], Pedro A. Algarabel[2], M. Ricardo Ibarra[2], Zdenek Arnold[3] and Clemens Ritter[4]

[1] CEMES-CNRS,
29, rue Jeanne Marvig, 31055 Toulouse, France
magen@cemes.fr
[2] Instituto de Ciencia de Materiales de Aragon, Universidad de Zaragoza-CSIC,
Pedro Cerbuna 12, 50009 Zaragoza, Spain
algarabe@unizar.es
[3] Institute of Physics, AS CR,
Na Slovance 2, 182 21 Prague 8, Czech Republic
arnold@fzu.cz
[4] Institute Laue-Langevin,
6, rue Jules Horowitz, 38042 Grenoble, France
ritter@ill.fr

Abstract. The rare earth intermetallic compounds $R_5(Si_xGe_{1-x})_4$ exhibit an extraordinary magnetostructural coupling which gives rise to an unprecedented variety of magnetoresponsive properties. The most exceptional is the discovery in $Gd_5(Si_xGe_{1-x})_4$ of the giant magnetocaloric effect, an extremely large magnetic-field-induced temperature and entropy change, which has promoted great advances in the field of the room temperature magnetic refrigeration based on these and other materials. Additionally, this potent coupling between the magnetic and crystallographic lattices opens the prospect of the modification of the physical properties of these alloys by externally modifying their volume cell/atomic distances, and thus the crystal structure and magnetic properties. Recent investigations have confirmed the strong impact of a moderate hydrostatic pressure on the magnetic and crystallographic properties of $R_5(Si_xGe_{1-x})_4$. In this work, we emphasize the newly discovered significance of pressure as a new control parameter of the physical properties of the 5:4 alloys by reviewing the most outstanding findings arising from the effect of pressure on the strong interplay between the crystallographic structure and magnetism in the compounds with R = Gd, Tb and Er.

1 Introduction to the $R_5(Si_xGe_{1-x})_4$ Alloys

The first $R_5(Si_xGe_{1-x})_4$ alloy (R = rare earth) was accidentally discovered by *Smith* et al. in 1966 as a remnant impurity phase during the synthesization of the compound Sm_5Ge_3 [1]. This was followed by a basic X-ray diffraction and magnetization characterization of the pure 5:4 silicides and germanides [2–4], which already revealed important details of their physical properties. Most of them presented a complex orthorhombic structure (space group *Pnma*), although some silicides presented monoclinic or tetragonal

R. Haug (Ed.): Advances in Solid State Physics,
Adv. in Solid State Phys. **46**, 231–243 (2008)

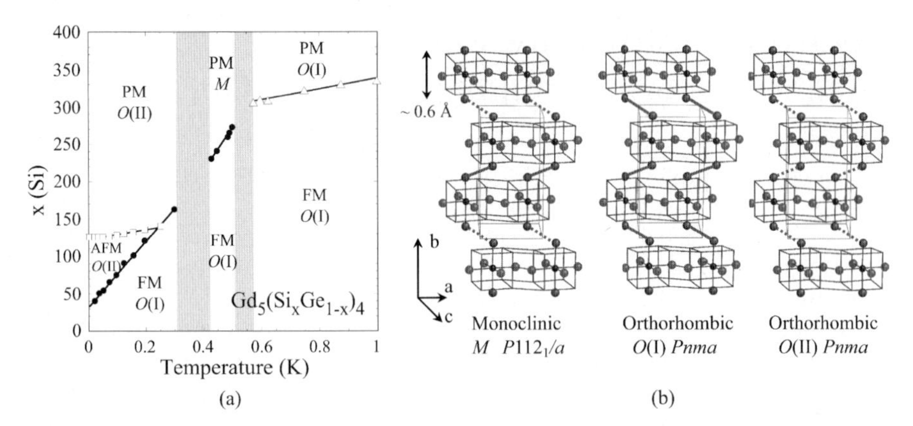

Fig. 1. The $Gd_5(Si_xGe_{1-x})_4$ series. **(a)** Magnetic-crystallographic phase diagram. **(b)** Representation of the possible crystal structures, where the formed X-X covalent bonds (*continuous red lines*) and broken pairs (*dashed red lines (online color)*) are indicated

distortions. *Holtzberg* et al. [4] studied the magnetic properties for R = Gd, Tb, Dy, Ho and Er and observed the clear trend of the silicides to be ferromagnets (FM) and the germanides antiferromagnets (AFM). This early work already indicated the incomplete solubility of the system Gd_5Si_4-Gd_5Ge_4, the different structures adopted by the silicides and germanides (otherwise presenting *Pnma* symmetry), and the existence of an unknown intermediate phase. Even more surprising was that the Curie temperature of Gd_5Si_4 was $T_C = 336$ K, more than 40 K higher that the undiluted pure metallic Gd ($T_C = 294$ K). Unfortunately, this system no longer attracted much attention in the forthcoming 30 years.

In 1997, *V. K. Pecharsky* and *K. A. Gschneidner* discovered the giant magnetocaloric effect (GMCE) in $Gd_5Si_2Ge_2$, renewing the interest in the 5:4 alloys [5]. Further studies revealed that this novel property could be tuned from 40 K to room temperature simply by modifying the Si/Ge ratio through the composition range $0 \leq x \leq 0.5$, see Fig. 1(a), opening a wide range of applications in the field of the magnetic refrigeration [6]. This promising technology is based on the change of temperature that a magnetic material experiments upon application of a magnetic field, a phenomenon called the Magnetocaloric Effect (MCE), and is considered as environmental friendly due to the absence of ozone depletion and greenhouse chemicals, and more energetically efficient than the conventional vapor compression refrigeration techniques [7]. The MCE in $Gd_5Si_2Ge_2$ was found to double the effect observed in the best candidate for a near room-temperature magnetic refrigerant in that time, which was metallic Gd, thus being labeled "giant".

The phenomenon of the GMCE in $Gd_5(Si_xGe_{1-x})_4$ is closely related to the existence of a reversible first-order coupled magnetic-crystallographic transformation between two sub-nanometer-thick layered structures [8]. A shear movement of these layers along the a axis, see Fig. 1(b), provokes a drastic volume change and the making (or cleavage) of interlayer covalent-like X-X (X = Si, Ge) bonds connecting consecutive layers [9]. In addition to the conventional RKKY indirect exchange between the $4f$ ions, the bonding of X-X pairs "switchs on" a new superexchange-like FM interaction mediated by this X-X pairs which strongly modifies the magnetic properties of the system [10,11], and produces a complex magnetic-crystallographic phase diagram, see Fig. 1(b), where magnetism and crystal structures are strongly correlated [12]. The system $Gd_5(Si_xGe_{1-x})_4$ presents three extended solid solution regions at room temperature: the Si-rich solid solution $0.575 \leq x \leq 1$ is FM and presents the orthorhombic Gd_5Si_4-type structure O(I) (space group $Pnma$) with all the slabs covalently connected; the intermediate phase $0.4 < x \leq 0.503$ is paramagnetic (PM) and crystallizes in a monoclinic M structure (space group $P112_1/a$) with only one half of the interlayer X-X pairs bonded; and the Ge-rich region $0 < x \leq 0.3$ is also PM and shows the Gd_5Ge_4-type structure O(II), also in the orthorhombic space group $Pnma$ with all the X-X bonds broken. A first-order magnetic-crystallographic transition on cooling in the composition range $0 < x \leq 0.504$ involves reforming the specific covalent X-X bonds and the onset of a common low-temperature O(I)-FM ground state for all compositions $0 < x \leq 1$ with all the layers covalently bonded [13–15]. This potent magnetostructural coupling enables this transformation to be reversibly induced by the application of a magnetic field, whose associated magnetic and structural entropy changes gives rise to the GMCE [8, 13, 14].

This strong interplay between magnetic and structural degrees of freedom has recently suggested the possibility of utilizing the hydrostatic pressure as a new control parameter of the physical properties of the 5:4 alloys. Due to this magnetostructural coupling, The pressure-induced reduction of the volume cell should induce remarkable modifications of the crystal structure and, by means of such a strong magnetostructural coupling, also of the magnetic interactions and magnetic structure, and in the transport properties. In the present work, we review our recent investigations on the pressure effects in the magnetostructural and magnetocaloric properties of the $R_5(Si_xGe_{1-x})_4$ alloys (R = Gd, Tb, Er).

2 Pressure Effects on the $Gd_5(Si_xGe_{1-x})_4$ Series

A comprehensive study of the effect of pressure on the magnetic and structural transition of the $Gd_5(Si_xGe_{1-x})_4$ alloys have yielded two general features. First, hydrostatic pressure induces the shift of the transition temperatures to higher values in all cases [13, 16–18], which is a remarkable fact itself, in contrast to the well-known tendency of pressure to destroy the FM interactions, especially among the $3d$ magnetic materials [19]. In addition, depending on the nature of the phase transition we are dealing with, a clear distinction can be done: the increase of the transition temperature upon pressure

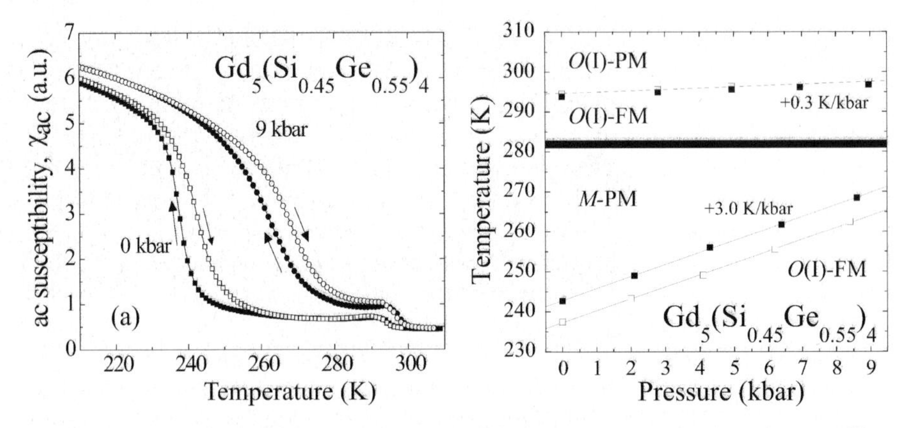

Fig. 2. (a) ac susceptibility as a function of hydrostatic pressure in polycrystalline $Gd_5(Si_{0.45}Ge_{0.55})_4$. **(b)** T-P phase diagram of the x = 0.45 compound and the impurity phase, x ≈ 0.55

is remarkably higher in the case of the magnetostructural transformations, whereas pure magnetic anomalies are poorly affected by pressure [16]. This fact evidences the importance of the existance of a crystal change with large volume contraction in amplifying the effect of pressure on the transformation. These phenomena are clearly illustrated in Fig. 2 by a magnetization experiment on a x = 0.45 polycrystalline specimen which summarizes both situations. The large jump in magnetization associated with the O(I)-FM↔M-PM phase transition of the compound x = 0.45 moves to higher temperatures at a rate of $dT_C/dP = +3.0\,K/kbar$. On the other hand, the small step at room-temperature signals the existance of a small amount of secondary phase, indexed as x ≈ 0.55, and experiments a pressure rate of +0.3 K/kbar. This result has been reproduced in single-phase Si-rich compounds. The PM-AFM second order transformation of the Ge-rich compounds also exhibits a low pressure rate ($dT_N/dP = +0.7\,K/kbar$). Further studies on single crystals (x = 0.5) have reproduced this general behavior for the magnetostructural anomaly, but with an even higher pressure rate (+4.7 K/kbar), revealing a strong dumping effect which might be caused by point defects and pinning at the grain boundaries of the polycrystals [18].

Recent studies in the parent compound Gd_5Ge_4 have revealed that volume cell reduction induced by hydrostatic pressure can, not only shift the magnetostructural transformation, but also can cause the onset of a new magnetic state and the nucleation of a crystallographic structure unexistent at ambient pressure [17]. This alloy is exceptional among the $Gd_5(Si_xGe_{1-x})_4$ system due to the absence of the O(I)-FM ground state in zero field and ambient pressure. Levin et al. demonstrated that the expected FM ground state can be reached by applying a magnetic field of ∼ 2 kOe in the O(II)-AFM state, through a first-order phase transition [20]; later the same was observed to occur with the O(I) crystal phase [21]. Furthermore, at very low temperatures this high

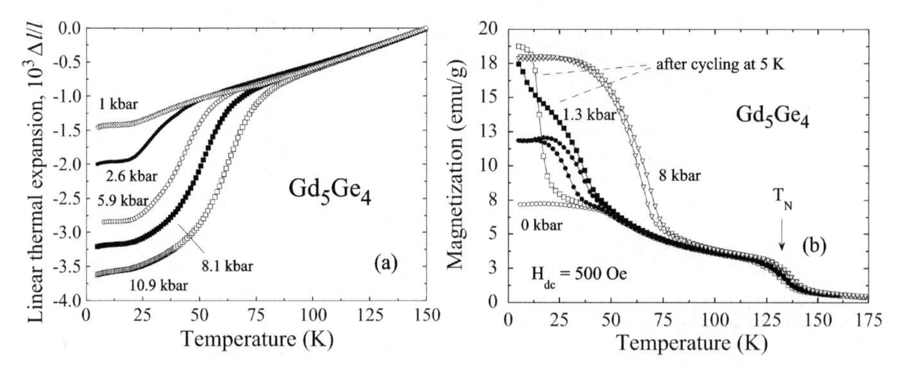

Fig. 3. **(a)** Linear thermal expansion collected on heating at several pressure values in Gd_5Ge_4. **(b)** Low-field magnetization of Gd_5Ge_4 as a function of pressure

field state O(I)-FM remained "arrested" after removal of the magnetic field due to freezing of the kinetics of the first-order phase transition.

We have demonstrated that the application of pressure at low temperature reduces the interlayer distances to a point in which the compound Gd_5Ge_4 experiments the reformation of the X-X bonds, the subsequent crystal structure transformation to the O(I) phase and the onset of three-dimensional FM state [17]. This pressure induced first-order O(II)-AFM→O(I)-FM phase takes place through an intermediate inhomogeneous state in which some regions of the sample volume in O(I)-FM state are nucleated within an O(II)-AFM matrix. This phase segregation scenario is illustrated in Fig. 3. The linear thermal expansion (LTE) experiment as a function of pressure depicted in Fig. 3(a) evidences that, from an ambient pressure curve with no anomaly, the increasing pressure induces a step-like anomaly. This jump signals the O(II)→O(I) transformation and progressively increases until saturation at ~ 11 kbar. This behavior is mimicked in the magnetization measurements shown in Figure 3(b), where the application of pressure gives rise to growing intermediate magnetization values between the virgin curve and the magnetic-field cycled and kinetically arrested saturation.

3 Hydrostatic Pressure Enhancement of the GMCE in the Alloy $Tb_5Si_2Ge_2$

The compound $Tb_5Si_2Ge_2$ has been reported to present the first deviation with respect to the Gd-like behavior among the 5:4 compounds. This compound was initially considered to experiment a first-order magnetic-structural transformation from a high temperature M-PM to a low temperature O(I)-FM state at $T_C = 105$ K [22], but a more detailed study of the transformation revealed that both transitions were not coupled; a difference of around 10 K exists between the magnetic ordering $T_C = 110$ K and the structural change

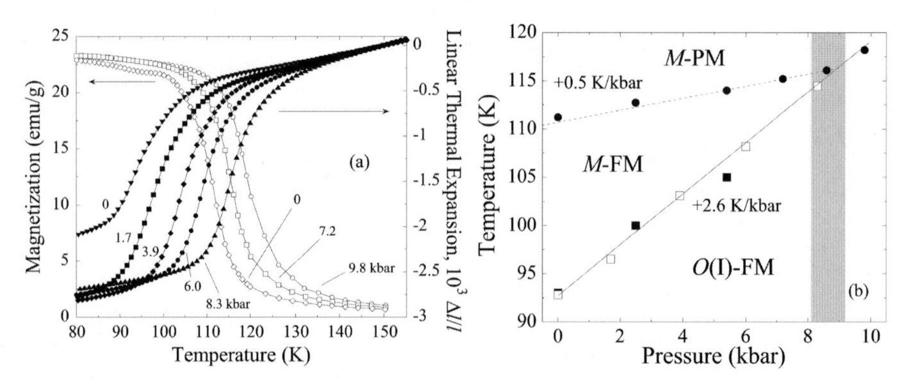

Fig. 4. (a) Pressure dependence of magnetization (*left axis*) and LTE (*right axis*) experiments on heating in $Tb_5Si_2Ge_2$. **(b)** T-P phase diagram elaborated from the magnetization (*solid symbols*) and the LTE (*open symbols*) temperature transitions. *Dashed* and *solid lines* depict the second-order and first-order phase boundaries, respectively. The gray area signals the region of uncertainty where both lines merge

$T_S = 100\,\mathrm{K}$, becoming the first experimental evidence of decoupling betwee the magnetic and structural transitions among the $R_5(Si_xGe_{1-x})_4$ [23].

Considering the pressure effects observed in the $Gd_5(Si_xGe_{1-x})_4$, it is reasonable to speculate about the influence of the different pressure rate of the magnetic and structural transformation in the decoupling of $Tb_5Si_2Ge_2$, and whether the application of pressure result in a recoupling of the anomalies. As can be seen in Fig. 4, magnetization and LTE experiments confirm that these transitions have a similar pressure rate as the Gd alloys, and a coincidence of the temperature transitions of the magnetic and the structural changes at around 8-9 kbar is predicted. The occurrence of this recoupling has been experimentally confirmed by neutron diffraction experiments upon hydrostatic pressure in the two-axis neutron powder diffractometer D20 at the Institute Laue-Langevin (ILL), in Grenoble, France. This experiment is presented in Fig. 5, where an angular range where pure magnetic reflections corresponding to the two crystallographic phases involved is shown, i.e. M and O(I) states. At 0 kbar, we observe a diffraction peak associated with the magnetic ordering of the M phase in the aproximate temperature range 110-100 K. However, this magnetic M reflection has dissapeared at 9 kbar and only magnetic peaks with an O(I) crystal structure remain, demonstrating that the application of pressure suppresses the intermediate M-FM state and enables a coupled magnetic-structural M-PM↔O(I)-FM transition [24].

As a consequence of this magnetostructural recoupling, the existence of a pressure enhancement of the MCE in $Tb_5Si_2Ge_2$ has been evidenced. One of the magnitudes that defines the MCE is the magnetic entropy change (ΔS_{mag}), which can be calculated by integrating the Maxwell relation

$$(dS/dH)_T = (dM/dT)_H \; .$$

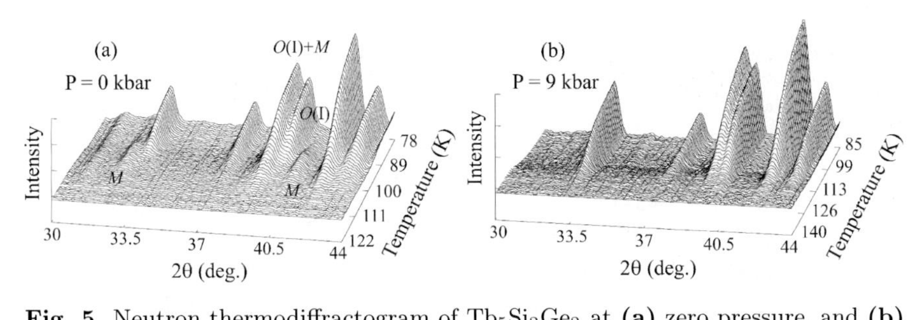

Fig. 5. Neutron thermodiffractogram of $Tb_5Si_2Ge_2$ at **(a)** zero pressure, and **(b)** 9 kbar in a selected angular range as measured on cooling in D20. The diffraction peaks labeled as O(I), M, and O(I)+M are purely magnetic in origin

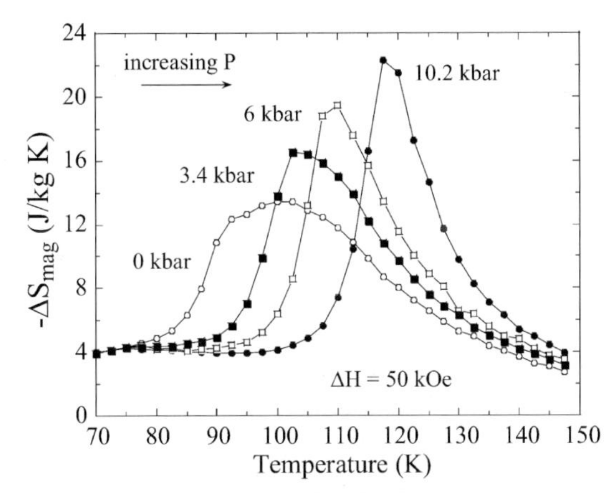

Fig. 6. MCE of $Tb_5Si_2Ge_2$ for a 50 kOe magnetic field change as a function of hydrostatic pressure (pressure values are at the maximum of the entropy change)

Figure 6 depicts the result of the MCE calculation from numerous magnetization isotherms measured in the vicinity of both transformations at several pressure values. This plot clearly demonstrates an enhancement of the MCE on increasing pressure from a low pressure situation with a small and broad peak related to the convolution of the partially overlapped entropy contributions of the two independent transitions to a large and narrow high pressure anomaly due to the merging of the entropy changes coming from both magnetic and crystallographic transformations. This exceptional result confirms the importance of a coupled first order magnetic-structural transformation in obtaining the optimal performance of a magnetocaloric material and points to a new procedure to improve the MCE capabilities of magnetic materials presenting crystallographic transformations and a strong magnetoelastic coupling.

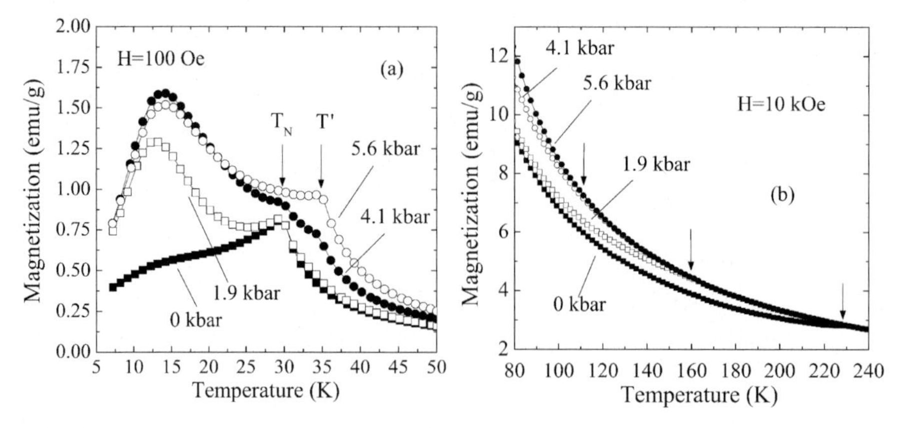

Fig. 7. Temperature dependence of the magnetization as a function of pressure upon heating in Er_5Si_4. **(a)** Low-field low-temperature magnetization. **(b)** High-field magnetization in the vicinity of the high-temperature O(I)↔M purely structural transition. The approximate transition temperatures are indicated with *arrows*

4 Novel Pressure Effects in Er_5Si_4

After the discovery of the decoupling phenomena in $Tb_5Si_2Ge_2$, a novel and more astonishing situation was found in the compound Er_5Si_4. Since its early characterization, this alloy was considered a simple FM with $T_C = 30$ K crystallizing in an O(I) structure [4]. However, a recent investigation of the temperature dependence of its crystal structure has revealed the occurrence of a reversible purely crystallographic transformation on cooling to a low-temperature M phase in the temperature range of 200-230 K [25, 26]. In addition to the enormous temperature span between the magnetic ($T_C = 30$ K) and crystallographic ($T_t = 200$-230 K) transitions, the order of the crystallographic phases is opposite to that conventionally observed below room temperature in other previously studied $R_5(Si_xGe_{1-x})_4$ compounds, where the room temperature M phase always transforms into the O(I) crystal structure on cooling.

In this situation, with a low volume phase [O(I)] at room temperature and a high volume state (M) at low temperature, the possibility of strongly affecting the behavior of the structural transformation by applying pressure and reducing the cell volume seems rather plausible [27].

The magnetization experiments performed in Er_5Si_4 are summarized in Fig. 7. At low temperatures, see Fig. 7(a), a complex pressure-induced process is observed. First, the growth of a low-temperature bump on increasing pressures is evident. Furthermore, in addition to the magnetic anomaly associated with the FM ordering at $T_C = 30$ K, the onset of a new magnetic anomaly is signalled by the appearance of an additional peak at around 35 K. Surprisingly, the size of this anomaly increases with pressure, whereas the relative

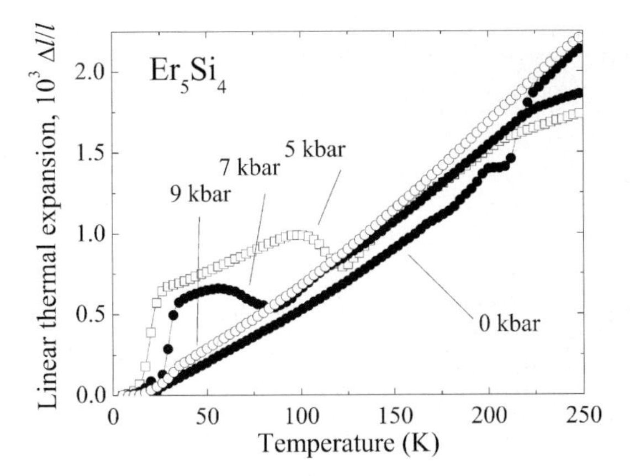

Fig. 8. LTE as a function of temperature under selected values of hydrostatic pressure (room temperature values)

weight of the low pressure peak at 30 K decreases. Eventually, at 5.6 kbar the initial peak has nearly disappeared in favor of the 35 K spike. This behavior suggests a progressive change of the magnetic state. Taking into account the coexistence of both peaks at intermediate pressure values, a first-order process might be consistent. The high-field magnetization measuremens depicted in Fig. 7(b) reveals that the effect of pressure in the crystallographic transition is also remarkable. As can be seen in the 0 kbar curve, T_t is signaled by a kink clearly observable in magnetization and, on increasing pressures, this rapidly decreases at an enormous pressure rate, $\mathrm{d}T_t/\mathrm{dP} \approx -23\,\mathrm{K/kbar}$.

The phenomenon found in the high-field magnetization is in perfect agreement with the LTE experiments shown in Fig. 8. In this plot, a clear and rapid decrease of the high-temperature anomaly associated to the O(I)↔M transition at an impressive rate of $\mathrm{d}T_t/\mathrm{dP} \approx -30\,\mathrm{K/kbar}$ is confirmed. The difference with respect to the value extracted from magnetization curves is surely associated with the clear uncertainty in the determination of the latter. Unexpectedly, a second and inverse first-order anomaly is induced by pressure at low temperatures, and shifts to higher temperatures at $\mathrm{d}T_{t2}/\mathrm{dP} = +6\,\mathrm{K/kbar}$. The opposite sign and similar size of both anomalies suggest that the low temperature anomaly may correspond to a O(I) reentrance. Eventually, the opposite sign of their pressure rates causes that the curve measured at a pressure value (at room temperature) of 9 kbar presents no anomaly, due to the cancelation of the high temperature and the low temperature anomalies. This behavior is perfectly verified by the neutron thermodiffractograms collected in D20 as a function of pressure and depicted in Fig. 9. In this case, whereas at 0 kbar we observe a single O(I)→M crystallographic transformation on cooling at ~200 K, the 4 kbar experiment reveals the strong decrease of T_t in addition to the appearance of new O(I) reflections at very low temperatures. At 10 kbar, the reflections coming from the M phase have vanished

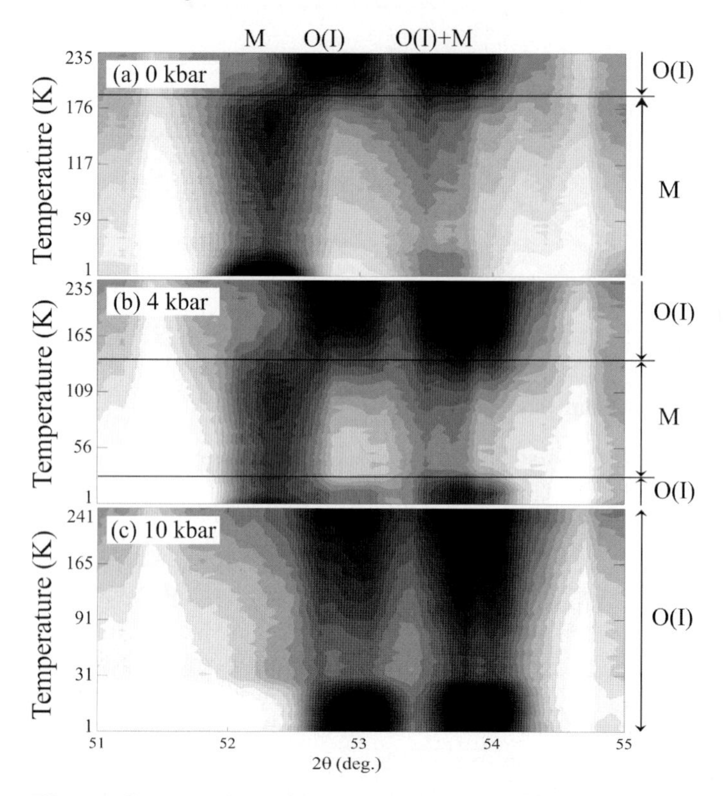

Fig. 9. Contour plots of the neutron thermodiffractograms of Er_5Si_4 collected in D20 at selected pressures (values at room-temperature): **(a)** ambient pressure, **(b)** 4 kbar, and **(c)** 10 kbar (note the non-linear scale of temperature). *Darker color* represents stronger intensity, whereas *lighter color* depicts weaker intensity areas

as both anomalies have collapsed and we only observe O(I) reflections in the whole temperature range. Further neutron experiments (not shown here) have demonstrated that the new magnetic anomaly at 35 K corresponds to the magnetic ordering of the newly nucleated O(I) phase at low temperatures. This idea perfectly matches with the phenomenology observed in magnetization in the way of a first order transformation which enables an intermediate pressure range of coexisting M and O(I) phases presenting different ordering temperatures, and also agrees with previous knowledge on the 5:4 compounds, as it gives rise to a O(I)-FM ground state, in this case at high pressure, as in other members of the family. Additionaly, the higher T_C of the pressure-nucleated phase agrees with the idea of reinforced superexchange-like FM interactions in the case where all the covalent bonds are formed.

In the light of these results, we can elaborate a T-P crystallographic-magnetic phase diagram of Er_5Si_4 which we represent in Fig. 10. In addition to the high temperature purely crystallographic phase boundary, which

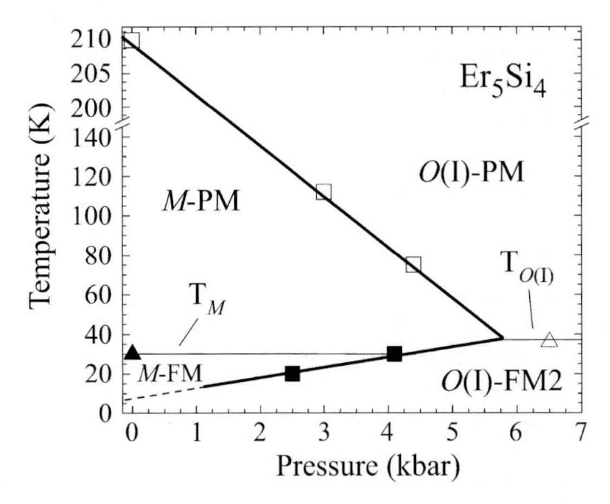

Fig. 10. Magnetic and crystallographic T-P phase diagram of Er_5Si_4. The temperatures of the suggested magnetic and structural transitions are indicated. *Thick lines* mark the crystallographic phase boundaries, and *narrow lines* are used for the magnetic transitions

rapidly decreases with pressure due to the pressure-induced favoring of the low volume phase [O(I)], and the low temperature FM ordering of the M state at $T_M = 30\,\mathrm{K}$, a second first-order anomaly grows in the low temperature regime giving rise to the onset of the O(I) phase also at low temperature. On increasing pressure, a phase coexistence scenario takes place in which the sample volume in divided in M-FM1 and O(I)-FM2 regions. Eventually, the opposite pressure rates of the high and low temperature inverse transitions gives rise to their collapse around $\mathrm{P} > 6\,\mathrm{kbar}$. At higher pressure, only the O(I) phase remains in the whole temperature range with an ordering temperature $T_{O(I)}$=35 K. This complex phase diagram underlines once more the paramount role of the interlayer distances, the chemical bonding and the subsequent change of the magnetic interactions in the macroscopic magnetic and structural properties of the 5:4 materials.

5 Conclusions

The aim of this work has been to present a brief overview of our investigations of the extraordinary effects of hydrostatic pressure on the interplay between the magnetism and the structure in the $R_5(Si_xGe_{1-x})_4$ alloys. The wealthy phenomenology displayed by the $R_5(Si_xGe_{1-x})_4$ systems upon application of pressure reveals the extraordinary importance of the hydrostatic pressure and the external control of the volume cell size in ruling the physical properties of the $R_5(Si_xGe_{1-x})_4$ alloys. The source of this outstanding magnetostructural coupling emerges from the strong correlations between the crystal structure, chemical bonding and magnetic properties in these compounds. This also opens a new possibility for enhancing or tuning the magnetocaloric properties of other systems which might present decoupled magnetic and structural changes, both among the 5:4 alloys and other magnetocaloric materials presenting first order crystallographic changes.

References

[1] G. S. Smith, Q. Johnson, A. G. Tharp: Acta Cryst. **22**, 269 (1967)

[2] G. S. Smith, A. G. Tharp, Q. Johnson: Nature (London) **210**, 1148–1149 (1966)

[3] G. S. Smith, Q. Johnson, A. G. Tharp: Acta Cryst. **22**, 940 (1967)

[4] F. Holtzberg, R. J. Gambino, T. R. McGuire: J. Phys. Chem. Solids **28**, 2283 (1967)

[5] V. K. Pecharsky, K. A. Gschneidner, Jr.: Phys. Rev. Lett. **78**, 4497 (1997)

[6] V. K. Pecharsky, K. A. Gschneidner, Jr.: Appl. Phys. Lett. **70**, 3299 (1997)

[7] V. K. Pecharsky, K. A. Gschneidner, Jr.: J. Magn. Magn. Mater. **200**, 44–56 (1999)

[8] V. K. Pecharsky, A. P. Holm, K. A. Gschneidner, Jr., R. Rink: Phys. Rev. Lett. **91**, 197204 (2003)

[9] W. Choe, V. K. Pecharsky, A. O. Pecharsky, K. A. Gschneidner, Jr., V. G. Young, Jr., G. J. Miller: Phys. Rev. Lett. **84**, 4617 (2000)

[10] V. K. Pecharsky, K. A. Gschneidner, Jr.: Adv. Mater. **13**, 683 (2001)

[11] E. M. Levin, V. K. Pecharsky, K. A. Gschneidner, Jr.: Phys. Rev. B **62**, R14625 (2000)

[12] A. O. Pecharsky, K. A. Gschneidner, Jr., V. K. Pecharsky, C. E. Schindler: J. Alloys and Compounds **338**, 126 (2002)

[13] L. Morellon, P. A. Algarabel, M. R. Ibarra, J. Blasco, B. Garcia-Landa, Z. Arnold, F. Albertini: Phys. Rev. B **58** (1998)

[14] L. Morellon, J. Blasco, P. A. Algarabel, M. R. Ibarra: Phys. Rev. B **62** (2000)

[15] Q. L. Liu, G. H. Rao, J. K. Liang: The Rigaku Journal **18** (2001)

[16] L. Morellon, Z. Arnold, P. A. Algarabel, C. Magen, M. R. Ibarra, Y. Skorokhod: J. Phys.: Condens. Matter **16**, 1623–1630 (2004)

[17] C. Magen, Z. Arnold, L. Morellon, Y. Skorokhod, P. A. Algarabel, M. R. Ibarra, J. Kamarad: Phys. Rev. Lett. **91** (2003)

[18] C. Magen, L. Morellon, P. A. Algarabel, M. R. Ibarra, Z. Arnold, J. Kamarad, T. A. Lograsso, D. L. Schlagel, V. K. Pecharsky, A. O. Pecharsky, K. A. Gschneidner, Jr.: Phys. Rev. B **72**, 024416 (2005)

[19] J. Kamarad: *Encyclopedia of Materials: Science and Technology* (Oxford: Elsevier Science 2001) p. 4976

[20] E. M. Levin, V. K. Pecharsky, K. A. Gschneidner, Jr.: Phys. Rev. B **63**, 064426 (2001)

[21] C. Magen, L. Morellon, P. A. Algarabel, C. Marquina, M. R. Ibarra: J. Phys.: Condens. Matter **15**, 1 (2003)

[22] C. Ritter, L. Morellon, P. A. Algarabel, C. Magen, M. R. Ibarra: Phys. Rev. B **65**, 094405 (2002)

[23] L. Morellon, C. Ritter, C. Magen, P. A. Algarabel, M. R. Ibarra: Phys. Rev. B **68**, 024417 (2003)

[24] L. Morellon, Z. Arnold, C. Magen, C. Ritter, O. Prokhnenko, Y. Skorokhod, P. A. Algarabel, M. R. Ibarra, J. Kamarad: Phys. Rev. Lett. **93** (2004)
[25] V. K. Pecharsky, A. O. Pecharsky, K. A. Gschneidner, Jr., G. J. Miller: Phys. Rev. Lett. **91**, 207205 (2003)
[26] Y. Mozharivskyj, A. O. Pecharsky, V. K. Pecharsky, G. J. Miller, K. A. Gschneidner, Jr.: Phys. Rev. B **69**, 144102 (2004)
[27] C. Magen, L. Morellon, Z. Arnold, P. A. Algarabel, C. Ritter, M. R. Ibarra, J. Kamarad, V. K. Pecharsky, A. O. Tsokol, K. A. G. Jr.: to be published

Magnetoelectric Correlations in Multiferroic Manganites Revealed by Nonlinear Optics

Manfred Fiebig[1,2]

[1] HISKP, Universität Bonn,
Nussallee 14–16, 53115 Bonn, Germany
fiebighiskp.uni-bonn.de
[2] Max-Born-Institut,
Max-Born-Straße 2A, 12489 Berlin, Germany

Abstract. A variety of magnetoelectric (ME) effects are observed in multiferroic hexagonal $R\mathrm{MnO}_3$ with R = Sc, Y, In, Ho–Lu. A passive ME effects occurs in the form of a pronounced but non-controllable coupling of magnetic and ferroelectric domains. An active ME effects is revealed in the form of a magnetic or electric field controling the magnetic structure. Local ME effects are detected within the magnetic domain walls. The effects are observed by linear and, in particular, nonlinear magnetooptical techniques and explained microscopically by the interplay of magnetic superexchange, domain-wall magnetization, and ferroelectric distortion.

1 Introduction

Systems with a spontaneous uniform magnetic, electrical, elastic, or toroidic orientation of all unit cells are termed "ferroic" [1, 2]. While ferromagnetism has already been known in China and Greece since pre-Christian times the discovery that a compound can carry different forms of ferroic (or anti-ferroic) ordering *simultaneously* has been made barely half a century ago [3]. Nowadays, some of these so-called "multiferroics" are already in widespread use. For example, ferroelastic ferroelectric (FEL) compounds, which combine a spontaneous and switchable strain and a spontaneous and switchable polarization, are used in piezoelectric transducers, and ferromagnetic ferroelastics, with their coupled magnetization and strain, find applications as magnetomechanical actuators [4]. Multiferroics with a coexistence of magnetic and electrical ordering are particularly interesting because they may allow one to control the magnetic properties by electric fields or the dielectric properties by magnetic fields [5]. The source of this coupling is the magnetoelectric (ME) effect which denominates the induction of a polarization by a magnetic field and of a magnetization by an electric field. The linear ME effect is described by the relations $\boldsymbol{P} \propto \hat{\alpha}\boldsymbol{H}$ and $\boldsymbol{M} \propto \hat{\alpha}\boldsymbol{E}$. The ME coupling is described by the second-rank axial c tensor $\hat{\alpha}$ and contributes a term $H_{\mathrm{ME}} = \hat{\alpha}\boldsymbol{EH}$ to the free energy [5]. In the case of long-range ordering it may be more convenient to write $H_{\mathrm{ME}} = \hat{\alpha}^*\boldsymbol{DB}$ with $\boldsymbol{B} = \mu_0(\boldsymbol{H}+\boldsymbol{M})$ and $\boldsymbol{D} = \epsilon_0(\boldsymbol{E}+\boldsymbol{P})$ as magnetic and electric fields in matter and $\hat{\alpha}^*$ as naked ME coupling tensor from which the electric and magnetic permittivities have been extracted. The ME

R. Haug (Ed.): Advances in Solid State Physics,
Adv. in Solid State Phys. **46**, 245–257 (2008)

coupling coefficient is limited by the relation $|\hat{\alpha}|^2 \leq |\hat{\chi}_e|\,|\hat{\chi}_m|$ [6] with $\hat{\chi_e}$ and $\hat{\chi_m}$ as electric and magnetic susceptibilities, respectively. The components of $\hat{\alpha}$ can be largest in ferromagnetic ferroelectrics because of $|\hat{\chi}_{e,m}| \gg 1$. Therefore, ME multiferroics indeed have the potential for unusually large ME effects.

Unfortunately the search for ME multiferroics has proven to be unexpectedly difficult. It was finally recognized that the most common mechanisms for ferroelectric and magnetic ordering require empty and partially filled $3d$ orbitals, respectively, and, thus, exclude each other [7]. Two strategies for surmounting this serious restriction are being pursued. On the one hand, the ME effect is generated as product property in composites consisting of a piezoelectric and magnetostrictive constituent which are not ME in themselves. A magnetic field induces a strain in the magnetostrictive constituent which is passed on to the piezoelectric constituent where it is the source of a polarization. On the other hand, alternative mechanisms for ferroelectricity that allow the coexistence with magnetic ordering as an intrinsic property are being exploited. Up to now, symmetry reduction by doping, geometrically driven ferroelectricity, lock-in ferroelectricity, competition of bond and site centering of valency electrons, and modification of magnetic superexchange by lattice distortion were identified as mechanisms [5]. The variety indicates that a lot more research is indispensable in order to gain a comprehensive picture of the microscopic origins of ME effects in multiferroics and, thereby, drive them closer towards practical use.

In this paper we show that multiferroic $R\mathrm{MnO_3}$ with $R \in$ {Sc, Y, In, Ho-Lu} displays a multitude of ME effects. A passive ME effects in the form of a pronounced but non-controllable coupling of magnetic and ferroelectric domains is observed in $\mathrm{YMnO_3}$. On the other hand, an active ME effects is found in $\mathrm{HoMnO_3}$ where a magnetic or electric field is used to control the magnetic structure and the direction of the magnetic order parameter. Finally, a local ME effect is detected within the magnetic domain walls. The prime basis for the experimental observation of all these ME correlations are nonlinear optical experiments which allows one simultaneous imaging of the magnetic *and* the electrical structure. The observed ME effects are explained microscopically on the basis of the complex interplay of magnetic exchange, domain-wall magnetization, and FEL distortion.

2 Multiferroic Manganites

The eight hexagonal $R\mathrm{MnO_3}$ compounds form a group of multiferroics with FEL ordering at $T_\mathrm{C} = 570 - 990$ K [8–10], antiferromagnetic (AFM) $\mathrm{Mn^{3+}}$ ordering at $T_\mathrm{N} = 73 - 124$ K [11], and ferro-, ferri, or AFM R^{3+} ordering of the compounds with partially filled $4f$ shell at $T_{4f} \sim 5$ K [12]. The broad variety of R^{3+} cations allows systematic studies of ME behavior as function of cation radius, $3d$ and $4f$ magnetism, and the interplay of various ordered

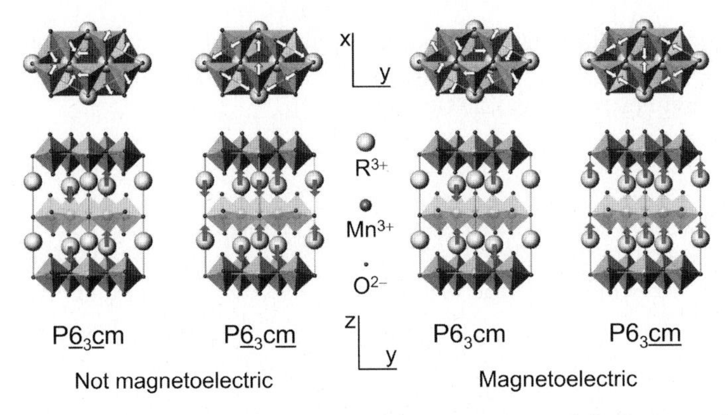

Fig. 1. Crystallographic and magnetic structure of the hexagonal manganites. The unit cell is indicated by *lines* and possesses the crystallographic symmetry $P6_3cm$. The four magnetic structures with the highest symmetry are shown. *Arrows* in the top view denote the Mn^{3+} magnetic moments. *Arrows* in the *side view* denote the R^{3+} magnetic moments (where applicable). In only two of the four structures the ME effect is symmetry allowed

sublattices. Investigation of the $RMnO_3$ system thus offers the possibility to reveal general criteria of ME behavior and multiferroicity.

The FEL phase of the multiferroic manganites possesses $P6_3cm$ symmetry and a polarization along the hexagonal z axis [10]. It is made up by three magnetic sublattices with $Mn^{3+}(3d^3)$ ions at 6c positions and, in the case of R = Ho–Yb, $R^{3+}(4f^{10\ldots13})$ ions at 2a and 4b positions. Anisotropy confines the Mn^{3+} spins to the basal xy plane where frustration leads to triangular AFM structures with strong intra-plane and two orders of magnitude weaker inter-plane coupling [11,13]. In contrast, the sublattices of Ho^{3+}, Er^{3+}, Tm^{3+}, Yb^{3+} are assumed to order Ising-like along z. Figure 1 shows the four magnetic structures with the highest magnetic symmetry. Both the ordering of the Mn^{3+} lattice and, if applicable, of the R^{3+} lattices are displayed. Note that the ME effect is allowed in only two out of the four magnetic symmetry groups displayed.

3 Experimental Techniques and Setup

Linear and nonlinear magneto-optical techniques were applied in order to investigate the magnetic and FEL structure in dependence of applied magnetic and electric fields. Faraday rotation couples to a magnetic field $\boldsymbol{H}$ or sublattice magnetization $\boldsymbol{M}$ by

$$P_i(\omega) = \epsilon_0 \, \gamma_{ijk} \, B_j(0) \, E_k(\omega), \tag{1}$$

in which the polarization $\boldsymbol{P}(\omega)$ induced by the static matter field $\boldsymbol{B}$ leads to a rotation of the polarization of the incident light with frequency ω. Here, the Faraday rotation from the Mn^{3+} $d-d$ transitions is employed to monitor the Ho^{3+}, Er^{3+}, Tm^{3+}, and Yb^{3+} ordering [14]. Further, second harmonic generation (SHG) couples to the magnetic order by

$$P_i(2\omega) = \epsilon_0 \, \chi_{ijk} \, E_j(\omega) \, E_k(\omega), \tag{2}$$

where $\boldsymbol{E}(\omega)$ and $\boldsymbol{P}(2\omega)$ represent the incident light at frequency ω and the polarization induced in the material at frequency 2ω. The set of nonzero tensor components χ_{ijk} is determined by the crystallographic and magnetic structure. For light incident along z and φ as angle between the Mn^{3+} magnetic moment and the local x axis [11,13] one gets $\chi_{xxx} \propto \sin\varphi$ and $\chi_{yyy} \propto \cos\varphi$ for spin structures with $P\underline{6}_3\,..$ symmetry and $\chi_{xxx} = \chi_{yyy} = 0$ for spin structures with $P6_3\,..$ symmetry [13,15].

The $RMnO_3$ crystals were flux-grown and prepared into polished z oriented platelets with a thickness of ~ 50 µm unless otherwise stated. For application of electric fields transparent indium-tin-oxide electrodes were evaporated onto the sample surfaces. Samples were excited with linearly polarized 5 ns light pulses from an optical parametric oscillator with photon energies up to 1.55 eV. The Faraday rotation was determined from the polarization of the linearly polarized light transmitted through the sample. Intensity and polarization of the SH signal were determined after suppressing the fundamental light wave with optical filters, using a liquid-nitrogen-cooled digital camera with a telephoto lens as spatially resolving detector.

4 Results and Discussion

4.1 Passive ME Effect

Figure 2a shows the SH spectra of $YMnO_3$ gained from χ_{yyy}, χ_{zyy}, χ_{yyz}, χ_{zzz} in (2). Observation of $\chi_{xxx} = 0$, $\chi_{yyy} \neq 0$ points to $P\underline{6}_3c\underline{m}$ symmetry. In Ref. 5 it was shown that χ_{yyy} couples to the bilinear product $\mathcal{P} \cdot \ell$ of FEL and AFM order parameters while χ_{zyy}, χ_{yyz}, and χ_{zzz} couple exclusively to $\mathcal{P}$. In Figs. 2b and 2c the spatially resolved SH light at 2.46 eV reveals the FEL and AFM domain structures. Dark and bright regions correspond to domains with opposite orientation of the order parameter $\mathcal{P}$ or ℓ. The FEL domain structure was visualized by interference of the SH signal wave from $\chi_{zyy}(\mathcal{P})$ with a planar reference light field [17,18]. Since regions with $\pm\mathcal{P}$ differ by 180° in the phase of the signal wave, the interference with the reference field can be constructive or destructive which leads to the different brightness of opposite domains. In a similar way the AFM domain structure was visualized by interference of the SH waves from $\chi_{yyy}(\mathcal{P}\ell)$ and $\chi_{zyy}(\mathcal{P})$. Depending on the orientation of ℓ the interference is constructive or destructive, this time leading to a different brightness for opposite AFM domains [16]. The comparison between Figs. 2b and 2c shows that any reversal of the FEL

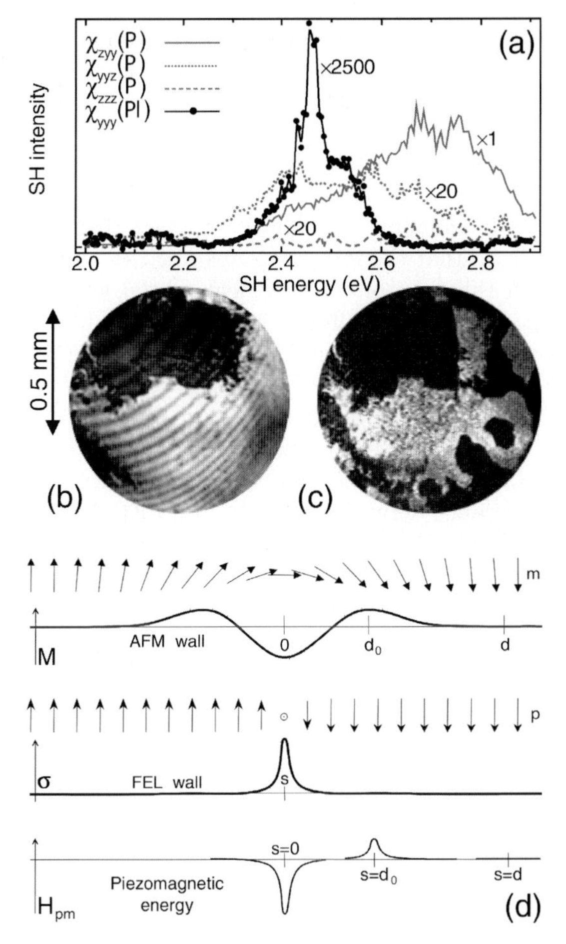

Fig. 2. Clamping of magnetic and electric domain walls in $YMnO_3$. **(a)** SH spectra of $YMnO_3$ (not normalized) at 6 K with light incident along the x axis. **(b)** FEL domains, and **(c)** AFM domains exposed with SH light at 2.46 eV. **(d)** Energy gain H_{pm} through interacting FEL and AFM domain walls for different distances s between the walls. m and p denote the magnetic moment of the Mn^{3+} ion and the electric-dipole moment per unit cell, respectively. M and σ denote magnetization and strain within the, respectively, AFM and FEL domain wall

order parameter is accompanied by a *simultaneous* reversal of the AFM order parameter. Consequently two types of AFM domain wall are found in Fig. 2c: "clamped" walls at any location of a FEL domain walls in Fig. 2b, and additional "free" walls within one FEL domain.

The pronounced coupling between FEL and AFM domains is quite surprising because according to Fig. 1 the $P\underline{6}_3c\underline{m}$ symmetry of $YMnO_3$ does not allow the ME effect. However, the coupling can be interpreted as *local* ME effect. Microscopically it is explained by a piezomagnetic interaction between the domain walls. The piezomagnetic effect is described by $H_{\mathrm{pm}} = q_{ijk} M_i \sigma_{jk}$ with H_{pm}, $\hat{q}$, $\boldsymbol{M}$, and $\hat{\sigma}$ as piezomagnetic contribution to the free energy, piezomagnetic tensor, magnetization, and strain tensor, respectively. In the present case, the magnetization is induced by the rotation of the Mn^{3+} spins in the AFM domain wall, and strain is induced by the opposite dipolar displacement of ions in opposite FEL domains. The AFM walls are therefore much broader than the FEL walls so that the former are regarded as perturbation to the magnetic lattice in the form of a topological soliton whereas the latter induce a local reduction of the crystallographic symmetry and are described by a step function with new tensor components σ_{jk} which are not allowed for the unperturbed FEL lattice [19].

As example, overlapping AFM and FEL domain walls in the xz plane at $y = 0$ are considered. Here the only contributions to H_{pm} are from M_y and σ_{yy} with $q_{yyy} \neq 0$. The corresponding piezomagnetic contribution to the free energy as a function of distance between the AFM and FEL domain walls is shown schematically in Fig. 2d. The largest energy gain is achieved by centrically overlapping AFM and FEL domain walls whereas walls at distance y_0 repel each other.

4.2 Active ME Effect

Figure 2 represents an example for a passive ME correlation between the magnetic and the electrical phases. The coupling between AFM and FEL domains is pronounced but cannot be controlled with respect to the generation of a specific domain structure or a desired orientation of the magnetic order parameter. In the following we therefore attempted to control the multiferroic state and the related order parameters actively via the ME effect.

Figure 3a shows the SH spectrum of $YbMnO_3$ at 6 K with no fields applied. The polarization of the SH signal reveals $\chi_{xxx} \neq 0$ and $\chi_{yyy} = 0$ and, thus, $P\underline{6}_3\underline{c}m$ as magnetic symmetry. The largest SH intensity is observed at 2.44 eV. Figure 3b shows the SH intensity at 2.44 eV in dependence of a magnetic field $H \parallel z$ for an $YbMnO_3$ crystal which was cooled to 1.4 K in zero magnetic field. Between 1.5 T and 3.0 T the SH intensity gradually decreases to zero in the whole investigated spectral range and does not recover upon field removal or reversal. According to [11] this points to a 90° rotation of all Mn^{3+} spins with a change of magnetic symmetry from $P\underline{6}_3\underline{c}m$ to $P6_3\underline{cm}$. Figure 3c shows the Faraday rotation (Φ) in $YbMnO_3$ at 6 K for a magnetic field $H \parallel z$. The rotation consists of a linear contribution $\Phi(\boldsymbol{H}) \propto \boldsymbol{H}$ from the applied field and a hysteresis $\Phi(\boldsymbol{M})$ which is due to an Ising-like magnetization of the Yb3+ sublattices along z. The slopes near 0 T and at 2.3 T correspond to single-domain ferri- and ferromagnetic orientation of the two Yb3+ sublattices, respectively. Note that magnetization along z is compatible with $P6_3\underline{cm}$ symmetry and, thus, field induced reordering of the Mn^{3+} lattice [20].

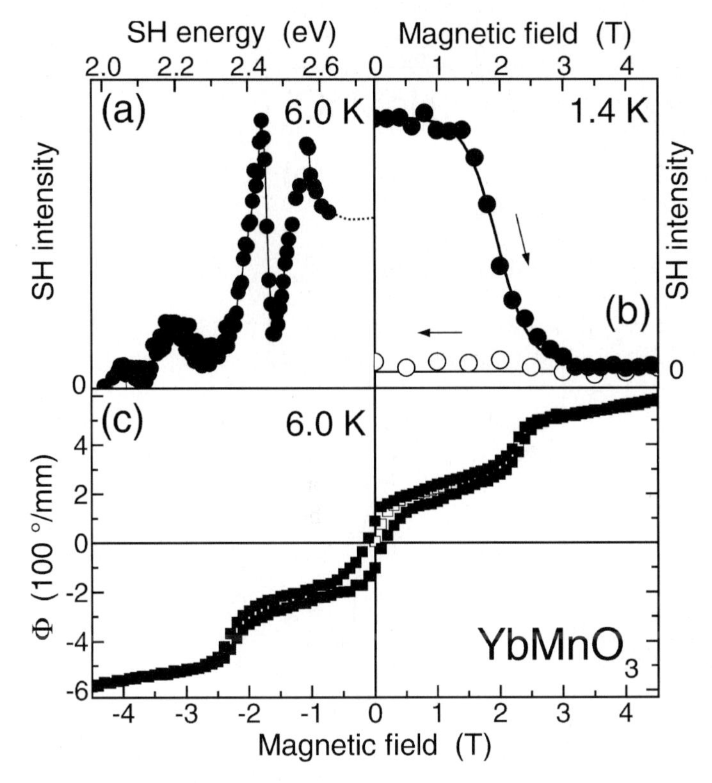

Fig. 3. Dependence of magneto-optical properties of $YbMnO_3$ on a static magnetic field $H \parallel z$ **(a)** SH spectrum at $H = 0$. Due to the large absorption the SH intensity was normalized to the intensity of the fundamental light transmitted through the sample. **(b)** SH intensity at 2.44 eV in dependence of H in field increasing and decreasing runs. **(c)** Hysteresis loop of Faraday rotation Φ at 1.23 eV in dependence of H

Figure 4a shows the SH spectrum of $HoMnO_3$ at 6 K and 50 K with no fields applied while Fig. 4b shows the temperature dependence of the SH intensity at 2.42 eV in the absence or presence of an electric field $E \parallel z$ with $E = E_0 = 10^5$ V/cm. At $E = 0$, x polarized SH light is observed below T_N, indicating $P\underline{6}_3\underline{c}m$ as magnetic symmetry [11]. It is followed at $T_R = 37$ K by 90° rotation of the Mn^{3+} spins and, thus, of the polarization of the SH signal. The spin reorientation changes the symmetry to $P\underline{6}_3c\underline{m}$ [11]. At 5 K the SH signal is quenched due to another 90° rotation of the Mn^{3+} spins which leads to $P6_3cm$ symmetry [11]. At $E = \pm E_0$ the temperature dependence of the magnetic SH signal is drastically different. Except from a small residual surface induced contribution [21] the SH signal is quenched in the whole temperature range below T_N. The nature of the phase induced by the electric field is revealed by Fig. 4c which shows the Faraday rotation induced in the $HoMnO_3$ sample by application of the field $\pm E_0$. Below T_N a

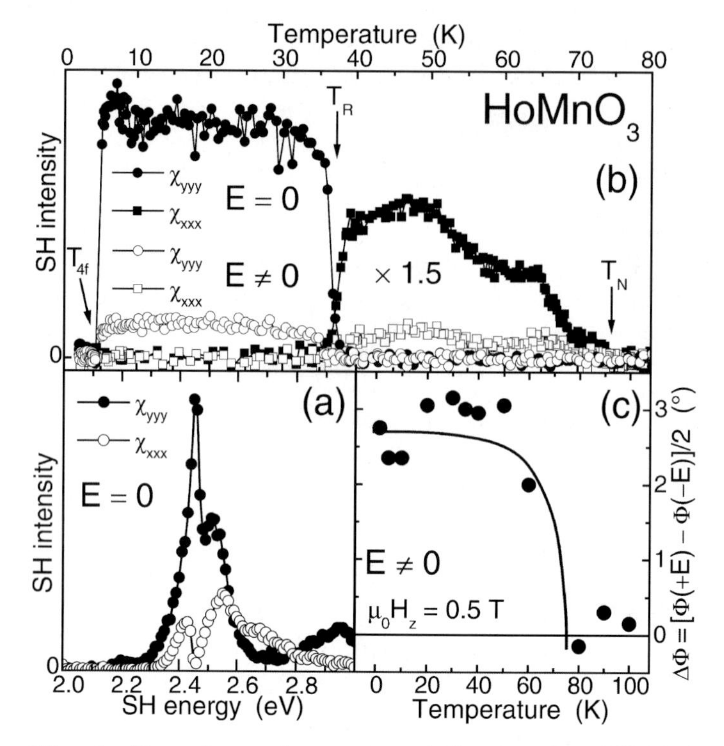

Fig. 4. Dependence of magneto-optical properties of $HoMnO_3$ on a static electric field $E \parallel z$. (**a**) SH spectrum at $E = 0$ (not normalized). (**b**) Temperature dependence of SHG at 2.42 eV in the absence and presence of an electric field. (**c**) Temperature dependence of the ferromagnetic contribution $\Delta\Phi$ to the Faraday rotation for $|E| = E_0$ at constant absorption with $\mu_0 H_z = 0.5$ T

ferromagnetic contribution $\Delta\Phi$ to the Faraday rotation is observed whose sign depends on the sign of the applied electric field. This contribution can only originate in ferromagnetic ordering of the Ho^{3+} sublattices: It is of the same order of magnitude as the rotation induced by intrinsic ferromagnetic Yb^{3+} ordering (Fig. 3) while it is *not* observed in compounds with paramagnetic R^{3+} sublattices ($R \in \{\mathrm{Sc, Y, In, Lu}\}$). According to Fig. 3 a quenched SH signal along with formation of a ferromagnetic moment indicates a change to $P6_3\underline{cm}$ symmetry.

The electric field therefore controls the magnetic order of the rare-earth sublattice in $HoMnO_3$ just as the magnetic field does in $YbMnO_3$. The origin of this unusual example of phase control expressed by Figs. 3 and 4 is the linear ME effect. In comparison to other compounds [22, 23] the ME energy $H_{\mathrm{ME}} = \hat{\alpha}^* \boldsymbol{D}\boldsymbol{B}$ of a magnetic FEL is enhanced by orders of magnitude by the presence of FEL and/or ferromagnetic ordering. When an external electric or magnetic field maximizes $\boldsymbol{D}$ or $\boldsymbol{B}$ by driving the system into a single-domain state, the microscopic processes detailed below mediate the reorientation of the Mn^{3+} spins to a state with $P6_3\underline{cm}$ symmetry and ferromagnetic interplane coupling between the R^{3+} spins. Only in this phase a macroscopic ME effect from $\alpha^*_{zz} \neq 0$ is active [22]. Therefore the field induced transitions into the $P6_3\underline{cm}$ phase cost small values of anisotropy and superexchange energy, but these are vastly overcompensated by the gain of ME energy from $\alpha^*_{zz} M_z P_z$.

Microscopically the Ho^{3+}–Mn^{3+} interaction leads to an energy contribution $H_{\text{ex}} = \sum \boldsymbol{S}^{\text{Ho}} \hat{A} \boldsymbol{S}^{\text{Mn}}$ with $\boldsymbol{S}^{\text{Ho,Mn}}$ as ordered magnetic moments, the summation including all Ho^{3+} and Mn^{3+} ions in the unit cell [24, 25]. $\hat{A}$ is the 3×3 interaction matrix which is made up by symmetric superexchange and asymmetric Dzyaloshinskii-Moriya contributions. If FEL distortions are neglected the sum is $H_{\text{ex}} = 0$ for all types of Mn^{3+} ordering in Fig. 1. FEL distortions, however, lead to a correction $\delta\hat{A} = \delta\hat{A}_0 P_z$ for which – in the case of $P6_3\underline{cm}$ symmetry – summation leads to the nonzero contribution $H_{\text{ex}} \equiv H_{\text{ME}} = \alpha^*_{zz} P_z M_z$ with $M_z = g\mu_{\text{B}} S_z^{\text{Ho}}$ and the ME susceptibility $\alpha^*_{zz} = (6/g\mu_{\text{B}}) S_y^{\text{Mn}} (\delta A_0^{2\text{a}} - \delta A_0^{4\text{b}})_{zy}$. Here superscripts 2a, 4b refer to the two Ho^{3+} sites (see Fig. 1). Note that although in the case of $P\underline{6}_3\underline{c}m$ and $P\underline{6}_3c\underline{m}$ symmetry the same Ho^{3+}–Mn^{3+} interactions are active, a macroscopic ME effect is not allowed since the microscopic ME contributions from the upper and the lower half of the magnetic unit cell cancel each other. Consequently $R\text{MnO}_3$ crystals with $P\underline{6}_3\,..$ symmtry can be called *antimagnetoelectric.*

4.3 Local ME Effect

Even in compounds where the ME is forbidden by symmetry, indications for ME behavior were reported. For instance, modifications of the dielectric constant at the Néel temperature [12, 26–28] and in the reentrant phase of the 90° spin rotation in $HoMnO_3$ (see Fig. 4) [29] were described. In the following the manifestation of a local ME effect will be discussed.

Figure 5 shows the temperature dependence and the spatial distribution of SHG in the course of the 90° spin reorientation in $HoMnO_3$. In [29] a drastic increase of the dielectric constant was reported for the reentrant phase interconnecting the $P\underline{6}_3\underline{c}m$ and $P\underline{6}_3c\underline{m}$ states. Observation of $\chi_{xxx} \neq 0$, $\chi_{yyy} \neq 0$ in Fig. 5a evidences that the reentrant phase possesses $P\underline{6}_3$ symmetry [11]. In the vicinity of T_{R} an anomaly of the SH signal is observed. In both temperature increasing and decreasing runs the SH component which is disappearing in the course of the phase transition (T increase: χ_{yyy}; T decrease: χ_{xxx}) decreases gradually with progress of the transition. However, the newly formed contribution (T increase: χ_{xxx}; T decrease: χ_{yyy}) passes through a local maximum with a width of $1 - 2$ K before the original value is restored.

Figures 5b and 5c depict the spatial distribution of SHG in the course of the spin reorientation. Curvy black lines distributed all over the sample indicate the position of domain walls which are visible due to an interference effect [21] and separate spin-reversal domains, that is, regions with opposite

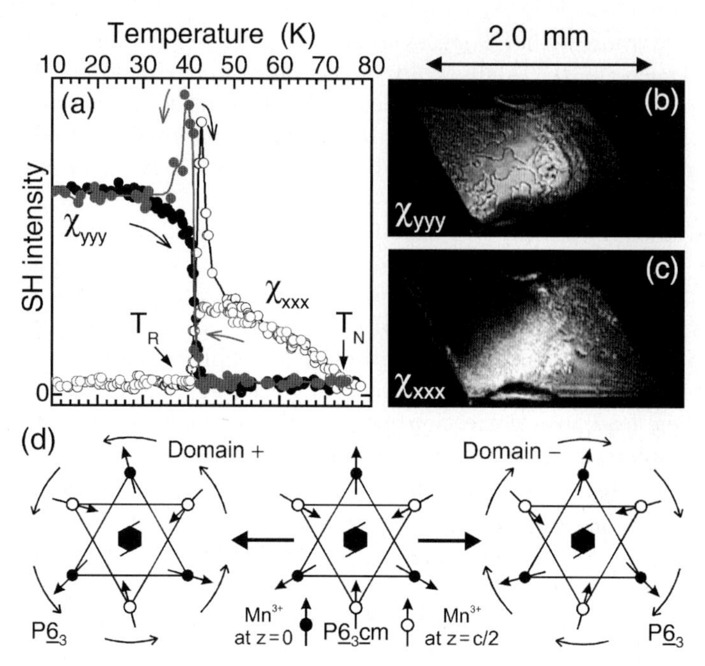

Fig. 5. Temperature dependence of SHG in $HoMnO_3$ at $E = H = 0$. **(a)** SHG from χ_{yyy} and χ_{xxx} in temperature increasing and decreasing runs. **(b, c)** Spatial distribution of SHG from χ_{yyy} and χ_{xxx} on the sample. *Curvy black lines* correspond to domain walls. **(d)** Model for the formation of spin-rotation domains in the reentrant $P\underline{6}_3$ phase at T_R in a temperature decreasing run

orientation of the AFM order parameter. Due to the gradual change of T_R across the sample which is caused by growth-induced inhomogeneities the unusual enhancement of SH intensity in the reentrant phase is clearly visible as bright region in the center of the image in Fig. 5c. In this region a grainy distribution of SH intensity without identifiable domain walls is observed in distinct contrast to the otherwise smooth distribution of SH intensity in which a network of domain walls creates a three-dimensional effect. A remarkable feature of the reentrant phase is that in spite of the featureless topography of the SH contribution from χ_{xxx} the contribution related to χ_{yyy} still displays a well pronounced distribution of domain walls. Figure 5d shows that the abnormal behavior of SH intensity from χ_{xxx} is caused by the nucleation of a tight network of spin-rotation domains which are distinguished by clockwise or counterclockwise rotation of all Mn^{3+} spins from the local y axis to the local x axis. Interference effects lead to the grainy distribution and the enhancement of the SH signal [21]. Upon further progress of the spin rotation the network of small rotation domains becomes unstable and collapses into the large spin-reversal domains of the $P\underline{6}_3c\underline{m}$ phase.

As shown in Fig. 4d the AFM walls carry an uncompensated magnetization in the basal xy plane which reduces the magnetic symmetry within the wall to $P\underline{2}$. Contrary to the *global* $P\underline{6}_3$ symmetry the low *local* $P\underline{2}$ symmetry in the walls allows ME contributions $P_z \propto \alpha^*_{zx} M_x$ and $P_z \propto \alpha^*_{zy} M_y$ to the electric polarization P_z by the wall magnetization $M_{x,y}$, and can therefore modify the dielectric function [30]. The high density of domain walls in the reentrant phase enhances this effect.

5 Conclusion

In summary, we have observed how multifold magnetic ordering in multiferroic manganites leads to a variety of ME coupling phenomena. In the compounds with partially filled $4f$ shell of the R^{3+} ion the linear ME effect was observed as bulk effect. The microscopic interplay of $R^{3+} - Mn^{3+}$ interaction and FEL distortion leads to a ME contribution to the free energy so that by application of external magnetic or electric fields the magnetic phase of the compounds can be controlled. In the compounds with filled or empty $4f$ shell of the R^{3+} ion the linear ME effect is forbidden as bulk effect, but *antimagnetoelectric* behavior was identified. In these compounds linear ME behavior is induced locally, in the domain walls. On the one hand ME effects manifest as piezomagnetic coupling between FEL and AFM domain walls. On the other hand the AFM walls with their lower symmetry are a source of linear ME behavior which modifies the dielectric function.

Based on the work presented here promising candidates for controlled ME switching are compounds with electronic states close to the ground state which are energetically lowered by ME contributions in an applied electric or magnetic field. Therefore frustrated systems or systems in the vicinity of phase boundaries or quantum critical points are prime candidates for ME phase control. Although the quest for multiferroics with a large magnetization ($> 1\ \mu_B$/unit cell) and polarization ($> 1\ \mu C/cm^{-2}$) at room temperature is yet far from being solved present advances in generating large ME effects and in understanding the origins of ME and multiferroic behavior are encouraging. In composite materials where the ME effect is generated as a product property of a mixture of piezoelectric and magnetostrictive constituents pronounced room-temperature ME effects at microwave frequencies are already a reality.

Acknowledgements

The author thanks D. Fröhlich, and Th. Elsässer for continuous support, R. V. Pisarev, Th. Lottermoser, and Th. Lonkai for participation in the experiments and helpful discussions, and DFG for subsidy.

References

[1] K. Aizu: Phys. Rev. B **2**, 754 (1970)
[2] H. Schmid: Ferroelectrics **252**, 253 (2001)
[3] G. A. Smolenskii, I. E. Chupis: Sov. Phys. Usp. **25**, 475 (1982)
[4] N. A. Spaldin, M. Fiebig: Science **309**, 391 (2005)
[5] M. Fiebig: J. Phys. D **38**, R123 (2005)
[6] W. F. Brown, R. M. Hornreich, S. Shtrikman: Phys. Rev. **168**, 574 (1968)
[7] N. A. Hill: J. Phys. Chem. B **104**, 6694 (2000)
[8] *Numerical Data and Functional Relationships*, Landolt-Börnstein, New Series, Group III, Vol. 16a (Springer-Verlag, Berlin 1981)
[9] P. Coeuré, F. Guinet, J. C. Peuzin, G. Buisson, E. F. Bertaut: ferroelectric Properties of Hexagonal Orthomanganites of Yttrium and Rare Earths, in: *Proceedings of the International Meeting on Ferroelectricity*; V. Dvorák, (Ed.), (Institute of Physics of the Czechoslovak Academy of Sciences, Prague 1966), pp. 332–340
[10] Th. Lonkai, D. G. Tomuta, U. Amann, J. Ihringer, R. W. A. Hendrikx, D. M. Többens, J. A. Mydosh: Phys. Rev. B **69**, 134108 (2004)
[11] M. Fiebig, Th. Lottermoser, R. V. Pisarev: J. Appl. Phys. **93**, 8194 (2003)
[12] N. Iwata, K. Kohn: J. Phys. Soc. Japan **67**, 3318 (1998) [Results for $ErMnO_3$ and $HoMnO_3$ have to be exchanged.]
[13] M. Fiebig, D. Fröhlich, K. Kohn, S. Leute, Th. Lottermoser, V. V. Pavlov, R. V. Pisarev: Phys. Rev. Lett. **84**, 5620 (2000)
[14] C. Degenhardt, M. Fiebig, D. Fröhlich, Th. Lottermoser, R. V. Pisarev: Appl. Phys. B **73**, 139 (2001)
[15] R. R. Birss, *Symmetry and Magnetism*, (North-Holland, Amsterdam 1966)
[16] M. Fiebig, Th. Lottermoser, D. Fröhlich, A. V. Goltsev, R. V. Pisarev: Nature **419**, 818 (2002)
[17] S. Leute, Th. Lottermoser, D. Fröhlich: Opt. Lett. **24**, 1520 (1999)
[18] M. Fiebig, D. Fröhlich, S. Kallenbach, Th. Lottermoser: Opt. Lett. **29**, 41 (2004)
[19] A. V. Goltsev, R. V. Pisarev, Th. Lottermoser, M. Fiebig: Phys. Rev. Lett. 90, 177204 (2003)
[20] Th. Lottermoser, Th. Lonkai, U. Amman, D. Hohlwein, J. Ihringer, M. Fiebig: Nature **430**, 541 (2004)
[21] M. Fiebig, D. Fröhlich, Th. Lottermoser, M. Maat: Phys. Rev. B **66**, 144102 (2002)
[22] T. H. O'Dell: *The Electrodynamics of Magneto-Electric Media*, (North-Holland, Amsterdam 1970)
[23] H. Schmid. Int. J. Magnetism **4**, 337 (1973)
[24] M. Fiebig, C. Degenhardt, R. V. Pisarev: Phys. Rev. Lett. **88**, 27203 (2002)
[25] K. Kritayakirana, P. Berger, R. V. Jones: Opt. Commun. **1**, 95 (1969)
[26] Z. J. Huang, Y. Cao, Y. Y. Sun, Y. Y. Xue, C. W. Chu: Phys. Rev. B **56**, 2623 (1997)
[27] D. G. Tomuta, S. Ramakrishnan, G. J. Nieuwenhuys, J. A. Mydosh: J. Phys.: Condens. Matter **13**, 4543 (2001)
[28] T. Katsufuji, S. Mori, M. Masaki, Y. Moritomo, N. Yamamoto, H. Takagi: Phys. Rev. B **64**, 4419 (2001)

[29] B. Lorenz, A. P. Litvinchuk, M. M. Gospodinov, C. W. Chu: Phys. Rev. Lett. **92**, 087204 (2004)
[30] Th. Lottermoser, M. Fiebig: Phys. Rev. B **70**, 220407(R) (2004)

Domain Wall Formation in Ferromagnetic Layers: An Ab Initio Study

Heike C. Herper

Theoretische Tieftemperaturphysik, Universität Duisburg-Essen,
Lotharstr. 1, 47048, Duisburg, Germany
heike@thp.uni-duisburg.de

Abstract. Domain walls are an inherent feature of ferromagnetic (FM) films consisting of layers with different magnetic orientations. Since FM films are used in electrical devices the question of the influence of domain walls on, e.g., the magnetoresistance has attracted much interest. Besides discussing the resistance contribution of domain walls, it is appropriate to study different types of domain walls and their energy of formation. The behaviour of domain walls is usually discussed within model calculations. In the present paper it is done within an ab initio Green's function technique for layered systems, i.e., the fully relativistic, spin-polarized screened Korringa-Kohn Rostoker method. Results are presented for fcc Co layers covered by two semi-infinite fcc Pt(001) bulk systems or by bulk fcc Co(001), respectively. The resistance, which is caused by the different types of domain walls is discussed within a Kubo-Greenwood approach considering Co(001)/Co_{24}/Co(001) as an example.

1 Introduction

Ferromagnetic (FM) thin films are widely used in electronic devices, e.g., in recording layers of storage media or in giant magnetoresistance (GMR) devices [1, 2]. All these devices have in common to work with FM domains or layers with different magnetic orientation. The type of the domains depends on the material, i.e., the ratio Q of anisotropy K and magnetostatic energy $2\pi M_s^2$. In case of a large $Q = K/2\pi M_s^2$ stripe domains are expected, whereas for small values of Q flux closure domains occur [3]. At the interface of two neighbouring magnetic domains with different magnetic orientation domain walls are expected, whereby the width of the domain walls depends on the magnetocrystalline anisotropy. Systems with large anisotropy, e.g., hcp Co form usually thinner domain walls as compared to materials with small magnetocrystalline anisotropy, e.g., bcc Fe [4]. During the last ten years numerous work has been done to investigate the influence of domain walls on the electronic properties. Especially, the contribution of the domain walls to the GMR has been widely discussed [4–8]. However, from an experimental point of view, it is quite difficult to clearly separate the resistance caused by the domain wall from other contributions, e.g., the anisotropic MR.

The discussion of the domain wall resistance is related to further, more basic questions concerning, e.g., the arrangement and the structure of domain walls in a FM film. Obviously, it is energetically favorable to build

R. Haug (Ed.): Advances in Solid State Physics,
Adv. in Solid State Phys. **46**, 259–269 (2008)

a domain wall at the interface of two magnetic domains with different orientation. However, domain walls are mostly treated within micromagnetic simulations [9, 10], which mainly focus on dynamic properties of the domain walls and model calculations [5]. Only few attempts have been made to discuss, for example, a Bloch wall from first principles [11, 12]. In the present paper the energy, which is needed to form different types of domain walls, is investigated by ab initio calculations following the method of Gallego et al. [12]. Results are shown for a Co layer of varying thickness sandwiched between two semi-infinite fcc Pt systems and a bulk like system, where the Co film is covered by fcc Co bulk.

Finally, we demonstrate the influence of domain walls on the electric properties of the Co film. The magnetoresistance is investigated by using a Kubo-Greenwood approach [13]. We assume a perpendicular geometry, for which the magnetic moments are in-plane oriented and the current is adjusted parallel to the surface normal (CPP geometry). This allows a direct calculation of the domain wall magnetoresistance (DMWR).

2 Theoretical Aspects

The energy of domain wall formation in fcc Co films has been calculated by means of the density functional theory. A fully-relativistic spin-polarized version of the screened Korringa-Kohn-Rostoker (SKKR) method for layered systems developed by Weinberger and Szunyogh [14, 15] has been used to investigate the electronic and magnetic properties of the FM Co films with different magnetic configurations. The Co films are embedded between two semi-infinite fcc Pt bulk systems. Some extra Pt layers are included to incorporate possible charge fluctuations at the Co-Pt interface [14], i.e., the systems are of the following form:

$$\mathrm{Pt}(001)/\mathrm{Pt}_m/\mathrm{Co}_n/\mathrm{Pt}_m/\mathrm{Pt}(001), \quad \text{with} \quad 2 \leq n \leq 54.$$

The number of buffer layers m varies between nine and eleven, because the total number of layers has to be a multiple of three, which is related to the special properties of the screened structure constants [14]. The three-dimensional lattice constant is kept fixed to $a_{\mathrm{Pt}} = 7.4164$ a.u., which corresponds to the experimental value of fcc Pt. In order to check the influence of the lattice spacing and bulk material on the magnetic properties of the Co film, additional calculations have been performed using Co bulk instead of Pt. In this case the lattice constant was $a_{\mathrm{Co}} = 6.6786$ a.u. being the experimental lattice constant of fcc Co.

2.1 Ab Initio Description of Domain Walls

The energy costs of a domain wall, which is placed in a FM film, can be obtained from the energy difference $\Delta E(C)$ between the energy of the domain wall configuration C and the energy of a magnetic reference state C_0 [14, 16]. The reference state is chosen to be an in-plane FM configuration with magnetic moments oriented in $\hat{x}$-direction,

$$C_0(S) = \{\underbrace{\hat{x}, ..., \hat{x}}_{S}\}, \tag{1}$$

with S being the total number of layers. Assuming that the domain wall is in-plane oriented (cf. also Fig. 1), the magnetic configuration of a 180° Bloch wall is given by

$$C(S) = \{\underbrace{\hat{x}, .., \hat{x}}_{S_1}, \underbrace{\hat{n}_1^{\phi}, ..., \hat{n}_L^{\phi}}_{L}, \underbrace{-\hat{x}, ..., -\hat{x}}_{S_2}\}, \qquad S = S_1 + S_2 + L, \tag{2}$$

where L denotes the width of the domain wall and $\hat{n}_p^{\phi}$ describes the orientation of the magnetic moments in layer p. In case of a 90° wall in (2), $-\hat{x}$ has to be replaced by $\hat{y}$. The energy needed to form a domain wall can be obtained from the difference between the energy of the reference configuration C_0 and the domain wall configuration C,

$$\Delta E(S, C) = E(S, C) - E(S, C_0). \tag{3}$$

Making use of the magnetic force theorem [17] the two energies correspond to grand canonical potentials at zero temperature and only the reference configuration is calculated self-consistently. The energy difference is then given by the sum over the layer-resolved energy differences,

$$\Delta E(S, C) = \sum_{p=1}^{S} \Delta E_p(S, C). \tag{4}$$

The energy difference in a layer p obtained for a given magnetic configuration C is then defined by

$$\Delta E_p(S, C) = \int_{\epsilon_b}^{\epsilon_F} \left(n_p(C, \epsilon) - n_p(C_0, \epsilon)\right)(\epsilon - \epsilon_F), \tag{5}$$

where $n_p(C, \epsilon)$ refers to the layer-projected density of states, which corresponds to the magnetic configuration C. The energy ϵ_b denotes the valence band bottom and ϵ_F is the Fermi energy of the Pt bulk system. This approach has been used to investigate three different types of Bloch walls: 90°, 180°, and split 180° walls. The latter type of domain wall consists of a 180° wall split into two 90° walls (Fig. 1, bottom).

3 Results and Discussion

Sandwich structures of $Pt(001)/Pt_m/Co_n/Pt_m/Pt(001)$ (the semi-infinite bulk systems not mentioned explicitely in the following) have been used to

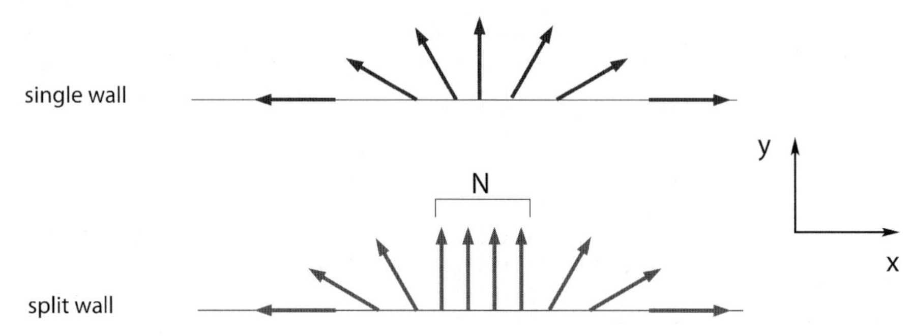

Fig. 1. Investigated arrangements of the magnetic moments in a 180° domain wall. *Top*: Single 180° domain wall. *Bottom*: Split domain wall. The two 90° walls are separated by N in-plane (but perpendicular to the reference configuration) oriented layers

investigate the formation of domain walls on the Co layer. In the present calculations the actual trilayers are covered by two semi-infinite bulk systems. This situation is different from the experimental setup, in which finite systems investigated. Hence, the magnetic structures and sizes of the magnetic moments, may differ when comparing theory and experiment.

3.1 Magnetic Moments

The magnetic moments of $Pt_9/Co_{30}/Pt_9$ are shown in Fig. 2 as a representative example. Except of the direct interface, the magnetic moments of the Co layers are mainly constant, whereby the magnetic moments at the interfaces are slightly larger as compared to the magnetic moments of the inner layers. The Pt layers are also magnetized having a magnetic moment of $m = 0.22\,\mu_B$ at the Pt-Co interface. However, the magnetic moments are induced by the neighboring Co layers and rapidly decrease with the distance from the interface. The magnetic moments of the Co layers are larger than the bulk value $m_{\text{bulk}} = 1.7\,\mu_B$ [18]. The spin moment amounts to $m_s = 1.87\,\mu_B$ and the orbital moments are about $m_l = 0.14\,\mu_B$. However, the enhancement of the spin magnetic moments is not caused by the confinement, but by the fact that the lattice constant of bulk Pt is larger compared to the lattice constant of bulk fcc Co. Ab initio investigations of bulk fcc Co have shown that for the lattice constant of Pt bulk, the spin magnetic moment results in an even larger value of $m_{\text{bulk}} = 1.95\,\mu_B$ [19]. If Pt is replaced by Co bulk the spin magnetic moment amounts to $m_s = 1.67\,\mu_B$ being quite close to the value, determined experimentally.

Additionally, the magnetic moment of Co also depends on the thickness of the Co layer n. This can be seen in Fig. 3, where the average magnetic moments per Co layer are plotted for different thicknesses. With increasing thickness of the Co layer the average spin magnetic moments, m_s, slightly decrease from $2.0\,\mu_B$ for $n = 2$ to $1.88\,\mu_B$ for thick layers with $n = 54$.

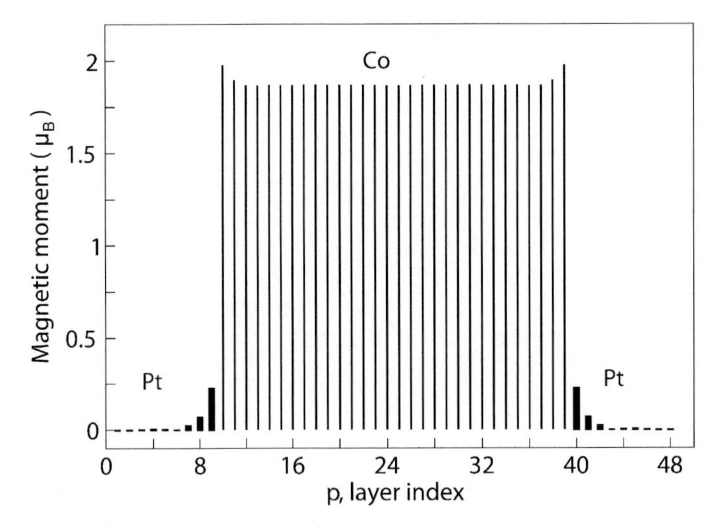

Fig. 2. Total magnetic moments of the $Pt_9/Co_{30}/Pt_9$ trilayer system. *Large, thin bars* belong to the magnetic moments of Co, which are enhanced as compared to the bulk magnetic moment of Co, see text

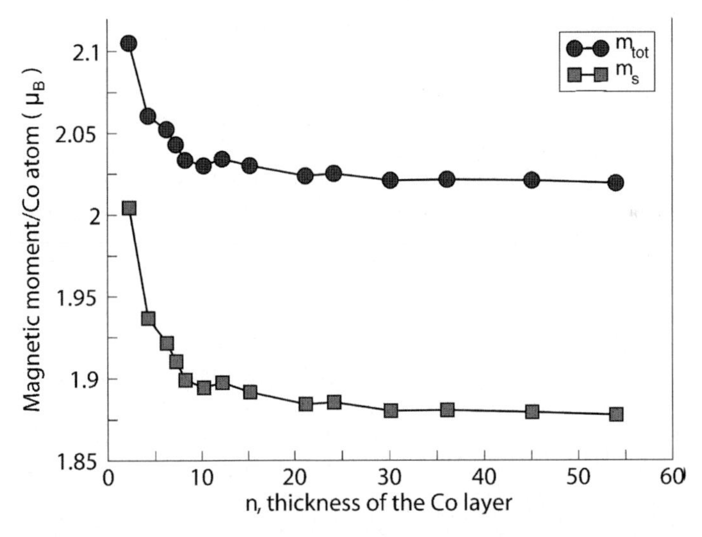

Fig. 3. Averaged magnetic moment of Co depending on the total thickness of the Co layer n. The spin magnetic moments are denoted by *squares*. *Circles* mark the total magnetic moments, i.e., including the orbital contributions

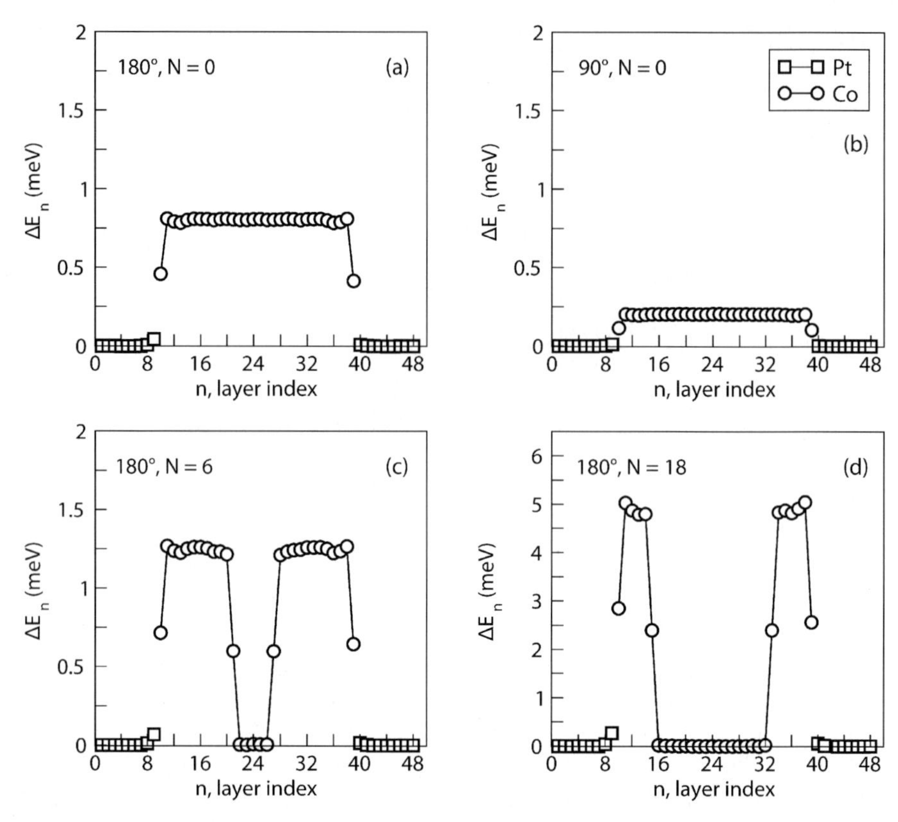

Fig. 4. Layer resolved energies of domain wall formation ΔE in $Pt_9/Co_{30}/Pt_9$ for: (**a**) a single 180° and (**b**) a 90° wall. In (**c**) and (**d**) results for a split 180° domain wall are shown. N marks the distance between the two partial walls

The total magnetic moments m_{tot} show similar behaviour, i.e., the orbital contributions lead to a more or less constant shift of the magnetic moments.

3.2 Energy of Formation

A domain wall inserted in a FM layer (single domain) will lead to an increase in energy. The actual costs can depend on the structure of the domain wall. Using the definition from (3), the energy difference ΔE between the FM reference state C_0 (see Fig. 1) and different configurations C containing Bloch walls, are discussed considering the $Pt_9/Co_{30}/Pt_9$ system as example. The results are displayed in Fig. 4. The formation of a the 180° Bloch wall leads to an increase of about 0.8 meV per Co layer at the inner part of the Co film and about half the size at the direct Pt-Co interface. Nearly no changes can be observed for the Pt layers and the energy difference is always smaller than 0.06 meV. This could be expected, because the Pt layers are mainly

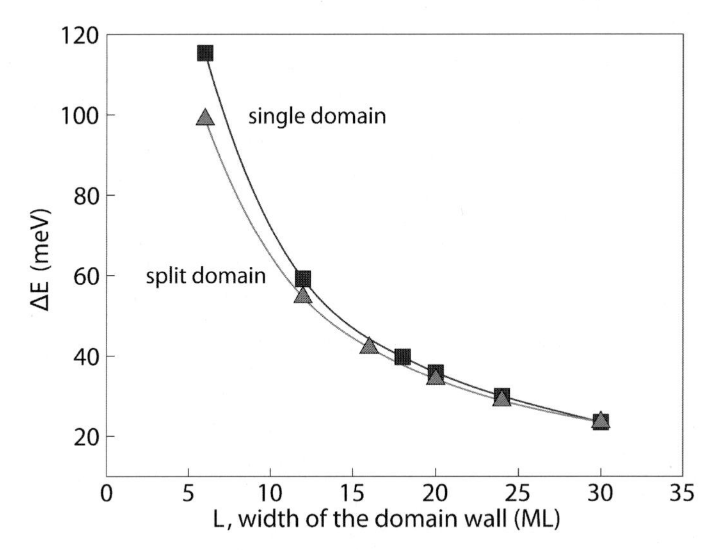

Fig. 5. The energy difference between single 180° walls and 180° walls split in two 90° walls as function of the width of the domain wall. This investigation has been done for a fixed Co thickness $n = 30$

non-magnetic and not affected by the magnetization reversal of the Co layer. In case of a 90° Bloch wall the energy differences ΔE are much smaller. Only 0.2 meV per Co layer (0.1 meV at the interfaces) have to be spend to establish the domain wall in the Co film. Accordingly, the energy, which is needed to form a 180° wall is four times larger as compared to the formation of a 90° domain wall. This suggests that the system could gain energy by splitting a single 180° wall into two 90° walls separated by a number N of layers with in-plane oriented magnetic moments, see bottom of Fig. 1. Two representative results are shown in Figs. 4(c) and (d) for $N = 6$ and $N = 18$, respectively. Obviously, in both cases the energy of formation per layer is large as compared to the case of a single 90° wall. Instead of 0.4 meV ($2\times\Delta E(90°)$) ΔE amounts to 1.25 meV ($N = 6$) or 5 meV in the case of $N = 18$. This can be understood from the fact that for finite N and fixed thickness of the Co layer the width of the domain walls has to decrease, which is equivalent to an increase of the rotation angle of the magnetic moments. Consequently, the split of domain walls is much more costly than formation of a single 180° wall, more precisely, a factor of two in the case of large N. However, the situation becomes different if the domain walls are of the same size, namely $L = 2 * N$. This has been proven by placing domain walls of different width in a Co film with $n = 30$, see Fig. 5. In the case of tiny domain walls consisting of six layers (i.e. $\approx$ 1 nm) one would indeed gain 16.5 meV using two 90° walls instead of one 180° wall. With increasing width of the domain wall the difference between single and split walls shrinks drastically (Fig. 5), e.g., to 1.67 meV in the case of domain wall covering 20 layers (3.9 nm). From the

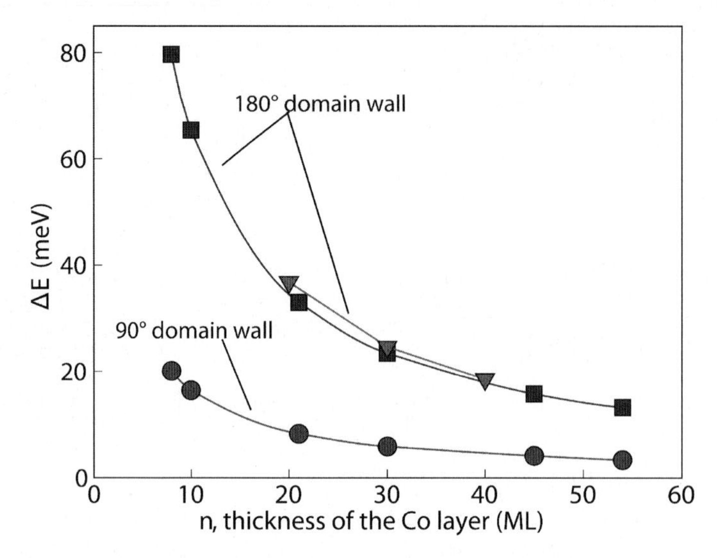

Fig. 6. Energy of domain wall formation as a function of the Co thickness. *Squares* mark the results for 180° walls and 90° walls are denoted by *circles*. The energy differences obtained for domain walls on fcc Co(001)/Co_{48}/Co(001) are given by *triangles*

results shown in the Figs. 4 and 5 it is obvious that the lowest energy of formation is always needed if a single 180° spans over the whole Co layer. However, these investigations have been performed for a relatively thin Co film with $n = 30$ (≈ 6 nm) and the results should not simply be assigned to thick layers. Investigations of bulk Permalloy have shown that the difference between single and split walls may be absent for thick films [12].

The energy of formation has been discussed so far for fixed width of the Co film n. In Fig. 6 the energy of formation is displayed depending on the thickness of the Co film. Here, the domain walls have always the same width as the Co film. With increasing width of the domain wall the energy of formation becomes smaller. From $L = 8$ to 54 the energy of formation decreases by a factor of six for both 180° and 90° walls. The results confirm what has already been found from the discussion of the layer-resolved energy differences for the Co_{30} film (see Figs. 4(a) and (b)), forming a 90° Bloch wall is always a factor of four less costly than a 180° wall. The formation energy does not change significantly if the Pt contacts are replaced by fcc Co, which is associated with a decrease in the lattice constant. The energy of formation of a Co(001)/Co_{48}/Co(001) system has been investigated for three different domain wall widths L marked by triangles in Fig. 6. The difference between ΔE of systems with and without Pt is in the range of 1 meV and is therefore tiny as compared to the absolute values of ΔE.

To avoid missunderstandings it should be added, that the formation of a domain wall costs energy as compared to a single domain state C_0, but it self-evidently lowers the energy in the case of two neighbouring domains or layers having different magnetic orientations. The costs of a direct spin flip would be one order of magnitude larger as compared to the domain wall formation.

3.3 Domain Wall Resistance

In layered systems, there are several sources of resistance, such as, the Lorentz force, AMR, and domain walls. The contributions from the Lorentz force are small and can be neglected. The resistance of the domain wall is given by [8]

$$R(S, C) = \frac{r(S, C) - r(S, C_0)}{r(S, C_0)}, \tag{6}$$

where $r(S, C)$ is the CPP sheet resistance of a system with width S and magnetic configuration C. To avoid contributions from the AMR one has to chose a perpendicular geometry. Suppose that the current flows perpendicular to the planes of atoms and the domain wall has in-plane magnetic orientation (Bloch wall). In order to determine the resistances one can make use of the Kubo-Greenwood equation [13], which provides the diagonal elements of the conductivity tensor. A detailed description of this method can be found in [12, 20, 21]

Assuming steady state conditions, the CPP resistances can be calculated within this approach for any magnetic arrangement C. For a film of S layers and a magnetic configuration C the sheet resistance can be written as

$$r(S, C, \delta) = \sum_{p,q}^{S} \rho_{pq}(S, C, \delta), \tag{7}$$

where the resistivity ρ_{pq} connects the current in layer p to the electric field in layer q. Due to numerical reasons, the Fermi energy has to be complex, $\epsilon_{\mathrm{F}} + i\delta$. The sheet resistance is calculated for a finite imaginary part δ. It has been shown in previous works [22, 23] that the sheet resistance varies linearly with δ. Therefore, the calculations have been carried out for several values of δ. The actual value for $\delta = 0$ is then determined by a linear fit.

The dependence of the DWMR on width, wall type and orientation of the system is discussed for thin domain walls on a Co_{24} layer embedded in two Co bulk systems. First of all, it can be seen from Table 1 that the DWMR decreases if the width of the wall increases, which agrees with the experimental findings [6] and theoretical model calculations [8]. Whether the decrease corresponds to the predicted $1/L^2$ behaviour could not be decided from these data. To this a much larger number of domain walls with different width L has to be investigated.

It has been shown in Sect. 3.2 that the formation of a 90° wall is less costly. This comes along with a tiny DWMR of 0.93 % being 6.5 times smaller

Table 1. Calculated magnetoresistance for domain walls of different type and width L on $Co(001)/Co_{24}/Co(001)$ and $Co(111)/Co_{24}/Co(111)$

growth direction	wall type	L	DWMR (%)
(001)	180°	6	6.08
(001)	180°	10	2.00
(111)	180°	6	5.43
(111)	90°	6	0.93

compared to the DWMR caused by a 180° wall (Table 1). Furthermore, the DWMR shows a slight dependence on the growth direction of the layer. In the case of Co, the DWMR of a layer oriented in (111)-direction is 0.65 % smaller as compared to the (001) oriented system, see Table 1.

4 Conclusion

Domain walls are an inherent part of FM films. The dynamical properties are intensively discussed within micromagnetic simulations. In the present paper the formation of Bloch domain walls on a Co film has been investigated on an atomistic level. This has been done by using a Green's functions technique, which provides formation energies of any noncollinear magnetic configuration. In the case of the Co film it could be shown that a 90° Bloch wall costs less energy than a 180° wall. One can save energy by splitting a 180° wall into two 90° walls provided that the walls are of the same width. It has been demonstrated, for some examples, that the difference in the formation energy comes along with different contributions to the magnetoresistance. The values obtained for the DWMR are quite large, but strongly decrease with increasing width of the domain wall.

Acknowledgements

The author would like to thank P. Weinberger for his kind support and L. Szunyogh for many helpful discussions. Financial support was provided from the Deutsche Forschungsgemeinschaft through the SFB 491 *Structure and Transport in Magnetic Heterostructures*.

References

[1] S. Maekawa, T. Shinjo (Eds.): *Spin dependent transport in magnetic nanostructures*, Advances in Condensed Matter Science (Taylor & Francis 2002)
[2] F. Jorgensen (Ed.): *The complete handbook of magnetic recording* (TAB Books, New York 1996)

[3] A. Hubert, R. Schaefer (Eds.): *Magnetic domains* (Springer, New York 1998)
[4] A. D. Kent, U. Rüdiger, S. Parkin: Domain wall resistivity in epitaxial thin film microstructures, J. Phys.: Condens. Matt. **13**, R461 (2001)
[5] L. Berger: Analysis of measured transport properties of domain walls in magnetic nanowires and films, Phys. Rev. B **73**, 014407 (2006)
[6] J. F. Gregg, W. Allen, K. Ounadjela, M. Viret, M. Hehn, S. M. Thompson, J. M. D. Coey: Giant magnetoresistive effects in a single element magetnic thin film, Phys. Rev. Lett. **77**, 1580 (1996)
[7] G. Dumpich, T. P. Krome, B. Hausmanns: Magnetresistance of a single Co wire, J. Magn. Magn. Mater. **248**, 241 (2002)
[8] P. M. Levy, S. Zhang: Resistivity due to domain wall scattering, Phys. Rev. Lett. **79**, 5110 (1997)
[9] R. Wieser, U. Nowak, K. D. Usadel: Domain wall mobility in nanowires: transverse versus vortex walls, Phys. Rev. B **69**, 64401 (2004)
[10] M. Bolte, G. Meier, C. Bayer: Spin-wave eigenmodes of Landau domain patterns, Phys. Rev. B **73**, 052406 (2006)
[11] J. Schwitalla, B. L. Györffy, L. Szunyogh: Electronic theory of bloch walls in ferromagnets, Phys. Rev. B **63**, 104423 (2001)
[12] S. Gallego, P. Weinberger, L. Szunyoggh, P. M. Levy, C. Sommers: *Ab initio* description of domain walls in permalloy: Energy of formation and resistivities, Phys. Rev. B **68**, 054406 (2003)
[13] R. Kubo: Statistical-mechanical theory of irreversible processes I. general theory and simple application to magnetic and conduction problems, J. Phys. Soc. Jpn. **12**, 570 (1957)
[14] P. Weinberger, L. Szunyogh: Perpendicular magnetism in magnetic multilayer systems, Comp. Mater. Sci. **17**, 414 (2000)
[15] L. Szunyogh, B. Újfalussy, P. Weinberger: Magnetic anisotropy of iron multilayers on Au(001): first-principles calculations in terms of the fully relativistic spin-polarized screened KKR method, Phys. Rev. B **51**, 9552 (1995)
[16] P. Weinberger: *Ab initio* theory of electric transport in solid systems with reduced dimensions, Physics Reports **377**, 281 (2003)
[17] H. J. F. Jansen: Magnetic anisotropy in density-functional thoery, Phys. Rev. B **59**, 4699 (1999)
[18] S. Hashimoto, Y. Ochiai: Co/Pt and Co/Pd multilayers as magneto-optical recording materials, J. Magn. Magn. Mater. **88**, 211 (1990)
[19] H. C. Herper: Calculated employing WIEN97 within the generalized gradient approximation, private communication
[20] H. C. Herper, P. Weinberger, L. Szunyogh, C. Sommers: Interlayer exchange coupling and perpendicular electric transport in Fe/Si/Fe trilayers, Phys. Rev. B **66**, 064426 (2002)
[21] H. C. Herper: Domain wall resistance in Co layers: An ab initio study, to be published in Phys. Rev. B
[22] P. Weinberger, L. Szunyogh, C. Blaas, C. Sommers: Perpendicular transport in Fe/Ge model heterostructures, Phys. Rev. B **64**, 184429 (2001)
[23] H. C. Herper, P. Weinberger, A. Vernes, L. Szunyogh, C. Sommers: Electric transport in Fe/ZnSe/Fe heterostructures, Phys. Rev. B **64**, 184442 (2001)

Domain Wall Spin Structures in 3d Metal Ferromagnetic Nanostructures

M. Laufenberg[1], M. Kläui[1], D. Backes[1,2], W. Bührer[1], H. Ehrke[1,5], D. Bedau[1], U. Rüdiger[1], F. Nolting[2], L. J. Heyderman[2], S. Cherifi[3], A. Locatelli[3], R. Belkhou[3], S. Heun[3], C. A. F. Vaz[4], J. A. C. Bland[4], T. Kasama[5], R. E. Dunin-Borkowski[5], A. Pavlovska[6], and E. Bauer[6]

[1] Fachbereich Physik, Universität Konstanz, 78457 Konstanz, Germany
[2] Paul Scherrer Institut, 5232 Villigen PSI, Switzerland
[3] Nanospectroscopy Beamline, Sincrotrone Trieste, 34012 Basovizza / Trieste, Italy
[4] Cavendish Laboratory, University of Cambridge, Cambridge CB3 0HE, U. K.
[5] Department of Materials Science and Metallurgy, University of Cambridge, Cambridge CB2 3QZ, U. K.
[6] Department of Physics and Astronomy, Arizona State University, Tempe, Arizona 85287-1504, USA

Abstract. In this article, a comprehensive study of head-to-head domain wall spin structures in $Ni_{80}Fe_{20}$ and Co nanostructures is presented. Quantitative domain wall type phase diagrams for NiFe and Co are obtained and compared with available theoretical predictions and micromagnetic simulations. Differences to the experiment are explained taking into account thermal excitations. Thermally induced domain wall type transformations are observed from which a vortex core nucleation barrier height is obtained. The stray field of a domain wall is mapped directly with sub-10 nm resolution using off-axis electron holography, and the field intensity is found to decrease as $1/r$ with distance. The magnetic dipolar coupling of domain walls in NiFe and Co elements is studied using X-ray magnetic circular dicroism photoemission electron microscopy. We observe that the spin structures of interacting domain walls change from vortex to transverse walls, when the distance between the walls is reduced. Using the measured stray field values, the energy barrier height distribution for the nucleation of a vortex core is obtained.

1 Introduction

Domain walls in nanoscale ferromagnetic elements are in the focus of interest because of their potential for applications in a variety of fields such as magnetic logic [1] as well as data storage [2, 3] and due to their associated fundamental physical effects [4–10]. Magnetoresistance effects related with domain walls have been investigated in recent years [4–6, 11, 12]. Domain wall motion induced by external fields [7–9] has been studied and wall mobilities and depinning fields [10] have been determined. Current induced domain wall motion and the underlying spin torque effect have been the subject of rising interest recently [13–15] with investigation of critical current densities and of domain wall velocities. The resistivity of a domain wall, its mobility, the

R. Haug (Ed.): Advances in Solid State Physics,
Adv. in Solid State Phys. **46**, 271–283 (2008)

depinning fields and critical currents as well as the spin torque effect depend critically on the wall spin structure [1, 4–10, 13–15]. In this field of research, temperature and heating effects play a key role [16–19], in particular the wall spin structure was predicted to depend on the temperature [19], which calls for studies at variable temperatures. Ring elements have proven to be a useful geometry for the investigation of domain walls since they can easily be created and positioned by applying an external uniform magnetic field. Ferromagnetic rings can be in the flux closure vortex state or in the onion state characterized by 180° head-to-head and tail-to-tail domain walls [20], which can be of either vortex or transverse type [21] with the spin structures shown in Figs. 1(a) and (b), respectively.

A strong influence of domain wall interaction on the switching of magnetic elements was found recently, when interaction-induced collective switching of adjacent elements was observed for small spacings [22, 23]. Such switching is dominated by domain wall motion and can only be understood with a detailed knowledge of the interacting domain walls' spin structures. Theoretically, the energies of the two wall types are different when interacting with an external field. We therefore expect the dipolar coupling to affect the two wall types in different ways and coupling-induced transitions from one domain wall type to another may occur [21]. A deeper understanding of the energetics involved is only possible if the stray fields of domain walls are determined quantitatively.

In this paper, we review our recent work on domain wall spin structures in mesoscopic NiFe and Co ring elements. For investigation of the spin structure of domain walls, we use high-resolution imaging by X-ray magnetic circular dichroism photoemission electron microscopy (XMCD-PEEM) as well as off-axis electron holography providing sub-10 nm resolution. In particular, we systematically study the spin structure of head-to-head domain walls at variable temperature. We present the phase diagrams and compare the findings with available theoretical calculations and the results of micromagnetic simulations. We derive a qualitative temperature dependence of the phase boundary and extract the energy barrier height for the vortex nucleation. Having comprehensively explored the spin structures of isolated domain walls, we study the interaction of adjacent walls and their dipolar coupling. By correlating the domain wall spin structure changes due to increasing edge-to-edge spacing between neighboring elements with a quantitative measurement of the domain wall stray field, we obtain a direct measure of the energy barrier distribution for the vortex core nucleation in our samples.

Arrays of 5×5 polycrystalline Co and NiFe rings with different thicknesses and widths were fabricated as described in [24, 25]. For the first set of samples, the edge-to-edge spacing between adjacent rings was more than twice the diameter to prevent dipolar interactions which might otherwise influence the domain wall type. To determine the spin structure of the domain walls as a function of the ring geometry, the samples were imaged using XMCD-PEEM [26]. Arrays of 350 nm wide rings with edge-to-edge spacings down to 10 nm were fabricated to investigate different dipolar coupling strengths

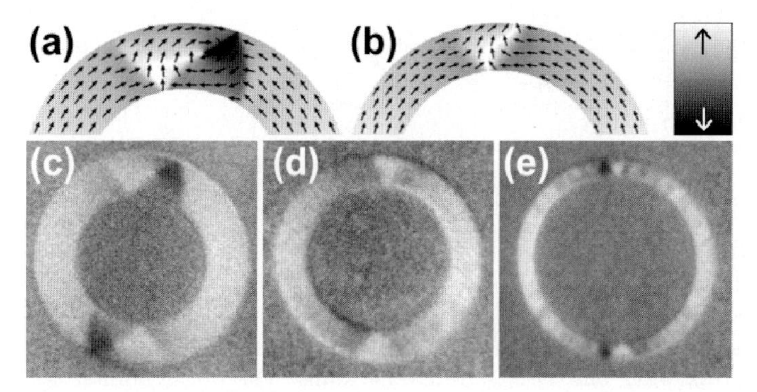

Fig. 1. (from [30]) Spin structure of (**a**) a vortex and (**b**) a transverse wall simulated using OOMMF. PEEM images of (**c**) 30 nm thick and 530 nm wide ($D = 2.7\,\mu$m), (**d**) 10 nm thick and 260 nm wide ($D = 1.64\,\mu$m), and (e) 3 nm thick and 730 nm wide ($D = 10\,\mu$m) NiFe rings. The *gray scale* indicates the direction of magnetic contrast

between domain walls in adjacent rings and resulting domain wall types. For the transmission off-axis electron holography experiments, 3/4-rings were patterned from 27 nm thick Co films on 50 nm thick SiN membranes as detailed in [27]. Open rings rather than full rings were grown on the fragile membranes in order to facilitate the lift-off process which cannot be assisted by ultrasound. In order to obtain quantitative information about stray fields, Co samples were investigated by off-axis electron holography [28, 29]. Co was chosen rather than NiFe for this investigation due to its higher saturation magnetization and therefore higher stray field.

2 Domain Wall Spin Structure

2.1 Domain Wall Phase Diagrams

In Fig. 1, we present PEEM images of (c) a thick and wide NiFe ring, (d) a thin and narrow ring, and (e) an ultrathin ring measured at room temperature. The domain wall type was systematically determined from PEEM images for more than 50 combinations of ring thickness and width for both NiFe and Co and the quantitative phase diagrams shown in Figs. 2(a) and (c) were extracted. The phase diagrams exhibit two phase boundaries indicated by solid lines between vortex walls (thick and wide rings, squares), transverse walls (thin and narrow rings, circles), and again vortex walls for ultrathin rings.

Now we first discuss the upper boundary shown in Figs. 2(a, c). Theoretically this phase boundary was investigated by *McMichael* and *Donahue* [21]. They calculated the energies for a vortex and a transverse wall and determined the phase boundary by equating these two energies. The calculated

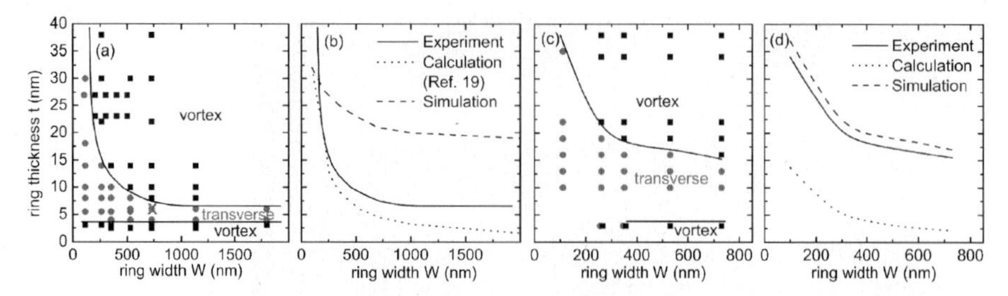

Fig. 2. (Color online, partly from [30, 31]) Experimental phase diagrams for head-to-head domain walls in (**a**) NiFe and (**c**) Co rings at room temperature. *Black squares* indicate vortex walls and *circles* transverse walls. The phase boundaries are shown as *solid lines (online color)*. (**b, d**) Comparison of the upper experimental phase boundary (*solid lines*) with results from calculations (*dotted lines*) and micromagnetic simulations (*dashed lines*). The thermally activated wall transitions shown were observed for the ring geometry marked with a *red cross (online color)* in (a) (W=730 nm, t=7 nm)

phase boundary (dotted lines in Figs. 2(b, d)) is of the form $t \cdot W = C \cdot \delta^2$ where δ is the exchange length and C a universal constant. It is shifted to lower thickness and smaller width compared to the experimental boundary (solid lines in Figs. 2(b, d)). This discrepancy can be understood by taking into account the following: The calculations [21] compare total energies and therefore determine the wall type with the absolute minimum energy as being favorable. In the experiment, the wall type was investigated after saturation of the ring in a magnetic field and relaxing the field to zero. During relaxation, first a transverse wall is formed reversibly [32]. For the formation of a vortex wall, an energy barrier has to be overcome to nucleate the vortex core. So the observed spin structure does not necessarily constitute the absolute minimum energy, but transverse walls can be observed for combinations of thickness and width where they constitute local energy minima even if a vortex wall has a lower energy for this geometry. To shed further light onto this, we have simulated the experiment by calculating the domain wall spin structure after reducing an externally applied field stepwise using the OOMMF code [33] (for NiFe: $M_s = 800 \times 10^3$ A/m, $A = 1.3 \times 10^{-11}$ J/m; for Co: $M_s = 1424 \times 10^3$ A/m, $A = 3.3 \times 10^{-11}$ J/m; for both: damping constant $\alpha = 0.01$, cell size 2–5 nm). The simulated boundary is shifted to higher thickness and larger width compared to the experiment. This we attribute to the fact that thermal excitations help to overcome the energy barrier between transverse and vortex walls in case of the room temperature experiment, while they are not taken into account in the 0 K simulation. So we can expect that for temperatures above room temperature the upper phase boundary is shifted to lower thickness, in other words, that transverse walls formed at room temperature change to vortex walls with rising temperature. This means that with rising temperature the experimental phase boundary

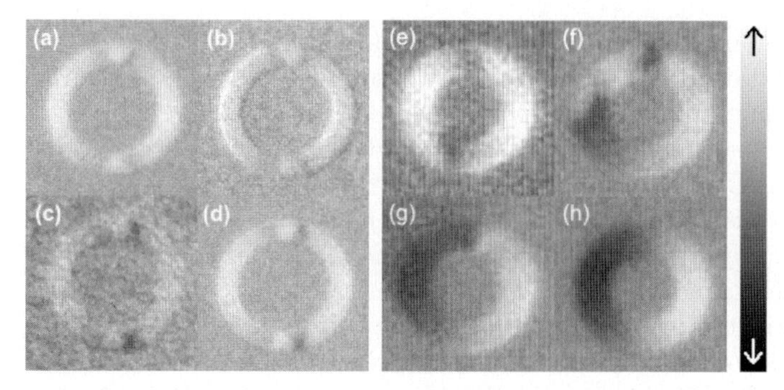

Fig. 3. (partly from [30]) PEEM images of a 7 nm thick and 730 nm wide ring imaged during a heating cycle at temperatures of (**a**, **d**) $T = 20°C$ (before and after heating, respectively), (**b**) $T = 260°C$, and (**c**) $T = 310°C$ (estimated errors are ± 10 K). Due to heating, rings (here (**e**): 7 nm thick, 1135 nm wide with two vortex walls) can attain either (**g**) a vortex state with a 360° domain wall or (**h**) the vortex state. The intermediate state, where one wall is displaced, is shown in (**f**). The *gray scale* indicates the direction of magnetic contrast

approaches the theoretical one since the walls attain the energetically lower spin structure.

2.2 Thermally Activated Domain Wall Transformations

In order to corroborate this explanation for the difference between the experiment on the one hand and calculations and simulations on the other hand, we have performed temperature dependent XMCD-PEEM studies. Figure 3 shows an image series of a 7 nm thick and 730 nm wide NiFe ring (geometry marked by a cross in Fig. 2(a)) for different temperatures of (a, d) $T = 20°C$ (before and after heating), (b) $T = 260°C$, and (c) $T = 310°C$. Transverse walls are formed (a) during saturation in a magnetic field and relaxation before imaging. At first, heating does not influence the spin structure of the domain walls as shown in (b), only the image contrast becomes weaker because imaging is more difficult at higher temperatures due to drift problems and decreasing magnetization. At a transition temperature between $T = 260°C$ and $T = 310°C$ corresponding to a thermal energy between 6.7×10^{-21} J and 8.0×10^{-21} J, the transverse walls change to vortex walls (c), so that a domain wall spin structure was created which is not accessible for the same ring geometry by only applying uniform magnetic fields. Figure 3(d) confirms that the vortex wall is stable during cooling down. This means that both domain wall types are (meta-)stable spin configurations and therefore constitute local energy minima at room temperature for this geometry. These PEEM experiments directly show that the position of the upper experimental phase boundary is temperature dependent and is shifted to lower thickness and width with increasing temperature. These results thus confirm the hypothesis about the discrepancy between experiment and theory put forward before: Both domain wall types constitute local energy minima, with the transverse wall attained due to the magnetization process, even if a vortex wall has a lower energy. Experimentally, we directly observe thermally activated crossing of the energy barrier between high energy transverse and low energy vortex walls.

It should be mentioned however, that the flux closure vortex state of the ring without any domain walls and with the magnetization aligned everywhere along the ring perimeter is the energetically most favorable state. Many rings attain this state when the temperature is increased as shown in Figs. 3(e–h). In order to observe the wall type transformations shown in Figs. 3(a–d), it is therefore necessary, that the energy barrier between transverse and vortex walls is lower than the barrier for the transition to the vortex state of the ring. This critically depends on imperfections of the ring microstructure which can serve as pinning centers and stabilize a domain wall.

It can be seen by comparing the boundaries for NiFe and Co in Fig. 2, that for NiFe the calculations [21] fit the experiment better than the simulations while for Co the opposite is true. The energy barrier between a transverse and a vortex wall can be overcome more easily in the case of NiFe rather than Co, so that transverse walls created are more likely to be retained at a certain temperature in a Co ring than in a NiFe ring with analogous dimensions. This is consistent with the observation that in NiFe there is a more abrupt change between transverse and vortex walls with varying geometry than in Co.

2.3 Walls in Thin and Wide Structures – Limits of the Description

We turn now to the discussion of the low thickness regime of the phase diagrams shown in Fig. 2, where a second phase boundary between 3 and 4 nm is found both for NiFe and for Co. In terms of energetics, this is not expected because the calculations [21] show that a transverse wall has a lower energy than a vortex wall in this thickness regime. But these calculations assume a perfect microstructure and do not take into account morphological defects such as the surface roughness. Holes, which might serve as nucleation centers for the vortex wall formation, were not observed in atomic force microscopy images. However, this does not exclude a spatial modulation of magnetic properties [37] such as the exchange or the saturation magnetization, which could locally allow for a stronger twisting of adjacent spins. Thus a vortex wall would be energetically more favorable in this thickness regime only due to imperfections of the microstructure or the morphology.

In the thin samples investigated, a ripple domain formation [38] is observed as shown in Figs. 4(a, b) (see also Fig. 1(e)). This can be attributed to statistical variations of the anisotropy of individual grains. Consequently, this phenomenon is more pronounced in the polycrystalline Co structures, in

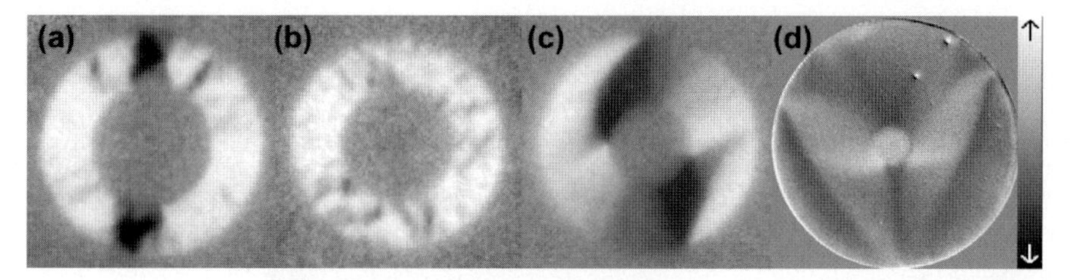

Fig. 4. Limiting cases of the structures investigated: 3 nm thick and 1.8 µm wide (**a**) NiFe and (**b**) Co rings showing ripple domain formation, (**c**) 6 nm thick and 3 µm wide ring with distorted transverse walls, and (**d**) 10 nm thick and 2.1 µm wide disc-like NiFe ring with 700 nm inner diameter in the triangle state (for detailed explanations of the contrast see [34–36]). The *gray scale* indicates the direction of magnetic contrast

which individual grains exhibit a non-negligible anisotropy compared to the weak anisotropy in NiFe.

The description in the frame of these phase diagrams is however limited by the width and thickness of the structure. In rings wider than ≈ 1.5 µm, we observe more complicated domain wall spin structures like distorted transverse walls (Fig. 4(c)). Wide rings with a hole in the center exhibit a disc-like behavior with a triangle state as shown in Fig. 4(d). This type of structure is discussed in more detail in [34–36]. In very thick elements double vortex walls were reported recently [39], which can also constitute (meta-)stable configurations in the thickness regime investigated here, as it turned out from experiments on current-induced domain wall transformations [40]. Recent theoretical calculations [41] propose to distinguish between symmetric and asymmetric transverse walls which is not done here, because both types are difficult to distinguish experimentally and sample irregularities can influence the detailed wall spin structure.

3 Domain Wall Coupling Energetics

3.1 Coupling Between Adjacent Domain Walls

After saturating with a magnetic field and relaxing the field, rings attain the onion state characterized by two head-to-head domain walls as shown before (see also Fig. 5(a)). An array of 25 rings in the onion state exhibits 50 walls in total. The domain walls inside the array interact with adjacent walls via their stray fields. Only 10 walls, which are located at the two opposite edges (top and bottom edges in Figs. 5(b) and (c)) of the array and therefore have no neighboring rings, are not influenced by stray fields of an adjacent wall. For all experiments, ring thickness and width were chosen such that isolated rings of this geometry exhibit vortex walls according to the phase diagram presented above.

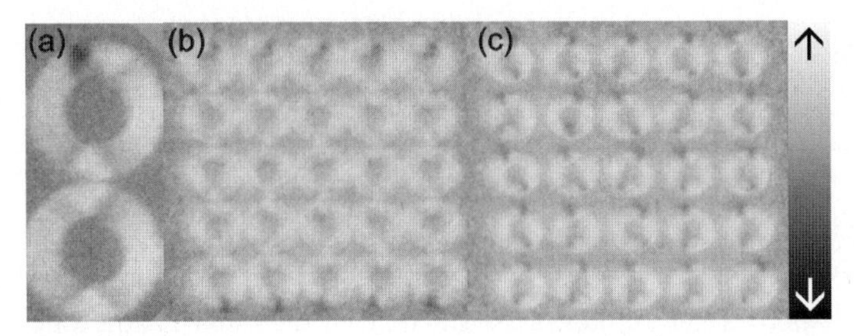

Fig. 5. (from [42]) (**a**) A high-resolution XMCD-PEEM image of two rings in the onion state, after saturation with an external field in the vertical direction and relaxation. White and black contrasts correspond to the magnetization pointing up and down, respectively. A non-interacting vortex wall (*top*) and three interacting transverse walls are visible. Overview images of an array of 27 nm thick and 350 nm wide NiFe rings with an edge-to-edge spacing of (**b**) 40 nm and (**c**) 500 nm, respectively. The transition from 100 % transverse walls inside the array for narrow spacings (b) to close to 0 % for large spacings (c) can be clearly seen. Since domain walls at the *top* and *bottom* edges of the array do not interact with adjacent walls, they are vortex walls for all spacings investigated

Figure 5 shows XMCD-PEEM images of arrays of 27 nm thick NiFe rings with (b) 40 nm and (c) 500 nm edge-to-edge spacing, respectively, as well as a high-resolution image (a) presenting both wall types. Vortex walls can be easily identified by black and white contrast which occurs because all magnetization directions corresponding to the full grey scale are present in a vortex (Fig. 5(a), top). In contrast, transverse walls exhibit the characteristic grey-white-grey contrast with the triangular spin structure in their center (Fig. 5(a), bottom).

In Fig. 6(a), we show the percentage of transverse walls inside the array as function of the edge-to-edge spacing for 27 nm thick NiFe rings (black squares) extracted from images of the type shown in Fig. 5. A decreasing number of transverse walls is found with increasing spacing. Domain walls at the edges of the arrays are vortex walls irrespective of the spacing due to the absence of dipolar coupling with adjacent walls. The data points for infinite spacings in Fig. 6(a) result from these domain walls. The transverse to vortex transition is characterized by a (10–90 %)-width of the switching distribution of $w = (65 \pm 9)$ nm and a center at $r_c = (77 \pm 5)$ nm. In Fig. 6(a), red triangles show a similar transition for 30 nm thick Co rings with $w = (328 \pm 130)$ nm and $r_c = (224 \pm 65)$ nm.

In order to explain these results, we first consider the process of domain wall formation in an isolated ring. When relaxing the applied external field from saturation, transverse walls are initially formed. In order to create a vortex wall, a vortex core has to be nucleated. This hysteretic transition from one wall type to the other involves overcoming a local energy barrier [32], since the nucleation of the vortex core is associated with a strong twisting of the

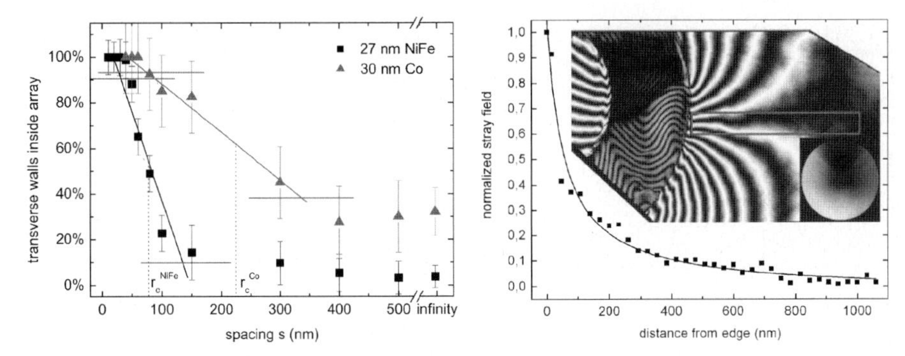

Fig. 6. (Color online, from [42]) (**a**) Percentage of transverse walls inside a ring array as function of edge-to-edge spacing. *Black squares* are for 27 nm NiFe, and *red triangles (online color)* for 30 nm Co, respectively. The error bars represent the absolute statistical error $1/\sqrt{n}$ due to the finite number n of domain walls investigated. The *horizontal lines* show the 10–90 %-levels of the transition from a transverse to a vortex domain wall. (**b**) The *inset* shows an off-axis electron holography image of a transverse wall in a 27 nm thick Co 3/4-ring. The color code (online color) indicates the direction of the in-plane magnetization and the *black lines* represent directly the stray field. The stray field strength was measured at several distances inside the marked area. The data points show the stray field normalized to the saturation magnetization as a function of the distance r from the ring edge for the wall shown in the *inset*. The line is a $1/r$-fit

spins in the core region [43]. In arrays of interacting rings, the edge-to-edge spacing dependent stray field stabilizes transverse walls so that for small spacings (corresponding to a strong stray field from the adjacent domain wall) transverse walls are favored (Fig. 5(b)). For increasing spacing, the influence of the stray field from an adjacent wall is reduced, until vortex walls are formed in the rings with the lowest energy barrier for the vortex core nucleation. The further the spacing increases the more rings nucleate vortex walls (Fig. 5(c)). Thus the spacing at which a wall switches from transverse to vortex is related to the nucleation barrier, which depends on local imperfections such as the edge roughness. So the number of domain walls that have switched from transverse to vortex as a function of the edge-to-edge spacing is a measure of the distribution of energy barriers for the vortex core nucleation.

For NiFe, a relatively sharp transition occurs from all walls being transverse to all walls being vortex walls. This corresponds to a narrow energy barrier distribution, while the domain walls in Co rings exhibit a much wider transition. This difference is thought to result from the different polycrystalline microstructures of the NiFe (magnetically soft fcc crystallites with negligible anisotropy) and the Co (hcp crystallites with strong uniaxial anisotropy leading to a larger number of pinning sites). Furthermore, this results in the presence of transverse walls in our Co sample even at infinite spacings.

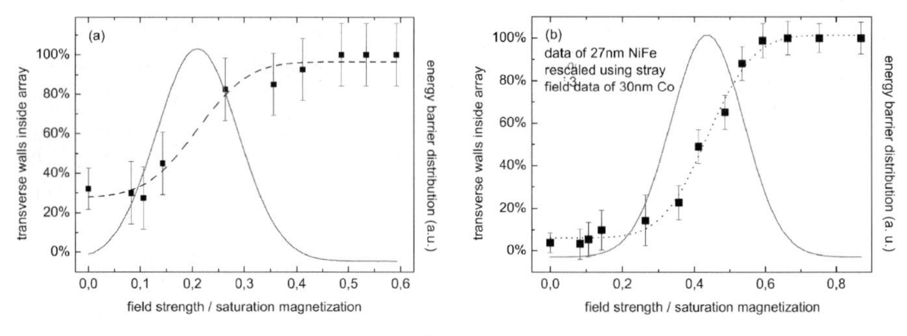

Fig. 7. (Color online, partly from [42]) (**a**) *Black squares* represent the same data points as shown in Fig. 6(a) for the 30 nm thick Co sample, but as a function of the normalized field strength. The *black dashed line* shows a fit with the error function. The corresponding Gaussian distribution of the energy barriers is shown as a *full red line (online color)*. (**b**) Corresponding data for the 27 nm thick NiFe sample rescaled using the stray field measurement of a domain wall in 30 nm thick Co

Thus, we chose Co for the electron holography measurements rather than NiFe in order to be able to observe a transverse wall and its stray field in an isolated structure.

3.2 Direct Observation of the Domain Wall Stray Field

This spacing-dependent distribution for the vortex core nucleation needs to be transformed to a distribution as a function of the stray field strength, which is in a first approximation proportional to the energy. To do this, the stray field as well as the magnetization of the domain wall was imaged using off-axis electron holography. The inset of Fig. 6(b) shows an image of the in-plane magnetic induction integrated in the electron beam direction, obtained from a transverse wall in a 27 nm thick isolated Co 3/4-ring designed with the same width as that of the rings imaged by XMCD-PEEM. No significant difference between the functional dependence of the stray field on the spacing is expected for a 27 nm and a 30 nm thick sample. The stray field was measured along the length of the region indicated in the image, and is shown as a function of the distance r from the ring edge in Fig. 6(b), normalized to the saturation magnetization of Co. The line is a $1/r$-fit which can be expected for the distance dependence of the stray field created by an area of magnetic poles for small r [44]. This dependence also confirms earlier results from indirect Kerr effect measurements [23]. In order to obtain the stray field of one single domain wall acting on an adjacent wall, the stray field of an isolated wall was imaged.

The spacing-dependent energy barrier distribution is now rescaled to a field-dependent distribution using the measured stray field decay of Fig. 6(b) and presented in Fig. 7(a). The rescaled data can be fitted with the error function $\mathrm{erf}(x)$, which is the integral of a Gaussian distribution. The error function is not the only possible fit covered by the error bars in Fig. 7(a), but it is consistent with our data. Assuming a similar dependence of the stray field for NiFe like measured for Co, a Gaussian distribution is also obtained for the energy barrier height distribution for NiFe as presented in Fig. 7(b). Thus a Gaussian distribution for the energy barriers is found, which is in agreement with the presence of independent local pinning centers at the particular wall position that determine the nucleation barrier. The position of the maximum is $H_{\mathrm{max}}/M_{\mathrm{s}} = 0.21 \pm 0.10$ and the full width at half maximum $w/M_{\mathrm{s}} = 0.16 \pm 0.05$, where M_{s} is the saturation magnetization. Using $E_{\mathrm{max}} = \frac{1}{2}\mu_0 M_{\mathrm{s}} H_{\mathrm{max}}$, an energy density of $E_{\mathrm{max}} = (8.4 \pm 4.0) \times 10^4\,\mathrm{J/m^3}$ equivalent to the field H_{max} can be obtained for the 30 nm thick Co sample.

Figure 6(a) shows that the transition for the Co sample saturates at a finite value for large spacings. In terms of the model described above, which explains how vortex walls are formed during relaxation from saturation, this means that an additional effective field would be needed to overcome the pinning of the remaining transverse walls at structural imperfections and to allow the vortex core nucleation and formation of an energetically favorable vortex wall. Since the pinning is much stronger in our Co sample than in the NiFe sample, this occurs here only for Co.

4 Conclusion

In conclusion, we have determined the spin structure of domain walls in NiFe and Co and extracted the corresponding room temperature phase diagrams with two phase boundaries between vortex walls for thick and wide as well as ultrathin rings and transverse walls for thin and narrow rings.

Using temperature dependent XMCD-PEEM imaging, we have observed a thermally activated switching from transverse walls established at room temperature to vortex walls at a transition temperature between 260°C and 310°C for NiFe rings. This gives direct experimental evidence for the fact that transverse and vortex walls are separated by an energy barrier, which can be overcome thermally. The low thickness regime of the phase diagrams revealed a second phase boundary which we attribute to spatial modulations of the magnetic properties in our thinnest samples.

Furthermore, we have mapped the stray field of a domain wall directly using off-axis electron holography with sub-10 nm resolution, and we find that the field strength falls off with a $1/r$-dependence. For interacting domain walls in 350 nm wide ring structures we observe a transition from a transverse to a vortex spin structure with increasing edge-to-edge spacing. By correlating this transition with the measured stray field, we are able to obtain the energy barrier height distribution for vortex core nucleation in our Co samples. This distribution has a Gaussian profile with the mean value being equivalent to an energy density of $(8.4 \pm 4.0) \times 10^4\,\mathrm{J/m^3}$.

Acknowledgements

The authors acknowledge support by the Deutsche Forschungsgemeinschaft through SFB 513, the Landesstiftung Baden-Württemberg, the EU through the European Regional Development Fund (Interreg III A Program), the EPSRC (U. K.), the Royal Society (U. K.), and by the EC through the 6th Framework Program. Part of this work was carried out at the Swiss Light Source, Villigen (Switzerland) and at Elettra, Trieste (Italy).

References

[1] D. A. Allwood et al., Science **309**, 1688 (2005)
[2] G. A. Prinz, J. Magn. Magn. Mater. **200**, 57 (1999)
[3] S. S. P. Parkin: U.S. patent 6,834,005 and patent application 10/984,055 (2004)
[4] P. M. Levy and S. Zhang, Phys. Rev. Lett. **79**, 5110 (1997)
[5] U. Ebels et al., Phys. Rev. Lett. **84**, 983 (2000)
[6] M. Kläui et al., Phys. Rev. Lett. **90**, 097202 (2003)
[7] D. Atkinson et al., Nature Mat. **2**, 85 (2003)
[8] Y. Nakatani, A. Thiaville, and J. Miltat, Nature Mat. **2**, 521 (2003)
[9] R. Wieser, U. Nowak, and K. D. Usadel, Phys. Rev. B **69**, 064401 (2004)
[10] M. Kläui et al., Appl. Phys. Lett. **87**, 102509 (2005)
[11] U. Rüdiger et al., Phys. Rev. Lett. **80**, 5639 (1998)
[12] A. D. Kent et al., J. Phys.: Cond. Mat. **13**, R461 (2001)
[13] A. Yamaguchi et al., Phys. Rev. Lett. **92**, 077205 (2004)
[14] M. Kläui et al., Phys. Rev. Lett. **94**, 106601 (2005)
[15] M. Kläui et al., Phys. Rev. Lett. **95**, 026601 (2005)
[16] D. Atkinson and R. P. Cowburn, Appl. Phys. Lett. **85**, 1386 (2004)
[17] D. Lacour et al., Appl. Phys. Lett. **85**, 4681 (2004)
[18] A. Yamaguchi et al., Appl. Phys. Lett. **86**, 012511 (2005)
[19] N. Kazantseva, R. Wieser, and U. Nowak, Phys. Rev. Lett. **94**, 037206 (2005)
[20] J. Rothman et al., Phys. Rev. Lett. **86**, 1098 (2001)
[21] R. D. McMichael and M. J. Donahue, IEEE Trans. Magn. **33**, 4167 (1997)
[22] X. Zhu et al., J. Appl. Phys. **93**, 8540 (2003)
[23] M. Kläui et al., Appl. Phys. Lett. **86**, 032504 (2005)
[24] L. J. Heyderman et al., J. Appl. Phys. **93**, 10011 (2003)
[25] Y. G. Yoo et al., Appl. Phys. Lett. **82**, 2470 (2003)
[26] J. Stöhr et al., Science **259**, 658 (1993)
[27] L. J. Heyderman et al., J. Magn. Magn. Mater. **290–291**, 86 (2005)
[28] A. Tonomura, Adv. Phys. **41**, 59 (1992)
[29] R. E. Dunin-Borkowski et al., J. Microsc. **200**, 187 (2000)
[30] M. Laufenberg et al., Appl. Phys. Lett. **88**, 052507 (2006)
[31] M. Kläui et al., Appl. Phys. Lett. **85**, 5637 (2004)
[32] M. Kläui et al., Physica B **343**, 343 (2004)
[33] The OOMMF package is available at http://math.nist.gov/oommf.
[34] C. A. F. Vaz et al., Phys. Rev. B **72**, 224426 (2005)
[35] C. A. F. Vaz et al., Nucl. Instrum. Meth. Phys. Res. B **246**, 13 (2006)
[36] M. Kläui et al., J. Appl. Phys. (in press) **99** (2006)

[37] I. Hashim, H. S. Joo, and H. A. Atwater, Surf. Rev. Lett. **2**, 427 (1994)
[38] A. Hubert and R. Schäfer: *Magnetic Domains – The Analysis of Magnetic Microstructures* (Springer, Berlin Heidelberg New York 1998)
[39] M. H. Park et al., Phys. Rev. B **73**, 094424 (2006)
[40] M. Kläui, et al., submitted
[41] Y. Nakatani et al., J. Magn. Magn. Mater. **290-291**, 750 (2005)
[42] M. Laufenberg et al., submitted
[43] R. P. Cowburn et al., Phys. Rev. Lett. **83**, 1042 (1999)
[44] M. McCaig: *Permanent Magnets in Theory and Practice*, 1st ed. (Pentech, London 1977)

Six Emerging Directions in Sculptured-Thin-Film Research

Akhlesh Lakhtakia, Melik C. Demirel, Mark W. Horn, and Jian Xu

Department of Engineering Science and Mechanics, Pennsylvania State University, 212 EES Building, University Park, PA 16802, USA
akhlesh@psu.edu, mcd18@psu.edu, mwh4@psu.edu, jianxu@psu.edu

Abstract. Sculptured thin films (STFs) are assemblies of shaped, parallel, identical nanowires generally grown by vapor deposition techniques on substrates. Their optical applications have advanced significantly during the last decade, and several new directions have begun to emerge in the past two years. These include: (a) STF light emitters; (b) STFs with optical gain; (c) electrically controlled STFs; (d) deposition of polymeric STFs by replamineform, multibeam lithographic, and mixed vapor deposition techniques; (e) growth of bioscaffolds on STFs; and (f) STFs with transverse architectures.

1 Introduction

Sculptured thin films (STFs) emerged during the period 1992–1995 [1], initially as a new class of optical nanomaterials that could underlie the *optical-circuit-in-a-chip* concept for possible integration with electronics, but also for applications that would exploit their optical response characteristics and/or their engineered nanoscale morphologies [2]. During the last decade, the optical response characteristics of STFs have been understood quite well and several optical applications have been developed [1, 3–6].

STFs are generally grown by physical vapor deposition or variations thereof [7]. The morphology of STFs comprises clusters of 3–5 nm diameter that are arranged to form parallel nanowires that are bent in some fanciful forms with feature size 30 nm or larger, as exemplified in Fig. 1 by images obtained from a scanning electron microscope (SEM). Accordingly, a STF is a unidirectionally nonhomogeneous continuum with direction-dependent properties at visible and infrared wavelengths that will affect the polarization state of light, depending on the frequency. Thus, STFs can be employed as optical filters for desired bandwidth and polarization state. The morphology of a STF can be changed sectionwise during deposition, and a multi-section STF can therefore be conceived of as an optical circuit that can be integrated with electronic circuitry on a chip. Being porous, a STF can act as a sensor of fluids and can be impregnated with liquid crystals for switching applications too. The nanowires of a STF can be thought of as a string of sausages that allows us the formulation of a nominal model to link the optical response characteristics of the STF to its nanoscale morphology [8].

R. Haug (Ed.): Advances in Solid State Physics,
Adv. in Solid State Phys. **46**, 285–296 (2008)

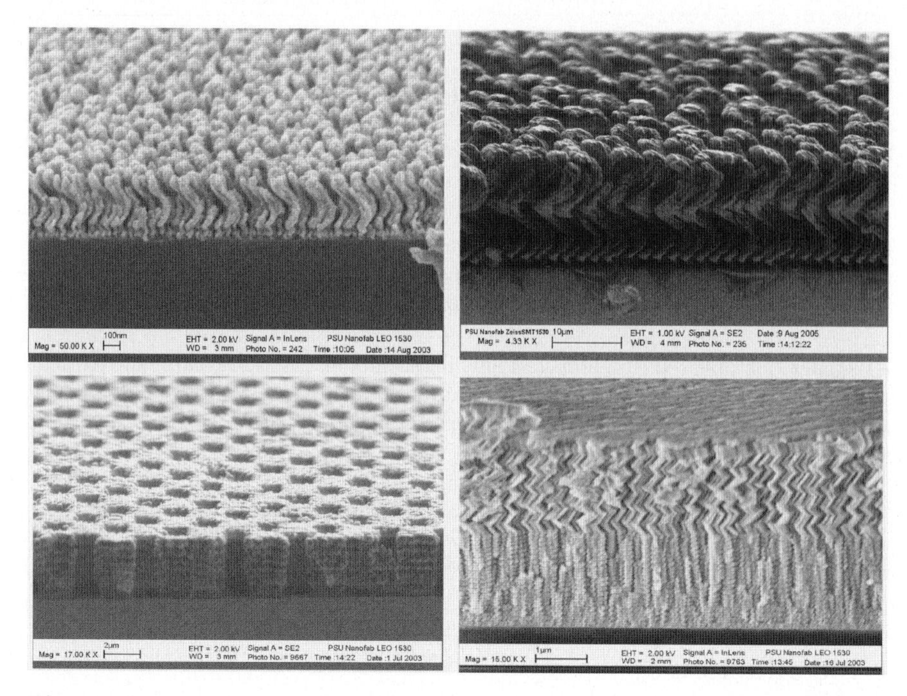

Fig. 1. Cross-sectional SEM micrographs of STFs. *Clockwise from upper left*: molybdenum STF, parylene STF, two-section STF of silicon oxide, and chiral STF of silicon oxide deposited on a topographic substrate

This paper is a survey of six directions in STF research that have emerged during the last two years. The first is light emission for genomic sensing as well as for monochromatic light generation of a specified polarized state. The second is the potential use of materials with optical gain to make STFs with enhanced remittances in quite narrow wavelength-regimes. The third is the exploitation of the (linear) Pockels effect in STFs made of electro-optic materials. Research in these three directions has been largely theoretical thus far. Experimental directions include the deposition of polymeric STFs by replamineform methods, mixed vapor deposition technique, and holographic lithography; the growth of bioscaffolds on STFs; and STFs with transverse architectures wherein the nanoscale and the microscale are blended. All of the theoretical topics summarized here involve chiral STFs that can function as chiral mirrors, because they exhibit the circular Bragg phenomenon: when circularly polarized light in a certain wavelength regime (called the Bragg regime) falls normally on a chiral STF, the reflectance is very high if the light and the STF are co-handed, but not otherwise [1, Ch. 9].

2 Light Emission

As any STF is porous, it can host luminophores that may be excited to radiate light. However, the polarization state and bandwidth of light emitted by the STF would be affected by the optical response characteristics of the host STF. This idea formed the basis of a thought experiment whereby chiral STFs, comprising helical nanowires, would host identical single-strand DNA molecules with a gene sequence specific to an organism (such as *E. coli*) [9]. When single-strand DNA fragments obtained from processing a sample of water, from a pond for instance, infected with the same type of organism would be introduced in the host STF, along with an appropriate ruthenium complex, those introduced fragments with the correct gene sequence would combine with the guest fragments to form double-strand DNA fragments. In the process, light would be emitted [10–12]. Since the host STF is chiral, by solving the Maxwell equations for a canonical problem it was found that the emitted light would have a specific circular or elliptical polarization state with a specific handedness, depending on the conformations of the double-strand DNA fragments [13], the structural handedness of the host STF [9], and the tilt angle of the helical nanowires [14]. The bandwidth of emission would also be affected by these factors, similarly to lasing in dye-doped cholesteric liquid crystals [15, 16].

The development of light-emitting devices (LEDs) of controlled circular polarization (CP) state is also of interest for biosensing. In the UV-to-NIR wavelength regime, many optically active biochemical materials, such as amino acids and sugars, show fingerprint anisotropic attenuation (circular dichroism) and dispersion (optical rotary dispersion) due to the molecular chirality [17–19]. With multispectral circular differential imaging using two CP light sources with orthogonal polarization states, it is possible to acquire detailed information on the protein or DNA composition of a biological sample as well as on associated secondary structures [20–22]. Such applications of CP light necessitate the development of LEDs of controllable CP state, tunable emission wavelength, high color purity, as well as compact size.

While a conventional CP light source is built by placing either a linear polarizer and a quarter-waveplate or a Fresnel Rhomb retarder in front of an unpolarized-light emitter, the implemented system is bulky and can hardly be integrated into chips of dense arrays. In addition, the polarization state that is not passed through the polarizer cannot be used, thereby degrading the overall efficiency of the LED. STF-based LEDs will potentially overcome these disadvantages by introducing chiral STFs as CP-selective chiral mirrors [23] to form a microcavity. The confinement applied by the CP-selective microcavity will alter the optical mode density within it, and spatially and spectrally redistribute the stimulated emission from an active medium occupying the microcavity, which would result in pure CP coherent radiation with high power-conversion efficiency.

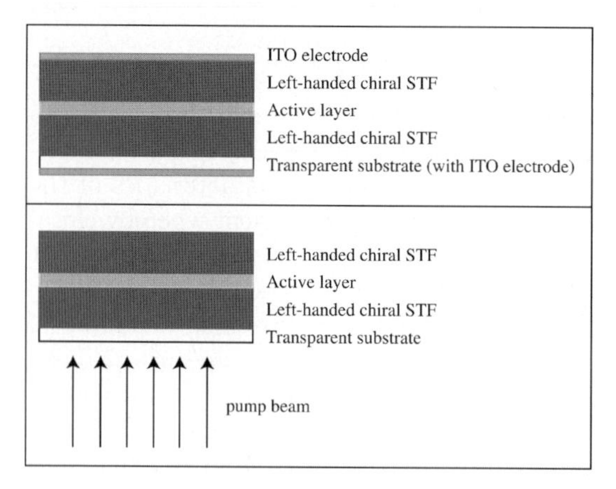

Fig. 2. Schematics of two device architectures for generation of coherent light of pure and controllable CP state. An active layer containing light-emitting NQDs is interposed between two identical chiral STFs. *Top*: electrical excitation. *Bottom*: optical excitation

Two possible device architectures are schematically shown in Fig. 2. The first architecture contains a pair of identical chiral STFs acting as chiral mirrors, and bottom and top indium-tin-oxide (ITO) electrodes to electrically excite an active layer of light-emitting nanocrystal-quantum-dots (NQDs); the second architecture does not contain the electrodes, and the active layer is to be optically excited. The active layer bounded by the chiral mirrors constitutes the microcavity. The optical thickness of the active layer and the ITO electrodes is fine-tuned to become half or integral multiples of the intended emission wavelength, which is determined by the size of NQDs in the active layer. In the Bragg regime, the CP-selective reflection from the structurally left-handed (right-handed) chiral STFs inhibits the presence of optical modes of the right (left) CP state in the active layer, which induces the resonance of left (right) CP state only.

Indeed, spectrally narrowed fluorescent emission of left CP light by sandwiching doped tris-aluminum-8-hydroxyquinolinate (Alq3) molecules between two structurally left-handed chiral STFs has been observed in preliminary experiments [24]. The emission bandwidth was 11 nm, which is much narrower than that of Alq3 fluorescence in free space. Right CP fluorescence, however, did not exhibit any spectral narrowing effects. The measured emission peak wavelength matched well with the resonance wavelength of the active layer. Since microcavity structures have been employed to fabricate organic semiconductor lasers, it appears possible to design and build an STF-based diode laser device to deliver coherent light of pure and controllable CP state.

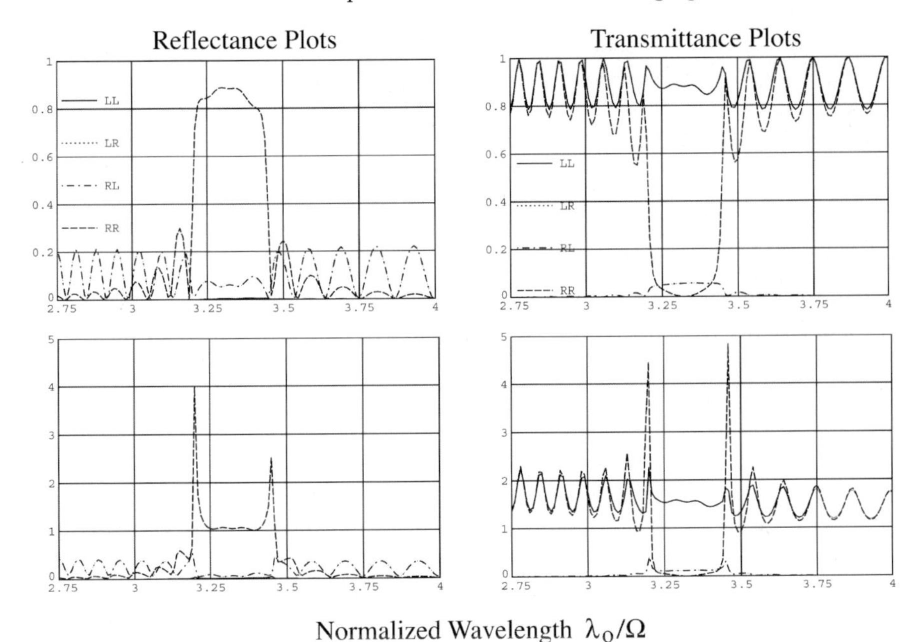

Fig. 3. Reflectances (R_{LL}, R_{RL}, R_{LR}, and R_{RR}) and transmittances (T_{LL}, T_{RL}, T_{LR}, and T_{RR}) of a structurally right-handed chiral STF computed as functions of λ_o/Ω, where λ_o is the free-space wavelength and 2Ω is the pitch of the chiral STF. T_{RL} is the fraction of incident power density **T**ransmitted into the **R**ight CP state when the incident plane wave has the **L**eft CP state. *Top row*: lossless chiral STF. *Bottom row*: chiral STF with gain

3 STFs with Optical Gain

Insofar as the linear optical response properties of STFs are concerned, attention thus far has been confined to lossless and passive (i.e., lossy) materials [1, 3–6]. However, dielectric materials with gain do exist. For instance, many direct semiconductors and rare-earth-doped dielectric crystals radiate light spontaneously on optical or electrical excitation [25, 26]. Upon the application of optical confinement, the stimulated-emission process clones photons that already exist in the material, which leads to the collapse of emission linewidth and enhanced radiation efficiency. Stimulated emission produces a net gain in the photon flux, and light amplification and lasing behavior can be achieved in mediums with gain. In addition, many organic polymers and low-molecular-weight molecules, such as rhodamine, exhibit optical gain when pumped with ultraviolet light [27, 28].

A theoretical study has shown that the remittances of chiral STFs with gain differ from those of lossless and lossy chiral STFs, chiefly at the edges of the Bragg regime, for normally incident plane waves [29], as can be deduced from Fig. 3. The edge differences can be explained in terms of the photonic density of states after invoking Fermi's golden rule, and the situation is quite similar to lasing in dye-doped cholesteric liquid crystals [15, 16]. Examination of a chiral STF with a central twist defect has shown that a narrowband feature exists to function as a reflection hole when the STF does not have optical gain and as an emitter when it does have gain, provided the film is not very thick [29]. As the thickness increases, the narrowband feature is replaced by an ultranarrowband feature which functions as a transmission hole if the STF is lossless; but the ultranarrowband feature does not appear in the presence of either gain or loss. The same understanding is expected to apply when the chiral STF with optical gain is chirped or endowed with many twist defects.

4 Electrically Controllable STFs

Electrical control of the optical response properties of a STF would not only impart tunability to a STF-based device but also the capability of switching between two or more optical modalities. The Pockels effect offers the possibility of electrical control, provided the STF is made of an electro-optic material [30].

This possibility has been theoretically demonstrated for chiral STFs possessing locally a $\bar{4}2m$ point group symmetry and subjected to a low–frequency (dc) electric field in the thickness direction. The Bragg regime could be widened by thus exploiting the Pockels effect [31], which concurrently implies that thinner optical devices are possible [32]. Quite a surprising result was that this effect can create a Bragg regime even when one is not there in the absence of the dc electric field [31], as shown in Fig. 4, offering thereby the possibility of electrically switchable optical filters. The underlying optical phenomenons can be explained in terms of increased optical birefringence due to the application of the dc electric field [32], and can be also be seen when the chiral STF is replaced by an ambichiral stack of achiral electro-optic layers [33]. Electrically controllable ultranarrowband filters are possible by inserting a central twist defect in an electro-optic chiral STF [34].

Electrical control does appear to require high dc voltages. These can be comparable with the half wave voltages of electro-optic materials [36, p. 420], but the dc electric field is much smaller than the characteristic atomic electric field strength [30, p. 3]. The photorefractive effect promises the delivery such high voltages [37]. The possibility of electric breakdown in the void regions of the STF exists, but it would significantly depend on the time that the dc voltage would be switched on for. This latter problem could be mitigated to some extent by filling in the void regions of an electro-optic STF with some appropriate material.

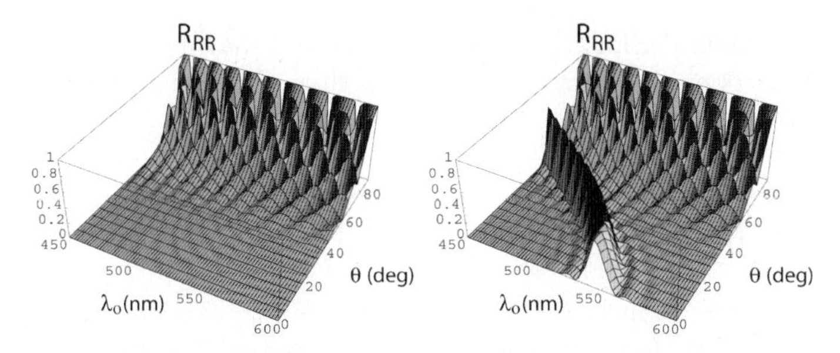

Fig. 4. Reflectance R_{RR} of a structurally right-handed chiral STF computed as a function of the free-space wavelength λ_o and the angle of incidence θ from the thickness direction. The STF is electro-optic and endowed with a local $\bar{4}3m$ point group symmetry [35]. *Left*: without dc voltage. *Right*: with dc voltage. The curved high-reflectance ridge indicates the occurrence of the circular Bragg phenomenon

5 Polymeric STFs

Although most STFs fabricated until recently were made from either inorganic dielectric materials or metals, polymeric STFs were considered attractive from the early days of STF research. Direct fabrication of polymeric STFs by physical vapor deposition being very difficult to implement successfully, a replamineform route was suggested [2]. This route has three stages: first, grow a STF from some inorganic material; second, fill in the void regions with a polymer; and third, etch out the inorganic skeleton. The result would be a STF comprising nanoholes rather than nanowires. Such a chiral STF was fabricated some years ago [38]; its Bragg regime is particularly suitable for being tuned piezoelectrically [39].

The replamineform route has been extended to realize a STF with polymeric nanowires [40]. In this five–stage route, the first three stages are the same as to fabricate a STF of nanoholes. The nanoholes are filled with the desired polymer in the fourth stage, followed by the etching out of the first polymer in the fifth stage.

A single-stage deposition route for polymeric STFs has recently emerged [41]. This route is a combination of chemical and physical vapor deposition processes. Parylene C, a polymer widely used for coating biomedical devices and biomaterials, is first pyrolized from the dimer form to the monomer form. Then it iss directed through a nozzle towards a rotating substrate in order to form a chiral STF, as shown in Fig. 1. This route is certainly less formidable than the two replamineform routes.

Holographic lithography provides the final route. Many optical beams at the same frequency are simultaneously launched in a photoresist, their interference pattern thus being recorded over a certain duration in the photoresist. After development by standard methods, a material with complicated morphology is obtained [42]. If at least one of the beams is elliptically polarized, the morphology is chiral [43, 44]. Holographic lithography appears very suitable for producing three-dimensional photonic crystals, but may also be used for making STFs that operate at lower frequencies than the optical beams used to produce them.

6 Bioscaffold Formation on STFs

Cell differentiation, proliferation and adhesion *in vitro* and *in vivo* are essential to successful research on and treatment of various diseases, disorders and injuries. Being porous, STFs may also serve as simulative tools for studying the interactions, transport and synthesis of biomolecules in confined environments as well as cell growth of tissue culture substrates. The use of STFs for cell-surface interactions is also appealing because they mimic the features of native extra-cellular matrix: clusters of 1-3 nm nanowires coalesce into 30-50 nm columns, and form bundles of 300-500 nm diameter. The key advantages of STFs for cell attachment are the increased specific surface area, multiscale and engineerable porosity, and high efficiency of conjugation due to nanostructure.

Preliminary experiments have shown that STFs made of parylene C can support cell growth, and differentiation in human kidney cells. This is an attractive development, because (i) the deposition process is robust, inexpensive and does not require any lithographic technique, and (ii) the deposited STFs have been shown to be highly bioactive and biocompatible.

In order to assess the adequacy of the deposited STFs of parylene C for the growth of human kidney cells, a scanning electron microscope as well as a confocal laser microscope and image processing software were used in conjunction with fluorescent probes. Human kidney HEK-293 cells were seeded on parylene-C STFs and bulk parylene-C films (control films), and both types of films were maintained under identical conditions. Figure 5 shows SEM micrographs of bulk films and STFs before and after incubation of kidney cells. Little or no cell growth and attachment occurred on bulk films, but three-dimensional scaffold formation on chiral STFs is evident in the SEM micrograph presented.

The cells were labeled with AlexaFluor 488-conjugated phalloidin (Molecular Probes) and DRAQ5 (Biostatus, UK) dyes for confocal laser microscopy. Figure 6 shows a confocal laser microscope image, wherein the fluorescent localization of F-actin and nucleic DNA is clearly evident. This figure unambiguously demonstrates the cell growth on, and thus the biocompatibility of, polymeric STFs. The extended time-course of cell growth after removal of unattached cells, also presented in Fig. 6, indicates an increasing area of cell coverage after 48, 72 and 96 hours in a serum-free environment. These conclusions have effectively been duplicated for fibroblast cells as well (not shown here), and are indicative of an extremely promising research direction.

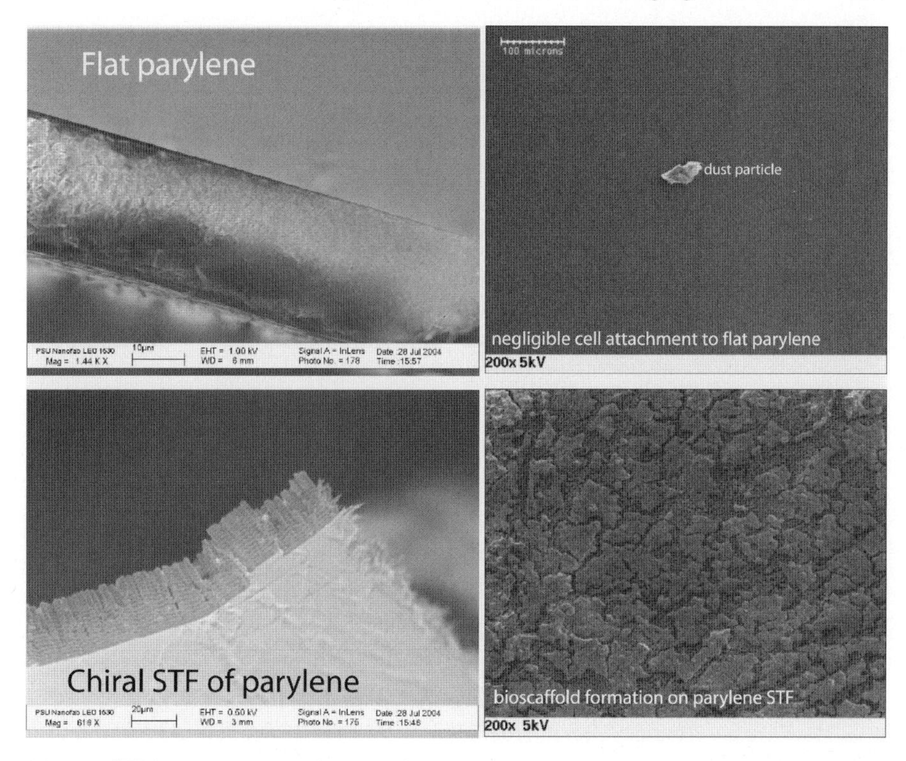

Fig. 5. SEM micrographs of a bulk film and STF, both made of parylene C, before and after incubation of human kidney cells. Little or no cell attachment occurs on the top surface of the bulk film, but a bioscaffold forms on the chiral STF

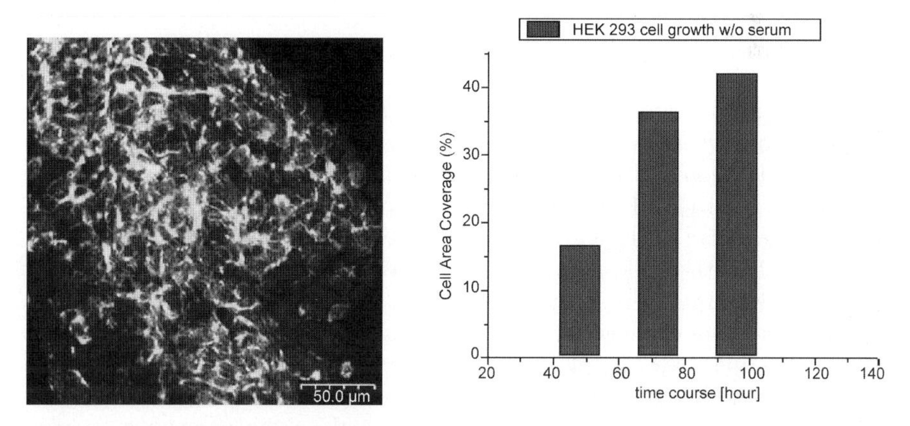

Fig. 6. *Left*: confocal laser microscope image of human kidney cells forming a bioscaffold on top of a parylene-C STF. The *red color* indicates nucleic DNA (stained with DRAQ5 dye), whereas the *green color (online color)* indicates F-actins (stained with phalloidins) in cellular cytoplasm. *Right*: time-course of cell coverage of a parylene-C STF's surface

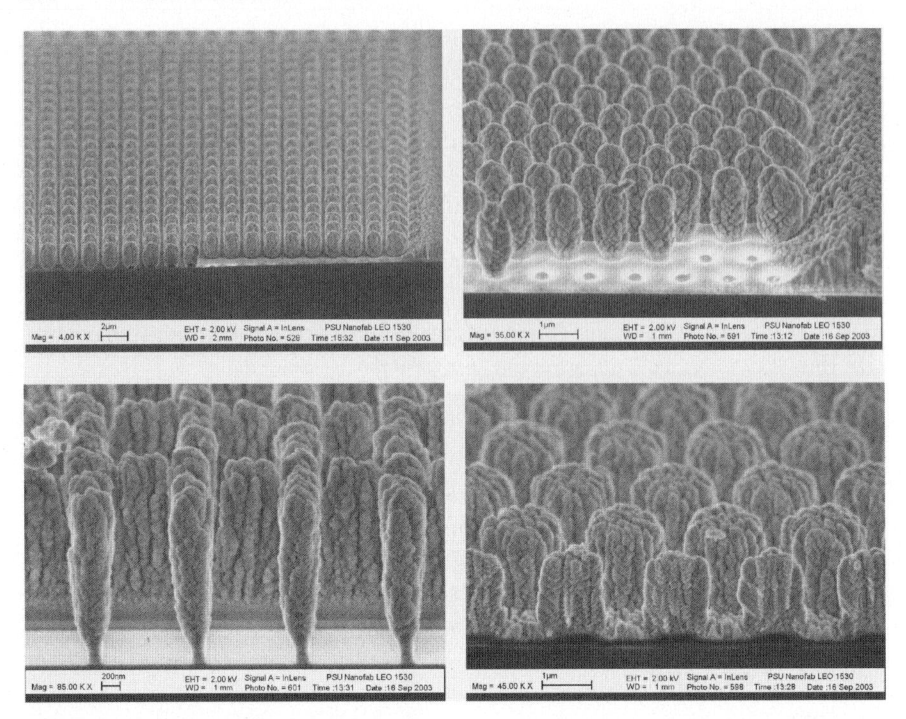

Fig. 7. Cross-sectional SEM micrographs of STFs with transverse architecture. Silicon oxide was deposited on micropatterned substrates. Note that the STFs are assemblies of nanowires; thus, the nanoscale and the microscale are blended in these films. See also the bottom right micrograph in Fig. 1

7 STFs with Transverse Architecture

Three major developments leading towards industrial-scale fabrication of STFs occurred simultaneously in 2003: the combination of large thickness (> 3 μm), large-area uniformity (75 mm diameter), and high growth rate (up to 0.4 μm/min) [45]. Furthermore, the deposition of assemblies of complex-shaped nanowires on lithographically defined patterns was achieved for the first time, and the nanoscale and the microscale were thus blended together in STFs with transverse architectures. SiO_x ($x \approx 2$) nanowires were grown by electron-beam evaporation onto silicon substrates both with and without photoresist lines (one-dimensional arrays) and checkerboard (two-dimensional arrays) patterns, as shown in Fig. 7. It was concluded that atomic self-shadowing due to oblique-angle deposition enables the nanowires to grow continuously, to change direction abruptly, and to maintain constant cross-sectional diameter. The selective growth of nanowire assemblies on the top surfaces of both types of arrays can be understood and predicted using simple geometrical shadowing equations.

8 Concluding Remarks

Sculptured thin films are a class of nanomaterials that have been developed and investigated mostly for optical applications and devices during the last decade. Spurred by developments in nanotechnology, during the last three years, new avenues for the use of STFs have begun to be explored. In this survey, six emerging directions have been discussed: STF light emitters, STFs with optical gain, electrically controlled STFs, polymeric STFs, bioscaffold formation on STFs, and STFs with transverse architectures. In the ensuing years, research in these six directions is expected to blossom and bear fruit, and more directions for research are expected to arise.

Acknowledgements

AL thanks J. Adrian Reyes of Universidad Nacional Autonoma de Mexico, and MCD and AL thank John Bylander and Judith S. Bond of the College of Medicine, Pennsylvania State University, for collaborations and discussions. MWH and AL gratefully acknowledge the US NSF-supported Penn State Materials Research Science and Engineering Center for partial funding.

References

[1] A. Lakhtakia, R. Messier: *Sculptured Thin Films: Nanoengineered Morphology and Optics.* (SPIE Press, Bellingham, WA, USA 2005)
[2] A. Lakhtakia, R. Messier, M. J. Brett, K. Robbie: Innovations Mater. Res. **1**, 165 (1996)
[3] I. J. Hodgkinson, Q. h. Wu: Adv. Mater. **13**, 889 (2001)
[4] J. A. Polo Jr.: In *Micromanufacturing and Nanotechnology* Ed. N. P. Mahalik. (Springer-Verlag, Heidelberg 2005)
[5] J. B. Geddes III: In *Frontiers in Optical Technology: Materials and Devices* Eds. P. K. Choudhury and O. N. Singh. (Nova Publishers, New York 2006)
[6] F. Wang: In *Surface Nanophotonics: Principles and Applications* Eds. D. L. Andrews and Z. Gaburro. (Springer-Verlag, New York 2006)
[7] D. M. Mattox: *The Foundations of Vacuum Coating Technology.* (Noyes Publications, Norwich, NY, USA 2003)
[8] J. A. Sherwin, A. Lakhtakia: Math. Comput. Model. **34**, 1499 (2001); corrections: **35**, 1355 (2002)
[9] A. Lakhtakia: Opt. Commun. **188**, 313 (2001)
[10] X.-H. Xu, A. J. Bard: J. Am. Chem. Soc. **117**, 2627 (1995)
[11] A. Islam, N. Ikeda, A. Yoshimura, T. Ohno: Inorg. Chem. **37**, 3093 (1998)
[12] E. S. Handy, A. J. Pal, M. F. Rubner: J. Am. Chem. Soc. **121**, 3525 (1999)
[13] A. Lakhtakia: Microw. Opt. Technol. Lett. **37**, 37 (2003)
[14] A. Lakhtakia: Opt Commun. **202**, 103 (2002); corrections: **203**, 447 (2002)
[15] M. Voigt, M. Chambers, M. Grell: Chem Phys Lett **347**, 173 (2001)
[16] M. Ozaki, M. Kasano, D. Ganzke, W. Haase, K. Yoshino: Adv. Mater. **14**, 306 (2002)

[17] W. Klyne, J. Buckingham: *Atlas of Stereochemistry.* (Oxford University Press, Oxford, United Kingdom 1978)
[18] E. Charney: *The Molecular Basis of Optical Activity.* (Krieger, Malabar, FL, USA 1985)
[19] A. Lakhtakia: *Beltrami Fields in Chiral Media.* (World Scientific, Singapore 1994)
[20] D. Keller, I. Tinoco: Proc. Nat. Acad. Sci. USA **82**, 401 (1985)
[21] T. W. King, G. L. Cote, R. McNichols, M.J. Goetz Jr.: Opt. Eng. **33**, 2746 (1994)
[22] T. Yamada, H. Onuki, M. Yuri, S. Ishikaza: Jpn. J. Appl. Phys. Pt. 1 **39**, 310 (2000)
[23] A. Lakhtakia, J. Xu: Microw. Opt. Technol. Lett. **47**, 63 (2005)
[24] J. Xu, A. Lakhtakia, J. Liou, A. Chen, I. J. Hodgkinson: Opt. Commun. (doi:10.1016/j.optcom.2006.02.025)
[25] B. E. A. Saleh, M. C. Teich: *Fundamentals of Photonics.* (Wiley, New York 1991)
[26] P. Bhattacharya: *Semiconductor Optoelectronic Devices, 2nd ed.* (Prentice-Hall, New York 1996)
[27] A. Dodabalapur, L. J. Rothberg, R. H. Jordan, T. M. Miller, R. E. Slusher, J. M. Phillips: J. Appl. Phys. **80**, 6954 (1996)
[28] X. Liu, D. Poitras, Y. Tao, C. Py: J. Vac. Sci. Technol. A **22**, 764 (2004)
[29] A. Lakhtakia, J. Xu: Optik (doi:10.1016/j.ijleo.2006.01.011)
[30] R. W. Boyd: *Nonlinear Optics.* (Academic Press, London 1992)
[31] J. A. Reyes, A. Lakhtakia: Opt. Commun. **259**, 164 (2006)
[32] A. Lakhtakia: Opt. Commun. (doi:10.1016/j.optcom.2005.12.031)
[33] A. Lakhtakia: Phys. Lett. A (doi:10.1016/j.physleta.2006.01.069)
[34] A. Lakhtakia: Asian J. Phys., at press (2006)
[35] A. Lakhtakia: Microw. Opt. Technol. Lett., submitted (2006)
[36] A. Yariv, P. Yeh: *Photonics: Optical Electronics in Modern Communications, 6th ed.* (Oxford University Press, New York 2007)
[37] N. Kukhtarev, T. Kukhtareva, M. E. Edwards, B. Penn, D. Frazier, H. Adeldayem, P. P. Banerjee, T. Hudson, W. A. Friday: J. Nonlin. Opt. Phys. Mater. **11**, 445 (2002)
[38] K. D. Harris, K. L. Westra, M. J. Brett: Electrochem. Solid State **4**, C39 (2001)
[39] F. Wang, A. Lakhtakia, R. Messier: J. Modern Opt. **50**, 239 (2003)
[40] A. L. Elias, K. D. Harris, C. W. M. Bastiaansen, D. J. Broer, M. J. Brett: J. Micromech. Microeng. **15**, 49 (2005)
[41] S. Pursel, M. W. Horn, M. C. Demirel, A. Lakhtakia: Polymer **46**, 9544 (2005)
[42] M. Farsari, G. Filippidis, G. Fotakis: Opt. Lett. **30**, 3180 (2005)
[43] E. R. Dedman, D. N. Sharp, A. J. Turberfield, C. F. Blanford, R. G. Denning: Photon. Nanostruct. Fund. Appl. **3**, 79 (2005)
[44] Y. K. Pang, J. C. W. Lee, H. F. Lee, W. Y. Tam, C. T. Chan, P. Sheng: Opt. Express **13**, 7615 (2005)
[45] M. W. Horn, M. D. Pickett, R. Messier, A. Lakhtakia: Nanotechnology **15**, 303 (2004)
[46] M. W. Horn, M. D. Pickett, R. Messier, A. Lakhtakia: J. Vac. Sci. Technol. B **22**, 3426 (2004)

Ion Beam Assisted Growth of Sculptured Thin Films: Structure Alignment and Optical Fingerprints

E. Schubert[1], F. Frost[1], H. Neumann[1], B. Rauschenbach[1] B. Fuhrmann[2], F. Heyroth[2], J. Rivory[3], E. Charron[3], B. Gallas[3], and M. Schubert[4]

[1] Leibniz-Institut für Oberflächenmodifizierung e.V., Permoserstraße 15, 04318 Leipzig, Germany
eva.schubert@iom-leipzig.de
[2] Interdisziplinäres Zentrum für Materialwissenschaften, Martin-Luther-Universität Halle-Wittenberg, Hoher Weg 8, 06120 Halle, Germany
[3] Institut des NanoSciences de Paris, Universitè Paris 6 et 7- UMR CNRS 75 88, Campus Boucicaut, 140, rue de Lourmel, 75015 Paris, France
[4] Electrical Engineering Department, and Center for Materials Research Analysis, University of Nebraska-Lincoln, 209N Walter Scott Engineering Building, 068588-0511 Lincoln, NE, U.S.A.

Abstract. Sculptured thin films from are grown by ion beam assisted deposition under conditions with very oblique angles of incidence for the particle flux. The nanodimensional structures within the sculptured thin films are designed in geometries of columns, chevrons, left-handed multi-fold and continuous screws, and comprise non-chiral and chiral properties. The growth is studied with emphasis on self-controlled process driven structure alignment across the substrate. Intriguing optical fingerprints from the various types of sculptured thin films are highlighted by reflection-type single-wavelength generalized Mueller matrix ellipsometry and spectrally-integrated diffracted light scattering intensity measurement scans. We suggest the ellipsometry approach for chirality assessment, and suggest possible applications of the sculptured thin films in sub-wavelength nanodiffractive structures, for example.

1 Introduction

Sculptured thin films (STF) with complex chiral or non-chiral geometries are promising candidates for a large variety of applications in the field of photonics, optics, sensors, and engineering mechanics, for example. Many materials are being explored for depositing three-dimensional structures on the nanoscale, because novel physical or chemical properties are expected due to confinement effects. Silicon is a primary candidate due to its extraordinary position in the field of microelectronic device fabrication. Recently, large efforts were undertaken for synthesis of silicon nanocomponents such as nanowires and nanoclusters. While known as an inefficient light emitter by virtue of the indirect electronic band gap property of the bulk silicon lattice,

R. Haug (Ed.): Advances in Solid State Physics,
Adv. in Solid State Phys. **46**, 297–308 (2008)

both porous silicon and silicon nanostructures reveal strong luminescence in the blue and red to near-infrared spectral regions, respectively. By integrating silicon into photonic materials electronic and photonic properties may be favorably complemented. The exploration of new approaches for fabrication of materials with three-dimensional structures on the nanoscale and sophisticated geometries is challenging many materials scientists. The glancing angle deposition (GLAD) approach is gaining attention due to the ability for depositing such nanostructures [1, 2]. In this technique, the naturally occurring particle beam shadowing is exploited for grow of high-aspect-ratioed columns with strong inclination away from the growing surface normal. These structures can be transformed into helices, chevrons, posts, multiple-sided screws, or networks thereof, by augmenting azimuthal substrate rotation, substrate tilt, and substrate prepattering [3–5]. During the initial growth process, seed structures with average dimension of approximately 20 nm are created [6, 7]. A highly-coherent growth of nanostructures emerging from these seed structures will occur if a lateral seeding ground selection was done by appropriate substrate patterning prior to the nanostructure deposition [8, 9]. Otherwise, the individual structures will compete during subsequent growth. Eventually, a certain number of adjacent columns will merge and form larger columns with average diameter of $\approx$ 150 nm. Here we summarize our recent findings on this self-controlled or process-driven structure alignment of silicon nanostructures deposited by GLAD. In particular, we employ for the first time ion-beam source technologies. The focus of our studies are non-chiral and chiral geometries. Specifically the nanostructures with chiral geometry present novel materials for which a wealth of new physical phenomena may be expected. Thermal expansion, mechanical [10], electronic, and optical properties [11–14] may vary at great length from their bulk crystal analogues due to the highly-rotational convoluted lattice planes of structures with nanodimensions. In this paper we focus on growth characteristics, and both polarization-optical and light-diffractive fingerprints initiating our studies of their physical properties. Specifically, we present and discuss atomic force and scanning electron microscopy investigations, light-diffraction, and generalized ellipsometry measurements.

2 Experimental

2.1 Growth

The growth experiments are performed in a ultra-high-vacuum chamber with a base pressure better than 8×10^{-9} mbar. The chamber is equipped with a double-grid high-frequency ion source, which consists of a focusing double-grid system of 40 mm diameter. The plasma within the ion source is generated by an inductively coupled high-frequency electric field (13.56 MHz), and the ion-current density is measured by a four-Faraday-cup arrangement at target position. A detailed account of the ion-source parameters is given elsewhere. Argon with a flux rate of $f_{\mathrm{Ar}} = 6.5$ sccm setting the deposition pressure of 1.3×10^{-4} mbar is used as the process gas, and a sintered polycrystalline silicon disc provides the target. Si (100) wafers are used as the substrates. The substrate holder is attached to a 2"-disc 3-axes $x - y - z$ manipulator. The sample polar angular movement about the substrate normal is controlled by a computerized stepper motor, and can be varied from 0.01 min^{-1} to 100 min^{-1}. The substrate tilt angle with respect to the incident particle direction can be varied manually over full 360. Here, the geometry between source, target, and substrate normals remain unchanged for all experiments. The ion beam is incident onto the target under 70° with respect to the target surface normal where maximum sputtered particle flux is obtained. The particle flux angular distribution variation is small in the direction towards the substrate, which leads to a quasi-directed particle flow. The substrate surface is tilted by 2° against the target normal. The ion energy is $E_{\mathrm{Ion}} = 1200$ eV, and all experiments are performed at room temperature.

2.2 Structural Characterization

Scanning electron (SEM) and atomic force microscopy (AFM) images are used to study structure alignment and surface morphology. Structure size and alignment are obtained from AFM images by calculation of the 2D Fourier-transform and the power-spectral-density function (PSD). The PSD is the Fourier decomposition of the surface height, and is here defined by

$$PSD(\mathbf{k}) = \frac{1}{A}\left(\frac{1}{2\pi}\int z(\mathbf{r})e^{-i\mathbf{kr}}d\mathbf{r}\right)^2, \tag{1}$$

where A is the scan area, z is the hight deviation in dependency of the lateral substrate surface position r, and k is the spatial frequency. Both PSD function and 2D Fourier image are obtained from surface-scan-size areas of 20 µm.

2.3 Optical Characterization

The STF samples are studied by generalized ellipsometry and unpolarized scattering intensity measurements. In both experiments, Mueller matrix elements are accessed, which represent the action of a sample onto an incident monochromatic polarized light beam described by the four Stokes parameters, which are defined in the p-s polarization mode system: $S_0 = I_p + I_s$, $S_1 = I_p - I_s$, $S_2 = I_{45} - I_{-45}$, $S_3 = I_{\sigma+} - I_{\sigma-}$, where I_p, I_s, I_{45}, I_{-45}, $I_{\sigma+}$, and $I_{\sigma-}$ denote the intensities for the p-, s-, +45°, -45°, right-, and left-handed circularly polarized light components, respectively [15–17]. Note that S_0 is proportional to the beam's total intensity. Arranging the Stokes parameters into a column vector, the Mueller matrix connects incident and emergent beam Stokes vector components at a given reflection or diffraction geometry

$$\begin{pmatrix} S_0 \\ S_1 \\ S_2 \\ S_3 \end{pmatrix}_{\text{emergent}} = \begin{pmatrix} M_{11} & M_{12} & M_{13} & M_{14} \\ M_{21} & M_{22} & M_{23} & M_{24} \\ M_{31} & M_{32} & M_{33} & M_{34} \\ M_{41} & M_{42} & M_{43} & M_{44} \end{pmatrix} \begin{pmatrix} S_0 \\ S_1 \\ S_2 \\ S_3 \end{pmatrix}_{\text{incident}} , \tag{2}$$

and contains all thereby accessible information from the sample. Note that all elements M_{ij} are real-valued (representing transfer of certain polarized intensities) and are bound by definition between -1 and 1 [17].

For the ellipsometry experiment, we measure for the reflected light the elements M_{ij}, except for those in row 4, normalized to the element M_{11}.[1] These ellipsometric data are acquired as a function of the angle of incidence Φ_a and the sample rotation φ, where the latter is an arbitrary but fixed sample azimuth (rotation) angle counted between a given but arbitrary line along the surface and the plane of incidence. A similar approach was used previously for characterization of uniaxial and biaxial anisotropic materials [19–21]. A rotating-analyzer ellipsometer with automated compensator function is employed (J.A.Woollam Co.). Measurements are made in reflection geometry with the incident beam set at $-\Phi_a$ and the reflected beam observed at Φ_a, where incident and reflected beams define the plane of incidence, and the *p*-*s* system. The elements M_{ij} are mapped at a single wavelength λ = 1550 nm, and as a function of Φ_a from 15° to 85° in steps of 15° and φ from $0\ldots$ to 360° in steps of 2°.

For the scatterometry experiment, we measure for the diffracted light the element M_{11} only, as a function of λ from 400 nm to 1000 nm. An in-house built scatterometry system is used, which employs a tungsten lamp as the source, and a spectrograph with CCD array as detector element. No polarizing components are intentionally involved. The incident light is directed normal to the sample surface, and the scattered light is measured under the angle Φ_s with respect to the surface normal, where the intensity was then recorded as a function of Φ_s and φ ($25° \ldots \Phi_a \ldots 85°$, $0 \ldots \varphi \ldots 360°$). The detected intensities are normalized to the source base line using a reflection standard (aluminum mirror), and are further divided by the effectively emitting surface area under the observation angle of Φ_s.

[1] For a detailed introduction to ellipsometry, applications, generalized ellipsometry and anisotropic materials see chapters by Jellison, Collins, Schubert and others within the recent Handbook compilation by Thompkins and Irene, [18].

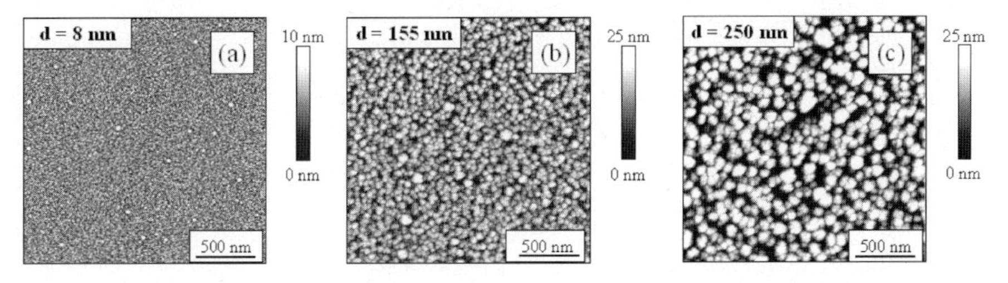

Fig. 1. AFM surface plots of columnar STF samples with similar growth conditions after different growth times: $T_G = 3$ min (**a**), $T_G = 60$ min (**b**), $T_G = 100$ min (**c**)

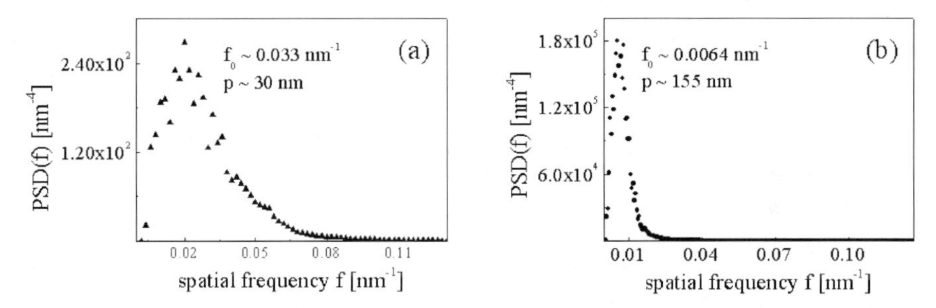

Fig. 2. Angular averaged PSD functions calculated from AFM images for columnar STF samples with 8 nm (**a**) and 250 nm (**b**) thickness. The pronounced maximum at f_0 reflects the average lateral structure size $\sim 1/f_0$, which increases from approximately 30 nm to 155 nm

3 Results and Discussion

3.1 Self-Controlled Sculptured Thin Film Growth

Figure 1 depicts AFM surface plots measured on columnar STF samples with different structure heights obtained by different deposition times T_G. At the initial stage of the STF growth self-organized seed structures with diameter of approximately 20 nm occur. Their lateral size increases during continued STF growth while the number of seeds reduces due to their coalescence. Figure 2 presents the angular averaged PSD functions for two samples with different thickness values reflecting their two different growth times. The PSD functions exhibit a distinct maximum at the spatial frequency f_0, which is relevant to the average structure seize. Here, the maximum frequency f_0 shifts to lower values for the STF sample with increasing T_G. The reduction of f_0 with structure height (T_G) is found to follow approximately an exponential law (Fig. 3a). Obviously, for large structure heights, the lateral size increase is slowing down, and f_0 may eventually stabilize during further increase in structure height. f_0 is also related to the average structure separation width. A reduction of f_0 is correlated with an increase in structure separation, as

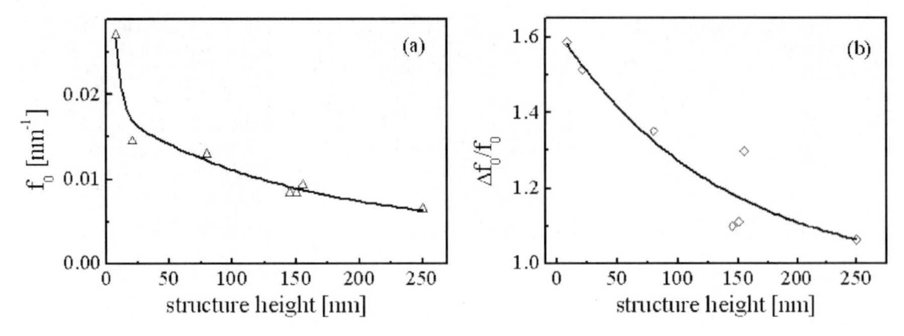

Fig. 3. Maximal spatial frequency f_0 (**a**) and correlation length $\Delta f_0/f_0$ (**b**) versus structure height of GLAD deposited columnar STF samples under equal growth conditions but varying growth times

can be inferred from Figs. 1a–c. The structure separation is explained by the geometric particle shadowing during deposition, which favors the growth of neighboring structures into combined larger structures thereby continuously reducing the numbers of individually surviving columns. Because the surviving structures increase in diameter as the particle net flux is kept constant, the space between them increases as well. The average structure separation may similarly stabilize when reaching separation lengths concordant with the particle shadowing length defined geometrically by the particle flux angle of incidence and the structure height. A measure for the correlation length between the structure building blocks forming a sculptured thin film can be estimated from the inverse ratio of f_0 and its peak full-with-at-half-maximum (FWHM) value Δf (Fig. 3b). For large structure heights $\Delta f/f_0$ approaches 1, indicative for correlation between next nearest neighboring structure elements only. This finding is in ideal agreement with the concurrent growth mechanism proposition, where geometric shadowing in the oblique angle deposition is responsible for self-organized three-dimensional structure formation. Figure 4 contains 2D Fourier image compositions of the AFM surface height scans from STF samples grown with different geometries, and reveal their preferential structure alignment. For these geometries, rotation of the substrates is controlled such that columns with alternating angles towards the substrate normal occur. Thereby *chevron* structures with two-fold (180° rotation steps, "zigzag"), and left-handed *tire-bouchon* structures or hollow screws with 4-fold (90° rotation steps, "corkscrew") symmetry are formed as building blocks. The structure alignment is isotropic and random for the columnar STF fabricated by continuous substrate rotation, indicative for homogeneous isotropic shadowing during growth. The chevron and tire-bouchon STF samples reveal their symmetry accordingly within the 2D Fourier images, which is now related to a preferential alignment of the structural building blocks due to the asymmetric shadow distribution during the growth.

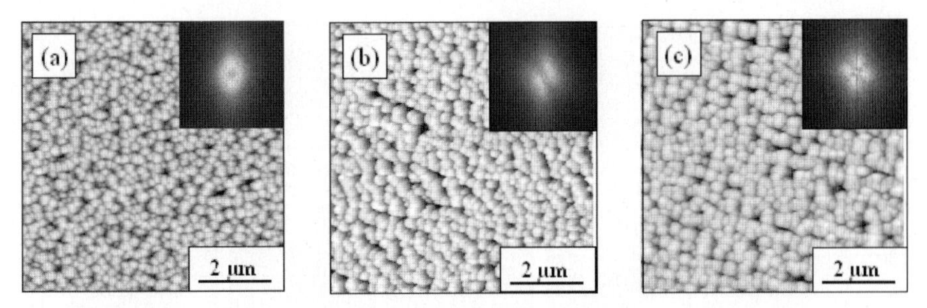

Fig. 4. 2D Fourier images of AFM surface scans from STF samples with isotropic columnar structures (**a**), chevron structures (**b**), and left-handed square-based tire-bouchon (hollow-screw) structures (**c**)

3.2 Light Scattering

Figure 5 summarizes scattering intensities together with the SEM images from the respective geometries of STF samples with chevron, left-handed tire-bouchon, and continuous helix structures. The integrated scattering intensities reveal distinct distributions according to the STF structures and identify their nanodimensional geometries. The scattering maxima occur at different angles Φ_s and φ, and represent sub-wavelength scale diffraction phenomena. For the columnar structure, the scattered light is isotropically distributed within a cone section centered at $\Phi_s \approx 70°$. The chevron structure causes distinct scattering in two preferential spatial directions, according to the plane of inclination of the alternating columns (Fig. 5b). The two maxima occur at $\Phi_s \approx 70°$, and the intersection between both points in Fig. 5g corresponds to the chevron plane. Note that the column inclination is approximately 45°, and hence the observed scattering is evidently a diffraction and not an oblique reflection phenomena. Three and four high-intensity scattering directions can be seen for the 3-fold and 4-fold tire-bouchon structures, respectively, where the maxima occur at $\Phi_s \approx 60°$, and their φ coordinates correspond to the individual column inclination planes (rotational positions of the substrate during deposition). The continuous helix structure causes a continuously varying distribution along a cone section centered at $\Phi_s \approx 70°$. Figures 5k–o present intensities recorded along Φ_s where the strongest scattering occurs, resolved within the three color regions. Short wavelengths are scattered more dominantly, as expected, but the maxima for the different structures occur at different azimuth angles, suggesting that wavelength-selective highly-directional diffractive 3D nanogratings may be devisable from STF structures.

3.3 Generalized Ellipsometry

Figures 6a–k, and 7a–k present gray-scale plots of Mueller matrix elements M_{ij}, normalized to the element M_{11}, versus sample rotation and angle of

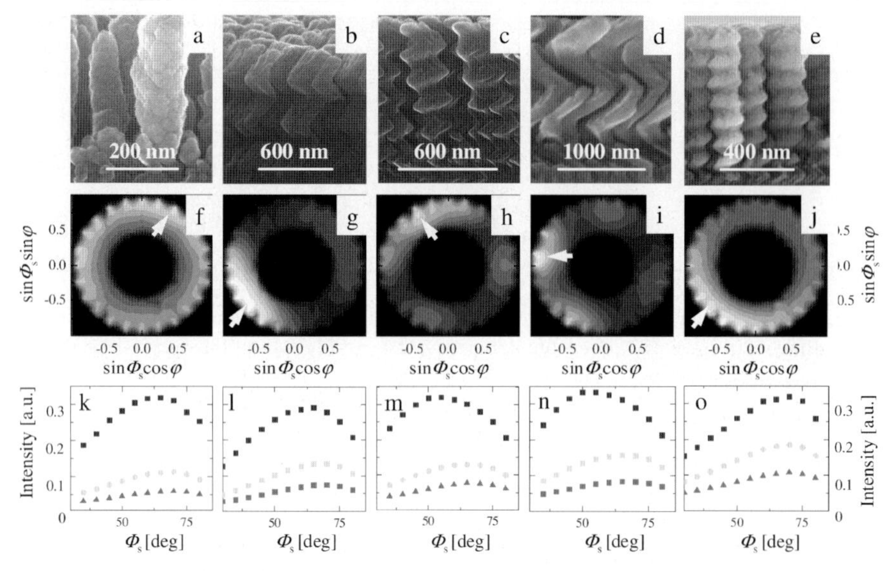

Fig. 5. SEM images of GLAD deposited STF samples with column (**a**), chevron (**b**), left-handed-triangular tire-bouchon (**c**), -fourfold tire-bouchon (**d**), and -continuous helix (**e**) structures. Figs. 5f–j depict 3D (x, y, z) gray-scale plots of the total scattering intensities (z-axis), or equivalently M_{11}, integrated over the wavelength range from 400 nm to 1000 nm for the respective STF samples in Figs. 5a–e, versus in-plane azimuth angle φ and scattering angle Φ_s in polar coordinates $(-1 \leq x = sin\Phi_s cos\varphi \leq 1, -1 \leq y = sin\Phi_s sin\varphi \leq 1)$. The z-coordinates of the measured data points are plotted from the upper hemisphere at their respective positions x, y according to Φ_s, φ in gray scale, where white corresponds to maximum intensity in arbitrary units, and black to 0. The coordinate origin corresponds to the position of the incident light. Figures 5k–o: Scattered intensities in the *blue* (*squares*), *green* (*circles*), and *red (triangles) (online color)* spectral regions along constant Φ_s as indicated by *arrows*

incidence for the STF with chevron structure, and the left-handed 3-fold tire-bouchon structure, respectively. Note that here all data points are acquired at oblique incidence. The elements M_{ij} reflect the highly optically anisotropic behavior of the STF samples, and serve as fingerprints to reveal the intrinsic birefringence of the inclined columns through elaborate model calculations [22]. While the latter will be subject of a forthcoming paper where a mored detailed account of the individual plots will be given, we point out here a very interesting observation regarding the symmetry of the 3D Mueller matrix plots. The plots of elements M_{12} and M_{21} can be transformed into each other by a mirror reflection at the line where $\Phi_a \approx 135°$. This line is equivalent with the plane of inclination in the chevron structure STF sample. The same holds for elements M_{13} and M_{31}, and M_{32} and M_{23}, and which appears to be an intrinsic property of arbitrary non-chiral anisotropic mediums. Inspecting the equivalent plots in Fig. 7 clearly reveals that no

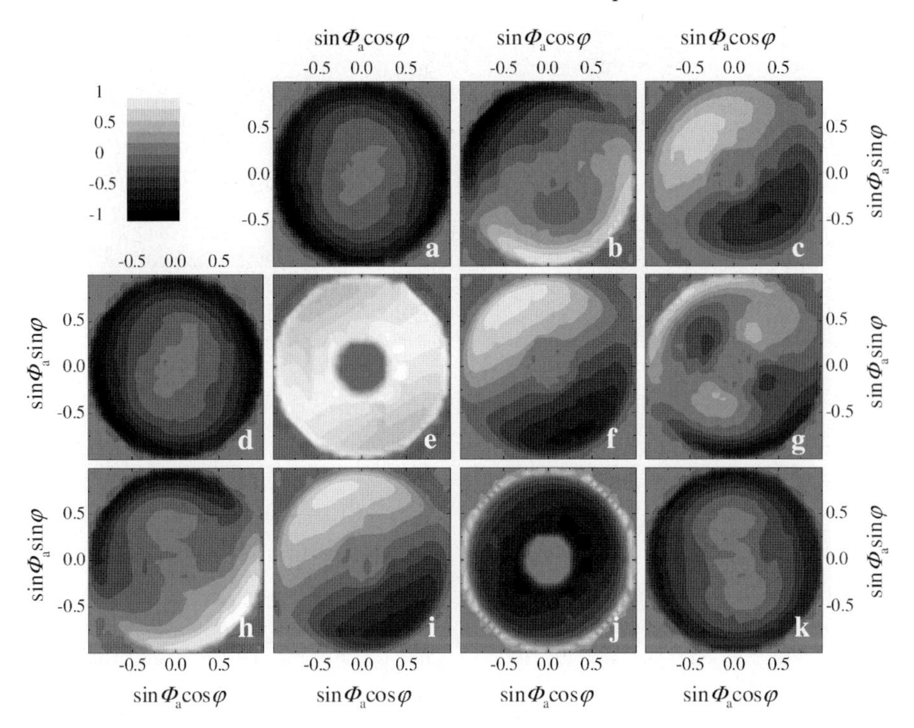

Fig. 6. Three-dimensional (x, y, z) gray-scale plots ($\lambda = 1550$ nm) of Mueller matrix elements $(z-axis)$ M_{12} (**a**), M_{13} (**b**), M_{14} (**c**), M_{21} (**d**), M_{22} (**e**), M_{23} (**f**), M_{24} (**g**), M_{31} (**h**), M_{22} (**i**), M_{33} (**j**), and M_{34} (**k**) from the STF sample with chevron structure (Fig. 5b). Data are plotted versus in-plane azimuth angle φ and angle of incidence Φ_a in polar coordinates $(-1 \leq x = sin\Phi_a cos\varphi \leq 1,\ -1 \leq y = sin\Phi_a sin\varphi \leq 1)$. The z-coordinates of the measured data points are plotted from the upper (lower) hemisphere at their respective positions x, y according to Φ_a, φ in *light (dark) gray* scale, where *white (black)* corresponds to $z = 1(-1)$

such mirror transformation exists in the case of the left-handed tire-bouchon structure, which seems the result of the handedness now being added to this sample extending its anisotropic attributes. Note that the individual building blocks, i.e., the inclined columns, are primarily similar to the ones used for the chevron structure, and possess similar intrinsic birefringence. For the tire-bouchon structure, as well as for any other chiral structure investigated by us so far, we observe this following symmetry property: The pair M_{12} - M_{21} requires rotation around the z-axis by π, whereas the pairs M_{13} - M_{31}, and M_{23} - M_{32} are congruent only after rotation around the z-axis by π and subsequent inversion at the $x-y$-plane. We therefore suggest inspection of Mueller-matrix ellipsometry versus sample rotation and angle of incidence for immediate chirality assessment of a given STF sample. Further results and discussions on chirality and anisotropy properties shall follow up soon.

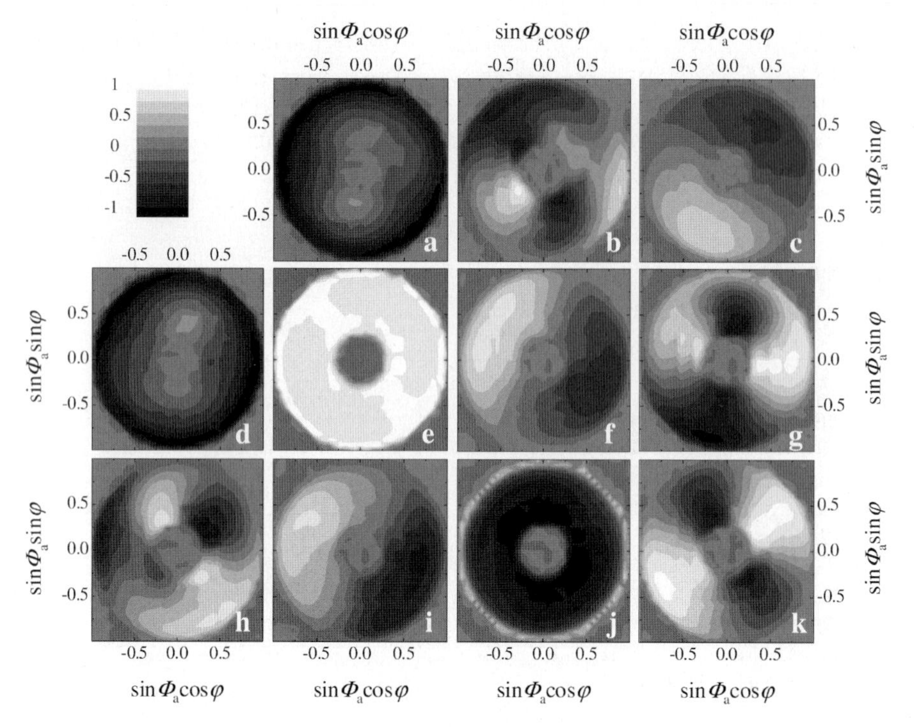

Fig. 7. Same as Fig. 6 for the STF sample with left-handed 3-fold tire-bouchon structure (Fig. 5c)

4 Summary

STF samples with different structures are deposited by ion beam assisted deposition under conditions with very oblique particle flux angle of incidence. The self-organized and -controlled growth process is dictated by the geometrical particle shadowing during deposition. We observe a competing growth of initial seed structures, which eventually evolve into larger structures with low lateral correlation, but stable lateral dimensions. The thereby obtained building blocks can be used to growth STF samples with column, chevrons, chiral multi-fold tire-bouchon and continuous helix structures. Light scattering measurements reveal intense sub-wavelength diffraction phenomena which are strongly influenced by the STF geometry. Generalized Mueller matrix ellipsometry scans versus sample rotation and angle of incidence reveal intriguing symmetry properties which suggest use for immediate chirality assessment of STF samples.

Acknowledgements

We acknowledge fruitful discussions with Kevin Robbie (Queen's University Kingston), Hans Arwin (Linkoping University), and Craig M. Herzinger

(J. A.Woollam Co.). M. S. acknowledges a research fellowship from the Universitè Paris, and support through startup funds from UNL and the JAWCo Foundation. E. S. acknowledges support within the DFG funded Research Group 522 "Architecture of nano- and microdimensional building blocks" under grant SCHU 1579/1-2.

References

[1] K. Robbie, L. J. Friedrich, S. K. Dew, T. Smy, M. J. Brett, J. Vac. Sci. Technol. A **13**, 1032 (1995).
[2] T. Karabacak, J. P. Singh, Y.-P. Zhao, G.-C. Wang, T.-M. Lu, Phys. Rev. B **68**, 125408 (2003).
[3] K. Robbie, G. Beydaghyan, T. Brown, C. Dean, J. Adams, C. Buzea, Rev. Sci. Instr. **75**, 1089 (2004)
[4] E. Schubert, Th. Höche, F. Frost, B. Rauschenbach, Appl. Phys. A **81**, 481 (2005).
[5] S. R. Kennedy, M. J. Brett, J. Vac. Sci. Technol. B **22**, 1184 (2004).
[6] E. Schubert, J. Fahlteich, B. Rauschenbach, M. Schubert, M. Lorenz, M. Grundmann, G. Wagner, J. Appl. Phys., (2006), in press.
[7] E. Schubert, J. Fahlteich, Th. Höche, G. Wagner, B. Rauschenbach, Nucl. Instr. Meth. B **244**, 40 (2006).
[8] B. Dick, M. J. Brett, T. J. Smy, M. Belov, M. R. Freeman, J. Vac. Sci. Technol. B **19**, 1813 (2001)
[9] B. Dick, M. J. Brett, T. J. Smy, M. R. Freeman, M. Malac, R. F. Egerton, J. Vac. Sci. Technol. A **18**, 1838 (2000).
[10] J. P. Singh, T. Karabacak, D.-X. Ye, D.-L. Liu, R. C. Picu, T.-M. Lu, G. C. Wang, J. Vac. Sci. Technol. B **23**, 2114 (2005).
[11] A. Lakhtakia and M. Messier: *Sculptured Thin Films*,(SPIE Press, Bellingham 2004).
[12] M. O. Jensen, M. J. Brett, Optics Express **13**, 3348 (2005).
[13] M. D. Arnold, I. J. Hodgkinson, Q. H. Wu, R. J. Blaikie, J. Vac. Sci. Technol. B **23**, 1398 (2005).
[14] N. J. Podraza, C. Chen, I. An, G. M. Ferreira, P. I. Rovira, R. Messier, and R. W. Collins, Thin Solid Films **455**–**456**, 571 (2004).
[15] R. M. A. Azzam, N. M. Bashara: *Ellipsometry and Polarized Light* (North-Holland Publ. Co., Amsterdam 1984)
[16] A. Röseler: *Infrared Spectroscopic Ellipsometry* (Akademie-Verlag, Berlin 1990)
[17] E. Hecht: *Optics* (Addison-Wesley, Reading MA 1987)
[18] H. Thompkins, E. A. Irene (Eds.): *Handbook of Ellipsometry* (William Andrew Publishing, Highland Mills 2004)
[19] M. Schubert, B. Rheinländer, J. A. Woollam, B. Johs, C. M. Herzinger: Extension of rotating-analyzer ellipsometry to generalized ellipsometry: determination of the dielectric function tensor from uniaxial TiO_2, J. Opt. Soc. Am. A **13**, 875–883 (1996)

[20] M. Schubert: Theory and Application of Generalized Ellipsometry, in W. S. Weiglhofer, A. Lakhtakia (Eds.): *Introduction to Complex Mediums for Optics and Electromagnetics* (SPIE, Bellingham, WA 2004) pp. 677–710
[21] M. Schubert, W. Dollase: Generalized ellipsometry for biaxial absorbing materials: determination of crystal orientation and optical constants of Sb_2S_3, Opt. Lett. **27**, 2073–2075 (2002)
[22] M. Schubert: Another century of ellipsometry, Ann. Phys. (submitted)

Organic Thin Film Devices for Displays and Lighting

Oliver J. Weiss[1,3], Ralf Krause[2,3], and Ralph Paetzold[3]

[1] Technical University of Darmstadt, Departement of Materials Science, Petersenstr. 23, 64287 Darmstadt, Germany
oliver.weiss.ext@erls.siemens.de
[2] University of Erlangen-Nuremberg, Departement of Materials Science, Martensstr. 7, 91058 Erlangen, Germany
[3] Siemens AG, CT MM 1, Guenther-Scharowski-Str. 1, 91058 Erlangen, Germany

Abstract. Organic materials can be used for fabrication of, e.g., electronic circuits, solar cells, light sensors, memory cells and light emitting diodes. Especially organic light emitting diodes (OLEDs) are increasingly attractive because of their huge market potential. The feasibility of efficient OLEDs was first shown in 1987 [3]. Only about ten years later the first product, a display for car radios, entered the market. Today monochrome and full colour OLED-displays can be found in many applications replacing established flat panel display technologies like TFT-LCDs. This substitution is a consequence of the outstanding attributes of OLED technology: Organic light emitting displays are self-emissive, thin, video capable and in addition they show a wide temperature operation range and allow a viewing angle of nearly 180 degree in conjunction with a low power consumption. As performance has steadily increased over the last years, today OLEDs are also under investigation as next generation light source. In contrast to inorganic LEDs, they can be built as flat 2-dimensional light sources that are lightweight, colour tunable, and potentially cheap. This will open up new degrees of freedom in design leading also to completely new applications. In this contribution we will have a brief view on the history of organic electroluminescent materials before we introduce the basic principles of OLEDs with a focus on the physical processes leading to light generation in thin organic films. Along with an overview of different concepts and technologies used to build OLEDs, the current status of OLED development will be illustrated. The last part focuses on the challenges that have to be overcome to enable a sustainable success in the display and lighting markets.

1 Organic Electroluminescent Materials and their History

The electroluminescence of organic materials was first demonstrated by A. Bernanose et al. in the early 1950s [1]. Their organic films were excited by a very high AC voltage of 500–2000 V. In 1963 *Pope* et al. [2] reported the electroluminescence of Anthracene single-crystals under DC voltage. However, the driving voltages were still higher than 400 Vbecause of the high sample thicknesses of 10–20 μm. The breakthrough in the development of commercially applicable OLED materials came when Tang and VanSlyke presented

R. Haug (Ed.): Advances in Solid State Physics,
Adv. in Solid State Phys. **46**, 309–320 (2008)

an electroluminescent thin film device with a driving voltage below 10 V in 1987 [3]. Their device consisted of two functional layers of small aromatic molecules (TPD and Alq_3) sandwiched between a transparent Indium-Tin-oxide (ITO) anode and a cathode made of an Mg-Ag-alloy. It reached a power efficiency of 1.5 lm/W. The electroluminescence of conjugated polymers was first reported by *Burroughes* et al. in 1990 [4]. The used organic material was Poly(p-Phenylene-Vinylene) (PPV) which emits – similar to Alq_3 – in the green-yellow wavelength region. Since these fundamental achievements the development of organic electronics proceeds rapidly. The two most important classes of organic electroluminescent materials can be distinguished by their molecular weight and therefore by the preferable deposition method. Lightweight small molecules (SM) can be thermally evaporated, which requires an expensive vacuum-chamber setup but enables multilayer stacks and guarantees good homogeneity and reproducibility. The molecular weight of polymer materials on the other hand is very high, so they can not be vapour phase deposited and decompose at high temperatures. Polymers are deposited in solution by wet chemical processes like spin coating, ink jet printing, dip coating or screen printing, enabling a possibility of low cost mass production. The fundamental physics is very similar in both material classes and will be shortly explained in the following section.

2 Basic Principles and Build-Up

In both molecular systems alternating single and double-bonds between carbon atoms are leading to delocalised π-electron systems. Charge carriers are captured via ionisation of the molecule and can move freely only inside the delocalised π-electron system of one molecule. Organic semiconductors have no band structure like inorganic semiconductors, thus the energetic position of the highest occupied (HOMO) and the lowest unoccupied molecular orbital (LUMO) is crucial for the electronic properties. Charge transport between the molecules occurs by hopping processes [5]. When "jumping" from one molecule to another the carriers have to overcome an energetic barrier, which depends strongly on the temperature and the applied electric field. The discussed organic materials are in fact insulators because they have nearly no intrinsic free charges. Their semiconducting properties arise when charge carriers are injected into the material from contact electrodes. The difference between the work function of the electrodes and the energetic position of the HOMO/LUMO level of the adjacent organic layers is the injection barrier. The charge carrier mobility in organic semiconductors is typically in the range of 10^{-2}–10^{-8} cm^2/Vs [6], which is very low compared to inorganic materials like silicon. In case of perfect injecting ohmic contacts [7] the current through the device is limited by the low mobility, leading to an accumulation of carriers near the electrode. The charge of these carriers creates a space

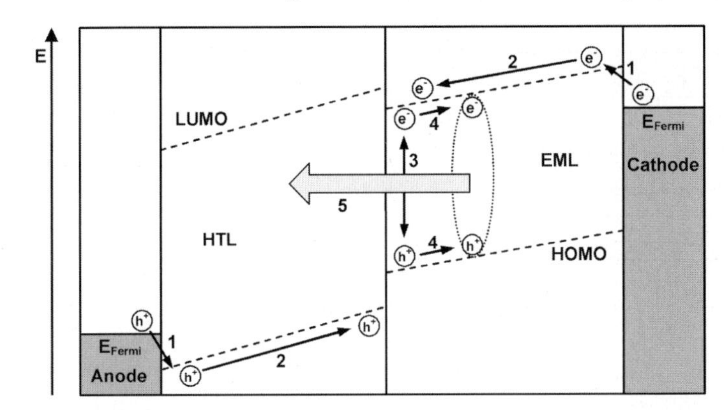

Fig. 1. *1*: Charge carrier injection, *2*: Charge carrier transport, *3*: Recombination of charges and exciton formation, *4*: Exciton diffusion, *5*: Light emission through radiative decay of excitons

charge zone, which lowers the effective electric field and limits the current (space charge limited current, SCLC) [8].

Figure 1 schematically shows the relevant processes in a two layer OLED during operation. After being injected (1) electron and holes move towards each other driven by the applied electric field (drift) and diffusion (2). When two carriers of opposite charge meet within their coulomb radius they can recombine and form excitons (3), which are $e^- - h^+$ pairs bound by coulomb interaction. During the lifetime of such an excited state it can diffuse through the material (4) until it decays radiatively (5) or nonradiatively. The emission colour depends on the materials energy gap between the HOMO and the LUMO.

Due to the random nature of spin orientation in electrically driven OLEDs 25% of the recombined carriers will form emissive fluorescent excitons (singlet states) and 75% non emissive triplet states [9]. By adding of heavy weight atoms – whose strong spin-orbit coupling favours the radiative decay of triplet excitons (phosphorescence) – it is possible to enhance the yield of emissive states η_{ST} up to 100%. Mainly metal-organic complexes with heavy transition metals are used as phosphorescent emitters. The lifetime of singlet excitons averages a few ns and that of triplet excitons up to several µs or even ms. Figure 2 shows that the fluorescence quantum efficiency η_{QE} (1) results additionally of the charge balance factor γ, the photoluminescence quantum efficiency η_{PL} and the outcoupling efficiency η_{EXT}.

$$\eta_{\mathrm{QE}} = \gamma \cdot \eta_{\mathrm{ST}} \cdot \eta_{\mathrm{PL}} \cdot \eta_{\mathrm{EXT}} \tag{1}$$

The basic device structure of an OLED is quite simple. A low number of different organic layers are sandwiched between two electrodes. For light extraction one of the electrodes has to be transparent. Because of unbalanced charge injection and transport devices with a single organic layer tend to have

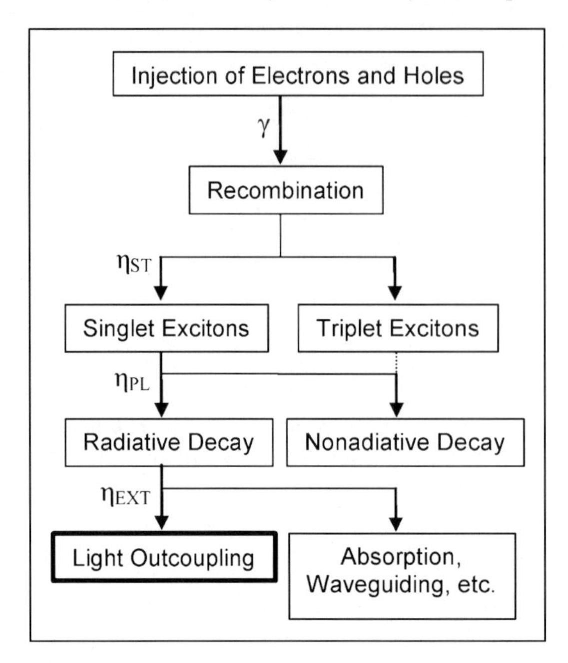

Fig. 2. Processes determining the fluorescence quantum efficiency of OLEDs

a low efficiency. To improve the charge carrier balance, standard polymer devices consist of two organic layers, usually a hole transport layer (HTL) and a light-emitting layer (EML). The additional HTL has a high hole mobility, lowers the injection barrier at the anode and flattens the rough surface of the ITO anode. An orbital offset occurs at the interface between the two organic layers, where the charge carriers accumulate and recombine. Because only two orthogonal (polar and unpolar) solvents are available and due to the fact that the upper layer must be applied onto the first one without dissolving it, polymer OLEDs (PLED) are mostly limited to a two layer stack. However, there are approaches to bypass this limitation, e.g., by the introduction of cross linkable polymers [10]. In contrast to PLEDs actual small molecule OLEDs (SMOLEDs) consist of three and more layers. On top of the anode a system of p-doped and/or intrinsic HTL is applied followed by at least one EML and a system of n-doped and/or intrinsic electron-transport layers (ETL) [11]. Since small molecules are deposited by thermal evaporation in vacuum (PVD) it is easily possible to build high efficient stacks of numerous functional organic layers, which is the most important advantage of SMOLEDs. For their SM device Tang and VanSlyke used Indium-tin-oxide (ITO) and an alloy of magnesium and silver as electrode materials. Due to its special properties like moderate electrical conduction (sheet resistance down to 10 Ω/square [12]), a high work function (4.7–5.1 eV) and visible transparency between 80% and 100% mainly ITO is used as anode. Cathode

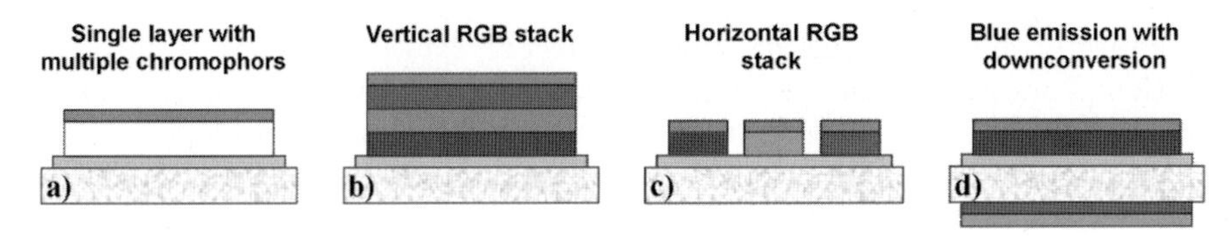

Fig. 3. Different concepts for white organic light emitting devices

materials should have a low work function, therefore alkali or earthy base metals like Barium (2.5 eV), Calcium (2.9 eV) or Magnesium (3.7 eV) are suitable. Unfortunately these low work function metals are very susceptible to oxidation by either water or oxygen. Thus bilayer cathodes are used: A very thin (5 nm) injecting layer is covered by a thicker layer (150–200 nm) of a more noble metal like Aluminium or Silver. This improves the conductivity and reduces corrosion. Despite using a cover layer, it is still necessary to process the OLED under inert atmosphere in a glove box system and encapsulate it afterwards to get high lifetimes anyway. Consequently it is desirable to completely avoid oxidation effects by using stable electrodes. For small molecule devices the combination of an ultra thin LiF layer (≤ 0.7 nm) and Al is often used as cathode [7]. Another approach is the usage of doped transport layers that permit noble contact electrodes [13]. In case of SMOLEDs these doped layers can be made by co-evaporation of the transport material (ETL or HTL) and donor or acceptor molecules. They enable ohmic injection almost independently of the electrode's work function. Thus for SMOLEDs it is possible to use noble metals (Ag, Al) for both electrodes. The total thickness of the organic layers in a typical OLED device is in the range of 200 nm. Therefore even single dust particles could create short circuits and lower the device performance and lifetime drastically. To avoid this kind of defects it is necessary to process OLEDs under cleanroom conditions.

3 White Organic Light Emitting Devices (WOLED)

For many applications especially for solid-state lighting purposes white emission is desired. Also for the use in displays it is of special interest as shown below in the corresponding section. In general white light is composed of all spectral colours. Since the emission spectra of organic devices are very broad the generation of white light requires the mixing of at least two complementary colours, e.g., turquoise and orange or three primary colours (blue, green and red). Depending on the material class (SM or polymers) different approaches to achieve white emission are used, an overview will be given in the following. The established device concepts for WOLEDs are presented in Fig. 3.

Approach a) has been realized for polymers and SM. In case of polymers the internal white emission of a single layer can be realized by applying a mixture of different emitting polymers in one solution (blend) or incorporation of several luminescent centres (chromophors) as co-monomers during polymer synthetisation [14]. It has been reported, that even blends of two blue materials can emit white light by emission of exciplexes [15] - these are excited states similar to excitons, but with charge carriers located on different molecules. In copolymers charge transport and blue emission could take place on the main chain (backbone) whereas green and red emission is generated by the chromophors. Excitation of the chromophors takes place either by transfer of excitons from the backbone or by trapping and recombination of charge carriers on the chromophors (direct recombination). For single emission layer devices copolymers are used preferably because of their easy processing, furthermore they have no drawbacks like phase separation occurring in polymer blends. White single layer devices could also be realised with SM, either by doping orange or red and green emitting materials into a blue emitting host or by using dopands of two or three[16] colours in a non emitting wide energy gap host material.

Because of the described difficulties in applying multilayer polymer stacks approach b) is mainly used for SM devices. In this concept the different colours are emitted from separated emitting layers stacked on top of each other. Besides the EML state of the art SMOLED stacks consist of several functional layers for injection, hole and electron transport, hole and electron blocking and exciton blocking. The number of layers is not limited by the preparation method, but the operation voltage gets higher with increasing device thickness. Orbital offsets at internal interfaces of two organic materials act as energetic barriers for charge carriers. To minimise the barriers it is necessary to arrange several organic materials in a way that their HOMO respectively LUMO levels increase cascading with small energetic steps. In order to maximise the device efficiency optimised systems of host and phosphorescent guest materials should be combined. Looking at three colour white SMOLEDs the lifetime and performance of red and green phosphorescent emitters is already satisfying as demonstrated in [17], [18] and [19]. Phosphorescent materials with saturated blue emission and sufficient operational stability are not yet known [20]. An additional difficulty is the fact that stable host materials possessing a high energy excited triplet level (T_1 state) are not yet available [20]. If the triplet level of the host material is too low excitons on emitter molecules could be quenched (forced to decay nonradiatively) by energy transfer to the matrix. In latest publications some groups are focusing on hybrid structures combining a fluorescent blue with phosphorescent red and green materials [21]. An elegant solution to achieve white illumination is the connection of individual single-colour OLED devices on top of each other. This so called tandem architecture seems to be attractive for getting highly efficient WOLEDs. The intermediate electrode layer connecting the emissive units has to separate the charges. For the lower unit this layer acts as cathode and for the upper as anode. The connection of two white OLEDs in series can lead to a significant improvement in lifetime by reducing the degradation induced by high current densities required for a high brightness in single-element OLEDs [22].

The principle of approach c) is based upon laterally structured single colour emitting materials. For the perception of white light the single light emitting areas must be small and closely arranged like pixels of a display. Only in case of indirect illumination the pixel area is less important. When the patterns are addressed separately it is possible to change the output spectrum during operation. Polymer materials in solution could be structured by ink jet or screen printing and by photolithographic techniques as presented in [10]. SM could be structured by evaporation trough shadow masks.

Approach d) is based on a blue emitting OLED with an outer layer of an inorganic phosphor. This so called colour changing media (CCM) partially absorbs the highly energetic blue light emitted by the OLED and converts it to a broad spectrum of lower energy (green and red) [23]. The resulting spectrum is a superposition of the original blue peak and the emission of the CCM layer. Either blue emitting polymers or SM could be used.

A disadvantage of approach b) could be that the emission colour changes with varying current densities. The colour shift accrues from the separated emission zones of these vertically stacked devices. At different current densities the density of excitons in each emission zone changes differently which affects the ratio between the single colours. Concerning the device concepts a), b) and c), one has to note that the degradation behaviour of several emitter materials is likely to be different. Thus colour shifts during the operational lifetime could occur when using devices with more than one organic emitter (differential aging). By the individual addressing of single coloured OLED elements – which works fine with approach c) – it is possible to compensate this colour shift. However the patterning is complex and increasing the current in one sort of the elements further accelerates the degradation process.

4 OLED Displays

The first commercially available monochrome OLED display with a configuration of 256×64 pixels [24] has been fabricated in 1997. It was a passive matrix (PM) SM display where the anode and cathode electrodes consist of parallel conducting paths orthogonal to each other (columns and rows); the crossing points define the pixels. To address an individual pixel a current is applied through the corresponding anode column and cathode row. Via PM switching it is not possible to address several pixels at the same time, therefore each column has to be controlled in a multiplex mode to realise video and picture applications. For example, in the 64 column display described above, each column is activated only a 1/64 time when driving the display in the multiplex mode. To get the same average brightness compared to a non-multiplexed display, each column (pixel) has to be driven 64 times brighter. The inertia of the human eye causes the perception of a stable and

Fig. 4. Picture depicted on a full colour OLED display (online color) based on a white emitting copolymer and colour filters. The display has 96 × RGB × 64 pixels

flicker-free picture of average brightness consisting of all sequentially activated pixels on the screen. The bigger the screen the faster the switching process and the brighter the pixels have to be. Therefore PM displays are limited in the size of their diagonals. In active matrix (AM) displays each pixel is switched independently by an own thin film transistor (TFT). As a matter of principle AM displays are not limited in the diagonal but due to the additional TFT technology they are of course more expensive and more complicated to fabricate. OLED displays, no matter if PM or AM could be made of either polymers or SM. Mainly the deposition and patterning of the organic layer(s) depend on the material class. In full colour displays one pixel usually consists of three (red, green and blue, RGB) sub pixels. A different approach - especially for PLED displays - is to use a single white emitting organic material in combination with RGB colour filters (CF). Unfortunately the filters could absorb about 60% of the emitted light. But compared to other kinds of full colour PLED displays realized, e.g., by ink jet printing the CF technology has several advantages. However, ink jet printing is a complex process. Besides the elimination of differential aging the main advantage of CF technology is the feasibility to apply only one white emitting (polymer) material without structuring. This could be done easily by spin coating or screen printing which is inexpensive and fast. The colour filters themselves are far developed and used for many years in the LCD technology. The first screen printed and colour filter based full colour polymer OLED display was presented 2004 by Siemens AG, CT MM 1 [25] and is shown in Fig. 4.

Fig. 5. Demonstrator of a bent OLED device on a 100 μm thick glass substrate

Besides ultra fast switching times (< 10μs) and low angular emission dependence the most striking advantages of OLED displays compared to established plasma screens or LCDs are their self-emission and flatness. OLEDs do not need background illumination, thus the active emission layer including electrodes is only a few hundred nanometres thick. It could be deposited on lots of different kinds of substrates. Combined with a thin film encapsulation lightweight and flexible OLED displays could be produced by applying the organic layers to thin glasses or metal- and plastic-foils [26]. Figure 5 shows an OLED device on a thin glass substrate. By encapsulation with another thin glass the device was fixed in a curved position.

Today a remarkable fraction of MP3 players features small OLED displays. They can also be found in lots of other applications like cell phones, shavers, and cameras. In 2004 Epson announced the development of a 40 inch full colour AM OLED display prototype [27], whereas the market entry of such large OLED displays still needs some time.

5 OLED Lighting

One of the previous sections dealt with the physical background of white light generation in organic materials, now applications will be discussed. White OLEDs are important for the use as solid state lighting source and will likely be able to enter completely new markets. Due to their unique combination of outstanding properties like flatness, flexibility and lightweight white OLEDs open up new degrees of freedom in design and shape of lighting applications, which are impossible to achieve with conventional techniques so far.

Fig. 6. OLED lighting demonstrator with an active area of $70\,\mathrm{cm}^2$, fabricated by Siemens AG, CT MM 1

For illumination purposes lighting tiles should have a brightness of about $1000\,\mathrm{cd/m}^2$, which is more than required for display applications. Taking a white display and increasing the area and driving voltage is not sufficient to get a large scale and high brightness lighting device. A lot of challenges arise scaling up the active area of OLEDs. On large active areas the probability of occurrence of small defects increases. They affect lifetime and performance on the one and usability on the other hand. Another very important criterion for a lighting device is the homogeneity in colour and brightness. Even differences in light intensity below 10% could be detected by the human eye. To avoid inhomogeneity of any kind it is necessary to achieve a uniform layer thickness and a homogeneous current distribution over the complete area. In large area devices the limited conductivity of ITO causes a voltage drop towards the middle of the anode. This results in a non-uniform current density and brightness distribution [28]. The difference between maximum and minimum brightness increases the steeper the IVL characteristics (current density and luminance vs. voltage) are. However, steep IVL curves are required for high brightness and power efficiency at low voltages. To reduce this brightness distribution it is necessary to increase the conductance of the electrode, e.g., by applying additional paths of metallization - so called bus bars - to the transparent anode. Another interesting approach to minimise the voltage drop are architectures consisting of several small lighting elements monolithically connected in series [29]. Although first demonstrators are already available (see Fig. 6) the development of OLED lighting applications is still facing the challenges discussed above.

6 Summary and Outlook

After a short excursus into the physics of organic light emitting thin film devices and a detailed look on their outstanding properties it is apparent that OLEDs are a very promising technology. As depicted the two most important OLED applications are displays and lighting. Small OLED displays are already for sale and meet with great success, up to day mainly in mobile applications. Large OLED displays are not yet brought onto the market. They will need some more time for further improvement in terms of lifetime and performance. White OLEDs will enable new degrees of freedom in design and shape of lighting devices. Several approaches have been proposed which could avoid potential difficulties scaling up the luminescent area of OLEDs. Which remains is the need of highly efficient and stable blue organic light emitting materials.

Acknowledgements

The authors want to thank C. Gaerditz and D. Buchhauser for support and helpful discussions on OLED lighting and display technology.

References

[1] A. Bernanose, M. Comte, P. Vouaux, J. Chim. Phys. **50**, 64–68 (1953)
[2] M. Pope, H. Kallmann, P. Magnante, J. Chem. Phys. **38**, 2042–2043 (1963)
[3] C. Tang, S. VanSlyke, Appl. Phys. Lett. **51**, 12, 913–915 (1987)
[4] J. Burroughes, D. Bradley, A. Brown, R. Marks, K. Mackay, R. Friend, P. Burn, A. Holmes, Nature **347**, 539–541 (1990)
[5] H. Baessler, Phys. Stat. Sol. (b) **175**, 15, 15–56 (1993)
[6] L. S. Hung, C. H. Chen, Mat. Sci. Eng. **R 39**, 143–222 (2002)
[7] M. Stoessel, Dissertation, University of Erlangen-Nuremberg (1999)
[8] M. Lampert, P. Mark: *Current Injection in Solids*, (Academic Press, New York, London 1970)
[9] M. A. Baldo, D. F. O'Brien, M. E. Thompson, S. R. Forrest, Phys. Rev. B **60**, 20, 14422–14428 (1999)
[10] K. Meerholz, Nature **421**, 829–833 (2003)
[11] M. Pfeiffer, S. R. Forrest, K. Leo, M. E. Thompson, Adv. Mater. **14**, 22 (2002)
[12] Cerac inc., http://www.cerac.com/pubs/proddata/ito.htm
[13] J. Blochwitz, M. Pfeiffer, T. Fritz , K. Leo, Appl. Phys. Lett. **73**, 729 (1998)
[14] C. Gaerditz, R. Paetzold, D. Buchhauser et al., Proc. of SPIE **5937**, 59370L (2005)
[15] M. Mazzeo, D. Pisignano, F. Della Sala, J. Thompson, R. I. R. Blyth, G. Gigli, R. Cingolani, G. Sotgiu,G. Barbarella, Appl. Phys. Lett. **82**, 334–336 (2003)
[16] B. W. D'Andrale, R. J. Holmes, R. Forrest, Adv. Mater. **16**, 7, 624–628 (2004)
[17] T. Tsuji, S. Kawami, S. Miyaguchi, T. Naijo, T. Yuki, S. Matsuo, H. Miyazaki, SID 04 Digest, 900–903 (2004)

[18] M. S. Weaver, V. Adamovich, M. Hack, R. Kwong, J. J. Beown, Proc. Int. Conf. Electrolum. Molec. Mat. Rel. Phen **O–35** (2003)
[19] H. Becker, H. Vestweber, A. Gerhard, P. Stoessel, R. Fortte, SID 04 Digest, 796–797 (2004)
[20] S.-W. Wen, M.-T. Lee, C. H. Chen, IEEE J. Display Tech. **1**, 1 (2005)
[21] Y. Sun, N. C. Giebink, H. Kanno, B. Ma, M. E. Thompson, S. R. Forrest, Nature **440**, 908–912 (2006)
[22] C. C. Chang, J. F. Chen, S. W. Hwang, C. H. Chen, Appl. Phys. Lett. **87**, 253501 (2005)
[23] A. R. Duggal, J. J. Shiang, C. M. Heller, and D. F. Foust, Appl. Phys. Lett. **80**, 19, 3470–3472 (2002)
[24] T. Wakimoto, R. Murayama, K. Nagayama, Y. Okuda, H. Nakada and T. Tohma, SID 96 Digest, 849 (1996)
[25] D. Buchhauser et al.: Proceedings SID Europe Chapter Spring Meeting, Frankfurt (2004)
[26] R. Paetzold, K. Heuser, D. Henseler, S. Roeger, G. Wittmann, A. Winnacker, Appl. Phys. Lett. **82**, 19 (2003)
[27] Epson, http://www.epson.co.jp/e/newsroom/news_2004_05_18.htm (2004)
[28] C. Gaerditz, R. Paetzold, D. Buchhauser, J. Wecker, A. Winnacker, SPIE Photonics Europe **06**, 6192–29 (2006)
[29] A. R. Duggal, D. F. Foust, W. F. Nealon, C. M. Heller, Appl. Phys. Lett. **82**, 16, 2580 (2003)

Index